The Technology Trap

The Technology Trap

Where Human Error and Malevolence Meet Powerful Technologies

Lloyd J. Dumas

 PRAEGER

AN IMPRINT OF ABC-CLIO, LLC
Santa Barbara, California • Denver, Colorado • Oxford, England

Library of Congress Cataloging-in-Publication Data

Dumas, Lloyd J.
   The technology trap : where human error and malevolence meet powerful technologies / Lloyd J. Dumas.
      p. cm.
   Includes bibliographical references and index.
   ISBN 978–0–313–37888–1 (hard copy : alk. paper) — ISBN 978–0–313–37889–8 (ebook)
1. Technology—Social aspects. 2. Accidents. 3. Terrorism. I. Title.
T14.5.D86   2010
303.48′3—dc22   2010021793

ISBN: 978–0–313–37888–1
EISBN: 978–0–313–37889–8

14 13 12 11 10   1 2 3 4 5

This book is also available on the World Wide Web as an eBook.
Visit www.abc-clio.com for details.

Praeger
An Imprint of ABC-CLIO, LLC

ABC-CLIO, LLC
130 Cremona Drive, P.O. Box 1911
Santa Barbara, California 93116-1911

This book is printed on acid-free paper ∞

Manufactured in the United States of America

Contents

List of Tables

Acknowledgments

Over the 35 years since I began exploring the connections among human fallibility, technical failure, and security, I have had so many discussions and debates, been given so many useful comments, and spent so many hours reading works drawn from so many literatures that I cannot even list all of those whose contributions to this work should be recognized. For this, I hope I will be forgiven.

Some whose contributions to this work are notable are still active and working to make this a better world. Others have since passed on, though their spirit is very much alive in the pages that follow. Among the physical and behavioral scientists, computer scientists, engineers, medical doctors, mental health professionals, and military officers whose comments and insights have contributed to the genesis of this work are Carl Sagan, Robert Lifton, Herbert Abrams, Warren Davis, Victor Sidel, Severo Ornstein, Lester Grinspoon, Bruce Blair, Herbert York, Bernard Feld, James Thompson, Margaret Ballard, Robert Karasek, Mikhail Milstein, James Bush, and Eugene Carroll. Among those who were important sources of support and encouragement through many years of researching and writing are Alice Barton, Lynne Dumas, Martha Hurley, Roger Kallenberg, and Abdi Yahya. I would also like to thank Tyler Ratliff, and especially Saheli Nath and Lindsay Bernsen, who gave freely of their time and energy, for their valuable research assistance.

I am grateful for the help my parents, Marcel and Edith Dumas, gave me in every project I ever undertook, and for the encouragement and inspiration of my friend and colleague, Seymour Melman. I miss them all. I am also thankful for my longtime intellectual co-conspirator Janine

Wedel and for the enthusiasm of my dearest friends and confidants, Dana Dunn and Yolanda Eisenstein. To be immersed for so long in such a difficult subject and still remain confident that these obstacles can be overcome, there must be a deeply felt sense of optimism, joy, and belief in the future. This has been the gift of my intellectual partner and soul mate, Terri Nelson Dumas.

Part I

Threatening Ourselves

1

Human Fallibility and the Proliferation of Dangerous Technologies

INTRODUCTION

We humans are creatures of paradox. When our talents and capabilities are put to positive use, we compose music that makes the spirit soar, create works of art and architecture of timeless beauty, and design and build devices that multiply our physical capabilities, extend the reach of our unaided senses, and connect us to the wider world. Yet we can and do also use those remarkable talents and capabilities to very different effect, despoiling our environment, debasing and degrading each other, and creating products and processes so destructive that we may yet become the first species responsible for its own extinction.

Of all the ways we affect the physical world for better or for worse, the development and application of technology is the most powerful. It has provided enormous benefits. The progress of technology has allowed us to make more and better goods available at prices low enough to put them within the reach of much of the earth's population. We can now travel faster, exchange information more easily, enjoy a wider variety of food and entertainment, have better protection from the elements, and access far better medical care than was available to even the richest monarchs of centuries gone by.

At the same time, the technologies we have developed and the ways we have used them have fouled the air, land, and water and driven thousands of living species into extinction. We have also put into our own hands the power to extinguish human society in a matter of hours and trigger processes that could eventually turn our bountiful planet into yet another lonely, lifeless sphere spinning aimlessly through space. This too is the legacy of our technological brilliance. Its two-sided nature is simply a reflection of the brighter and darker sides of the species that created it.

There has been more than a little rhapsodizing about the peaceful and prosperous Eden to which our technologies will carry us some day, a day that never quite seems to arrive. There has also been a great deal of ranting about the evils of technology by those who have urged us to return to a more natural lifestyle, but this book is neither a paean to the glories of technology nor a cautionary tale about the price we have paid for its bounty. It is instead, to use Arthur Schlesinger, Jr.'s term, "a passport to reality" and a guide to making wiser choices. It is also a warning—not about technology itself, but about technological hubris—about the trap inherent in the unthinking assumption that we can always control the technologies we create, no matter how powerful, no matter how dangerous, and permanently avoid disaster.

As a species, we humans are inherently fallible both in the sense of being prone to error—a characteristic we all share and demonstrate daily—and in the sense that, under the right circumstances, at least some of us are prone to malevolent behavior, criminal or terrorist. If we are to enjoy the vast bounty that technology provides without falling into the "technology trap," then we must learn to take the limitations imposed by both these dimensions of our fallibility more consciously and profoundly into account than we currently do in deciding what technological paths to follow. We cannot allow our fascination with the power of what we can do to blind us to how easily it can all go terribly wrong.

TECHNOLOGY AS MAGIC: THE TECHNOLOGICAL FIX

For all the mystique that surrounds it, technology is nothing more than the application of the biological, chemical, and physical laws of nature to the design and development of useful products and processes. Scientists extend and deepen the store of knowledge by applying the "scientific method" to the exploration of these laws. Engineers apply their knowledge to the design of actual products and processes.

Yet to many of us, technology is a kind of magic. Why it works is a mystery, but there is no doubt that it does remarkable things.

Flip a switch and a room lights up; push a few buttons and talk to a friend on the other side of town or the other side of the world; sit inside an enormous metal tube and roaring engines catapult you through the sky. It is all pretty amazing.

Some technologies are very transparent. It is easy to see how they work, easy to understand what is happening. It is obvious why a bicycle moves faster when you peddle harder or changes direction when you turn the handlebars, but most modern day technologies are much more opaque. Unlike a bicycle, you can drive a car for years without having the slightest idea why it moves faster when you press the accelerator pedal or turns when you turn the steering wheel. And driving the car teaches you little or nothing about how the underlying technology works. Neither does using a cell phone, listening to an iPod, or watching a DVD.

Engineers have worked hard to make increasingly sophisticated technology more "user friendly." Because of their efforts, it is not necessary to understand how cell phones, digital videodisc players, or laptop computers work in order to use them. And so, these technologies have become accessible to millions. But for all of its advantages, the "user friendly" approach has one unfortunate side effect. It allows the technology in common use to advance far beyond even the vague comprehension of the vast majority of those who use it. For most of us, more and more of the products of everyday life have become what engineers call "black boxes"—we know what goes in and what comes out, but not what goes on inside. This has reinforced the unconscious image of technology as magic, an image that has very serious negative consequences for making intelligent social decisions on which technological directions to explore and which technologies to develop and put to use.

The problem is that in magic, anything can happen. There are no limits, no tradeoffs, and no impossibilities. So, if technology is magic, then it has no limits, no tradeoffs, and no impossibilities. Whatever the problem, if scientists and engineers are clever enough and work hard enough, they can eventually find a way to solve it, they can find a technological fix.

In the 1950s, it was widely believed that the new agricultural technologies known as the "green revolution" would finally put an end to world hunger. These technologies were, in fact, enormously successful in increasing agricultural yields. Yet there were more people starving in the early years of the twenty-first century than there were 50 years before. The reason is simple: world hunger was not and is not the result of inadequate food production. It was and is mainly the result of economic, social, and political factors that affect food distribution and use.[1] The problem of world hunger is not solvable by technology because it is not a technological problem.

When nuclear weapons were first created, many believed they were so terrifying that they would finally put an end to the age-old scourge of war. Yet there have been well more than 150 wars since the dawning of the Atomic Age, taking the lives of more than 20 million people, two-thirds of them civilians.[2] This technological fix did not work either, and for the same reasons. Neither war itself nor the broader issue of security is a technological problem. Security and war are political, social, and economic problems. We will put an end to war and find more humane and reliable paths to security when we become committed to less threatening means of resolving even our most contentious conflicts. No brilliant technological innovation will save us from the difficult task of learning how to live with each other.

There is not always a technological fix even for problems that are mainly technological. In the real world, there *are* limits, tradeoffs, and impossibilities. Often enough, the technology that solves one problem creates others. As hard as it may be to believe today, in the late nineteenth and early twentieth centuries, the automobile was promoted in part as a technological fix for urban pollution. Horse-drawn transport had polluted the cities with an accumulation of animal waste that threatened public health. The automobile neatly solved that problem. Yet today, automobiles are one of the largest single sources of health-threatening urban air pollution, noise, and greenhouse gas emissions. More recently, the Montreal Protocol on Substances that Deplete the Ozone Layer—created in 1987 and ultimately signed by more than 180 nations—included a phase-out of a number of chemicals that deplete the ozone layer, among them hydrochlorofluorocarbons (HCFCs). This stimulated the development of other fluorocarbons that were much less of a threat to the ozone layer to replace those being phased-out. Among these were hydrofluorocarbons (HFCs). As it turns out, although HFCs have far less potential for depleting ozone, they are much more powerful greenhouse gases with a much higher potential for making global warming worse. History certainly has its ironies.[3]

The fact is, technology is not magic. It is the result of a systematic creative process of thinking and experimentation carried out by fallible human beings who inescapably imbed the essential imperfection that makes them human into every technology they develop. Not only does every particular technology have its own limits, technology itself is limited in what it can be expected to accomplish. It is very important and very powerful, but technology is not omnipotent.

SHAPING TECHNOLOGY

Technology reaches into the very fabric of society, reshaping the economic, social, political, and cultural world in which we live. At the same time it affects these dimensions of our lives, however, it is greatly affected by them as well. Technology is not self-propelled. It is important to take a brief look at the nature of the process by which technological choices are made in order to better understand the role of the society in shaping these choices.

The engineering design of any product begins with a central design objective and an explicit or implicit list of design criteria. The central design objective is related to the basic function that the product is supposed to perform. For example, for a refrigerator, the central design objective is providing a controlled and consistent environment inside the machine that is cooler than the external environment; for a truck, it is to effectively transport people and cargo; for a computer, it is to efficiently process data. The accompanying list of design criteria consists of product characteristics relevant to those who will buy and operate the product. Some of these criteria are virtually universal: purchase cost, operating cost, safety, ease of use, reliability, maintainability, and aesthetics. Others are more specific to the product at hand: for refrigerators, they include energy efficiency, ratio of usable internal space to external size, and convenience features such as movable shelves, icemakers, and cold water dispensers; for trucks, speed, range, acceleration, energy efficiency, and tailpipe emissions; for computers, speed, memory capacity, portability, and the like.

There are inevitably multiple tradeoffs among these design criteria; no single design will be best for all of them. A refrigerator that is more energy efficient may cost more to buy; a truck that pollutes less may have slower acceleration; a computer that is more portable may have a smaller keyboard, increasing the probability of keying errors. The critical point is that many different designs of any given product are possible, all of which achieve the central objective but differ in the degree to which the various design criteria are achieved. Therefore, whoever establishes the priorities attached to the different design criteria has enormous influence on the characteristics of the final product. This is actually true of any functional design situation, whether the designer is an engineer or an architect, whether the product is a toothbrush, an office building, or an urban transportation system. Those who set the priorities attached to the design criteria play a critical role in directing the application of existing technological knowledge. They shape the technology that shapes our society.

A very similar process also determines the direction of the development of new scientific and technological knowledge. Technological development is not unidirectional, and as in the case of design, the priorities followed have a great impact on the end result. If a high priority is attached to reducing energy use, for example, not only will existing technology be applied to produce more energy efficient cars and appliances, but more research will be done in areas that seem to have the most promise for achieving further gains in energy efficiency. In research, what is found is very strongly conditioned by what is sought, where the attention of the researchers is directed. Not everything is possible, and yet technology may be bent in many different directions.

Who is qualified to make this crucial choice of design or research priorities? The surprising answer is, all of us. Those trained in science and engineering are far more capable of developing and applying new scientific and technological knowledge than the rest of us, but they have no special knowledge that gives them a greater capacity—or a greater right—to determine the technological priorities that ultimately have such an enormous impact on the kind of world we live in. That is a matter of social choice; it is not a scientific prerogative.

In the case of products offered for sale, the priorities are established by managements strongly influenced by what sells. If car buyers are more interested in style than safety, carmakers will attach a higher priority to aesthetics and less to safety. If car buyers later become more interested in safety than style, it will be more profitable for car makers to shift emphasis to safety and pay less attention to aesthetics. Government can also exert strong influence on both design and research priorities—as a rule maker through regulation, as a source of research funding through the kinds of grants they make, and as a customer through their purchases.

It is worth emphasizing that someone must set those priorities for technology to progress. Whether they are set by the government or by the marketplace, all of us are capable of understanding enough about the choice of priorities to participate (through our purchases and our political activities) because social choice determines research and design criteria priorities, not scientific necessity. If we choose not to pay attention, these critical, society-shaping decisions will still be made.

But they will be made behind closed doors by a much smaller group of insiders, and we will all have to live with the consequences.

Although technological knowledge itself is neither good nor bad, it is not true that all technologies can be applied in ways that are beneficial. Some are extremely difficult if not impossible to put to constructive use. Given our fallibility and the darker side of the human psyche, there may even be some things we would be better off not knowing. There is simply too great a chance that they will be used in ways that turn out to be highly destructive, intentionally or unintentionally.

Despite the dangers, few of us would want to give up the benefits of modern technology. But the choice is not to embrace or reject at once the full range of technologies of modern society. We can *and should* pick and choose among existing technologies (and research directions) those that have the greatest net benefit, and we should firmly reject those that have the potential for doing us grave harm. While it is impossible to eliminate risk and danger from technologically advanced society, there is no reason why we cannot avoid falling into the technology trap and thus subjecting ourselves to the very real possibility of human-induced technological catastrophe.

DANGEROUS TECHNOLOGIES

"Dangerous technologies" are those capable of producing massive amounts of death, injury, and/or property damage within a short span of time. Some technologies are designed to be dangerous. Their intentional use, accidental use, or failure can do enormous damage. Nuclear weapons are a prime example. There is no doubt that deliberate nuclear attack is devastating, but so too is accidental nuclear war that is triggered by human error, miscalculation, or technical failure. Even the accidental explosion of a single nuclear weapon in the "right" place at the "right" time could take millions of lives.

Other technologies designed for benign purposes can still produce devastating damage if enough goes wrong. Nuclear power plants supply electricity to consumers, business, and industry. Yet the catastrophic human or technical failure of a nuclear plant has the potential of killing or injuring hundreds of thousands of people and contaminating huge amounts of the surrounding landscape. A so-called "maximum credible accident" at a single nuclear power plant could do billions of dollars worth of damage. Therefore, nuclear power must also be classified as a dangerous technology.

Weapons of Mass Destruction

Weapons of mass destruction are dangerous technologies by design. Although it technically possible to kill millions of people with conventional explosives (or even with rifle bullets or knives), what makes weapons of mass destruction different is that they can kill large numbers of people *with a single use.*

Nuclear weapons are the preeminent example of a weapon of mass destruction. They have enormous destructive power relative to bomb size or weight and a reach that extends far beyond their moment of use: people were still dying from the Hiroshima bombing decades after the attack. Used in large numbers, nuclear weapons assault the environment, tearing at the ecology and at the very fabric of life.[4]

Weapons that disperse nerve gas or other deadly chemicals are also capable of killing thousands or even millions of people. Biological weapons that release virulent infectious organisms can be still more destructive because the organisms are alive and can reproduce. Some pathogens (e.g., anthrax bacilli) are very resistant to external conditions and can remain dormant but infectious for many years.[5] As frequent flu epidemics clearly indicate, once a chain of infection begins, it can spread rapidly through large numbers of people. Wartime conditions of disruption and stress would make the epidemic worse.

Despite the advanced state of modern medicine, some deadly infections can be extremely difficult to treat, let alone cure. This is especially true when the nature of the infectious organism itself is unusual, as the AIDS epidemic has illustrated with frightening clarity.[6] Recent remarkable advances in biotechnology have made it far easier to intentionally produce genetically altered and therefore unusual versions of known pathogens as well as wholly artificial new infectious life forms.[7]

First cousins to nuclear weapons, radiological weapons (colloquially known as "dirty bombs") can kill by dispersing radioactive materials. They can contaminate large areas with high levels of radiation that can remain deadly for many centuries. Nuclear accidents like that at Chernobyl have already illustrated how easily radioactive materials can be dispersed over a wide area. In a sense, any nation that has nuclear power plants or waste storage sites within its borders already has "enemy" radiological weapons on its own soil. A military or terrorist attack against those facilities could unleash enormous amounts of dangerous radiation.

Given the ingenuity of weapons designers and military strategists, there will undoubtedly be more to come in the way of methods for killing large numbers of people. Perhaps we will come up with weapons that alter the climate, shifting rainfall patterns and temperatures so as to create widespread starvation or ecological devastation.[8] Computer simulations of nuclear war in the 1980s indicated that the large-scale use of nuclear weapons might well produce a "nuclear winter."[9] We have already tried small-scale, primitive ecological weapons such as Agent Orange, used to defoliate the jungles of Vietnam.[10] What about weapons that would produce hurricanes, tornados, or earthquakes on demand? At least as far back as the 1960s, research was already underway on the possibility of using underwater nuclear explosions to generate tsunamis (tidal waves).[11] On December 26, 2004, the world saw once again how extraordinarily devastating tsunamis could be when a powerful earthquake off the coast of Indonesia generated a huge tsunami that killed up to 150,000 people in one day and injured and displaced millions more.[12]

How about "ethnic weapons" designed to target one particular ethnic group? Small genetic variations in the gene pools of different ethnic groups (such as the predisposition of blacks to sickle cell anemia) might make it possible to find chemical agents, bacteria, or viruses to which that group would be particularly vulnerable. As grisly and bizarre as this sounds, it has long ago been discussed in the public military literature.[13] There is no doubt that the remarkably creative human mind directed to such dark purposes can add still unimagined means of annihilation to this already chilling list.

Nuclear Power and Nuclear Waste

Nuclear power is a dangerous technology not only because the consequences of a major malfunction are so potentially catastrophic, but also because it generates large amounts of highly toxic radioactive waste. Despite periodic announcements of breakthroughs, the problem of developing a safe, practical way to treat and/or store long-lived nuclear waste is still far from being solved.[14] The combination of inadequate nuclear waste treatment and storage technology and large amounts of very toxic, long-lived, nuclear waste is a nightmare, a disaster waiting to happen.

Nuclear waste is often corrosive and physically hot, as well as radioactively "hot." This makes it difficult to store safely and isolate from the biosphere, especially when it remains dangerous for extremely long periods of time. A single half-life of plutonium—a critical raw material

in nuclear weapons and a byproduct of nuclear power reactors—
is 24,000 years. That is far longer than all of recorded human history.
Considering all the political, social, economic, and cultural changes that
have occurred from a time more than one hundred and fifty centuries
before the pyramids of ancient Egypt up to the space age, what reason
is there to believe that we can safely contain, isolate, or even keep track
of materials that remain so deadly for so long? Until the problem of safe
storage is solved (if ever), nuclear waste storage sites are ecological time
bombs waiting to explode in the faces of this or future generations. This
is hardly a legacy of which we can be proud.

Highly Toxic Chemicals

The spectacular progress of chemicals technology during the twentieth
century, especially since World War II, has created its own set of problems. Some chemicals routinely used in industry are extremely toxic
themselves or are manufactured by processes that produce highly toxic
chemical waste. In fact, the worst industrial accident in history occurred
at a chemical plant—the Union Carbide pesticide plant at Bhopal, India.
On December 3, 1984, water entering a methyl isocyanate storage tank
triggered a reaction that resulted in the release of a cloud of deadly
gas that drifted over the city, killing at least 2,000 people and injuring
200,000 more.[15] The isocyanates used in industry are related to
phosgene, a chemical used as poison gas as far back as World War I.[16]

The Bhopal disaster may have been the worst to date, but there have
been many accidents during manufacturing, transportation, storage,
and use of highly toxic chemicals. Still, not all technologies involving
chemicals toxics are "dangerous." The chemicals must be toxic enough
and in quantities large enough for a major release to produce disaster.
Fortunately, most noxious chemicals in common use are not that toxic.
Unfortunately, all too many are.

Only in the last few decades have we begun to understand just how
much toxic chemical waste we have created as a result of our fascination
with the "magic of modern chemistry." Highly toxic chemical waste
may be less complicated to treat and store than nuclear waste, but it is
still a very serious problem. And it is very widespread. Toxic waste
dumps are scattered all over the United States. There are in fact few, if
any, parts of the world where this problem can be safely ignored.

Despite the fact that they kill or injure large numbers of people,
some technologies are not classified as "dangerous" because the

damage they do is not concentrated. Automobiles, for example, kill or maim tens of thousands every year. But they are injured in thousands of separate crashes, each involving no more than a relatively small number of people. Because it is dispersed in time and place, this carnage on the roadways has neither the same social impact nor the same social implications as concentrated disasters like Bhopal or Hiroshima.

PROLIFERATION

Dangerous technologies have spread all over the globe. Sometimes they were purposely propagated in pursuit of profit or other self-interested objectives; sometimes they spread despite determined efforts to prevent others from acquiring them. The laws of nature cannot be patented. Once it has been shown that a technology works, it is only a matter of time before other well-trained and talented people will figure out how to do it too.

Nuclear Weapons

It took years for a huge, well-orchestrated program led by an unprecedented collection of the world's most talented physical scientists to develop the first atomic bomb in 1945. Intensive efforts to keep the technology secret have not succeeded. The American monopoly on nuclear weapons did not last long. By 1949, the Soviet Union successfully tested an atomic bomb of its own. Britain followed in 1952, and France in early 1960. By the end of 1964, China too had joined the "nuclear club." A decade later, India detonated a "nuclear explosive device" but continued to insist that it was not a nuclear weapons state for almost 25 years. All such pretenses were blown away in May 1998 when India exploded a series of five nuclear weapons in three days.[17] Arch rival Pakistan then followed suit within a few weeks.[18] Then, on October 16, 2006, impoverished North Korea exploded a nuclear device of its own, followed by a second and apparently much larger bomb test on May 25, 2009.[19]

To date, eight nations have publicly declared themselves to be nuclear weapons states, but they are by no means the only countries capable of building nuclear weapons. All 20 countries listed in Table 1.1 have (or had) a significant nuclear weapons capability. Almost twenty countries in addition to those listed have the technical capacity to develop nuclear weapons, but they are not believed to have yet tried to do so

Table 1.1 Nations that have (or had) active nuclear weapons programs (April 2009)

Nation	Nuclear Weapons Status	Size of Arsenal
United States	Declared Nuclear Weapons	2,200 Strategic and 500 Tactical Operational Warheads; total inventory of 9,400 warheads
Russia	Declared Nuclear Weapons	2,790 Strategic and 2,050 Tactical Operational Warheads; total inventory of 13,000 warheads
United Kingdom	Declared Nuclear Weapons	160 Strategic Operational Warheads; total inventory of 185 warheads
France	Declared Nuclear Weapons	300 Operational Strategic Warheads
China	Declared Nuclear Weapons	180 Strategic Operational Warheads (Tactical Warheads Uncertain); total inventory of 240 warheads
India	Exploded "peaceful" nuclear device in 1974, and a series of admitted nuclear weapons tests in 1998	Estimated at more than 60 nuclear warheads
Pakistan	Exploded series of nuclear weapons in 1998	Estimated at 60 nuclear warheads
North Korea	Exploded small nuclear device in 2006 and larger, Hiroshima-sized bomb in 2009	Less than 10 nuclear warheads
Israel	Has not acknowledged possession of nuclear arsenal	Estimated at 80 nuclear warheads
Iran	Despite denials, Western intelligence sources believe active nuclear weapons program underway	Estimated to be making progress toward building nuclear weapons
Iraq	Unveiled in 1995 by defection of high-level official, covert nuclear program (operating since 1970s) was terminated well in advance of the US-led invasion in 2003.	None. No active nuclear weapons program

Country	Description	Nuclear Weapons
Algeria	Once suspected of having an embryonic program, Algeria renounced any attempt to develop nuclear weapons in 1995	None
Argentina	Suspected of having begun nuclear weapons program in 1970s, renounced any attempt to build them in the mid-1990s	None
Belarus	Part of former Soviet Union, all nuclear weapons on its territory transferred to Russia by mid-1990s	None
Brazil	Secret nuclear program run by military rulers in 1980s ended in 1990 by newly elected government	None
Kazakhstan	Part of former Soviet Union, all nuclear weapons on its territory transferred to Russia by April 1995	None
Libya	Admitted and dismantled covert nuclear weapons program in December 2003	None
Romania	Secret nuclear weapons program during Ceaucescu regime ended with 1989 overthrow by democracy	None
South Africa	Successful secret nuclear weapons program ended by 1989 order of President de Klerk	6 nuclear weapons built; weapons and related facilities dismantled by 1991
Ukraine	Part of former Soviet Union, all nuclear weapons on its territory transferred to Russia by June 1996	None

Sources: Federation of American Scientists, "Status of the World's Nuclear Forces 2009" (http://www.fas.org/programs/ssp/nukes/nuclearweapons/nukestatus.html, accessed June 1, 2009); Spector, Leonard S., Mark G. McDonough, and Evan S. Medeiros, *Tracking Nuclear Proliferation* (Washington, D.C.: Carnegie Endowment for International Peace, 1995) and accompanying "Tracking Nuclear Proliferation: Errata and Essential Updates" sheets (June 7, 1996); Albright, David, "The Shots Heard 'Round the World" and "Pakistan: The Other Shoes Drops," *Bulletin of the Atomic Scientists* (July/August 1998); Hamza, Khidhir, "Inside Saddam's Secret Nuclear Program," *Bulletin of the Atomic Scientists* (September/October 1998).

(The nations in this category include 14 European countries plus Australia, Canada, Japan, and South Korea.)

Two decades after the end of the Cold War, progress has been made in reducing global nuclear weapons arsenals. But there are still more than 23,000 nuclear weapons located at 111 sites in 14 countries, nearly half "active or operationally deployed."[20] It is possible that nuclear materials and important elements of weapons technology have already proliferated to revolutionary or sub-national terrorist groups (see Chapter 2). There is no hard public information on the extent of this proliferation nightmare, but it certainly must be taken seriously.

The longer government arsenals of nuclear weapons and government-sanctioned weapons development programs continue, the more likely both types of proliferation become. Renegade governments, revolutionaries, criminals, and terrorists will have more opportunities to steal the weapons themselves or acquire the critical materials, blueprints, and equipment needed to build them. Just as important, there are also more people with experience in designing, building, maintaining, and testing nuclear weapons with every passing day. In 2004, after two months of investigation, Pakistani officials concluded that Abdul Qadeer (A. Q.) Khan, the father of their nation's nuclear weapons program, was apparently involved in proliferating nuclear weapons technologies to both Iran and Libya.[21] It is simply not reasonable to assume that no other nuclear weapons expert will ever be convinced, coerced, or bought off by rogue governments, revolutionaries, criminals, or terrorists.

Testifying before Congress in January 1992, then CIA Director Robert Gates (later to become Secretary of Defense) estimated that some 1,000 to 2,000 of the nearly 1 million people involved in the former Soviet Union's military nuclear programs were skilled weapons designers.[22] Add to that the numbers in the American, French, British, Chinese, Pakistani, Indian, Israeli, South African, and now North Korean programs, and it is clear that many thousands of people have skills and experience critical to developing nuclear weapons.

Gates emphasized the grave danger posed by the possibility that the economic, social, and political distress in the former Soviet republics could lead to a large-scale emigration of nuclear, chemical, and biological weapons designers. In late 1998, it was reported that Iran had already recruited Russian biological weapons experts for its own germ warfare program.[23] The desperate scientists might also find work with the prosperous criminal underground in their own or other countries, or perhaps with a well-funded revolutionary or terrorist group. By the mid-1990s, the wealthy Japanese Aum Shinrikyo doomsday cult boasted of having thousands of members in Russia.

The vast improvement in Russia's economy during the early years of the twenty-first century have somewhat blunted these fears for the time being. But while Russia and the other the countries of the former Soviet Union have been the greatest concern, none of the countries that have nuclear weapons are immune to this problem. The United States today still has an extremist militia movement whose vicious, racist, anti-government rhetoric leaves plenty of room for more and more brutal acts of violence.[24] Britain and France have both been plagued by terrorist activities for years. In China, rising inequality has created tensions between those parts of the population swept forward by a rapidly growing economy and those left behind, as well as between the repressive old guard regime in Beijing and a rapidly modernizing urban society. These internal tensions, along with the frequently troubled relationship between mainland China and heavily armed Taiwan, could someday help to precipitate an explosion.

India and Pakistan have yet to settle the differences that have led them to war with each other three times since 1947 and nearly triggered a full-scale accidental war in January 1987 (see Chapter 5). Both countries have also had internal ethnic violence and secessionist movements. The region around Peshawar, Pakistan (a staging area for rebels during the Soviet-Afghan war) later became a hotbed of international terrorism. And in Pakistan's Northwest Frontier, radical Islamic fundamentalists such as the Taliban and Al Qaeda create continuing turbulence and instability for the Pakistani government. Israel has long been a favorite target of Arab terrorists and has seen the rise of extremist violence within its own citizenry. North Korea, the most recent declared nuclear weapons state, is a country whose population is in desperate economic straits. It remains in a pro forma state of war with South Korea and continues to pose a serious and vocal threat to peace and stability the region. As recently as May 27, 2009, North Korea warned that if any of its vessels were stopped by an American-led effort to intercept ships believed to be carrying nuclear weapons, "we will immediately respond with a powerful military strike."[25]

Nuclear Power

Nuclear plants are always subject to catastrophic accident, especially as they age. Even when operating normally, the reactors and the waste storage areas that service them are tempting targets for terrorists or hostile militaries bent on mass destruction. Yet, as shown in Table 1.2, this dangerous technology too has spread around the world. As of January 2009, there were 436 nuclear power reactors operating in 30 countries on five continents.

Along with the normal problems of aging equipment, nuclear reactors are subject to embrittlement caused by exposure to high levels of radiation over extended periods of time. Because this can seriously shorten their safe operating life, it is important to consider the age distribution of the world's nuclear power reactors. Table 1.3 gives the number of operating reactors of each indicated age category. By 2009, nearly half the reactors were more than 25 years old. Of these, more than 30 percent were 30 years old or older, and almost 12 percent were more than 35 years old.

Something on the order of 70,000 kilograms of plutonium are produced each year in the normal course of operation of these 400-plus power reactors. The technology required to extract plutonium from spent power plant fuel has been widely accessible for decades. Indeed, roughly 200,000 kilograms of plutonium has already been separated

Table 1.2   Nuclear power reactors in operation (Total = 436 as of May 2009)

Country	Number of Reactors	Country	Number of Reactors
United States	104	Switzerland	5
France	59	Finland	4
Japan	53	Hungary	4
Russia	31	Slovak Republic	4
Republic of Korea	20	Argentina	2
United Kingdom	19	Brazil	2
Canada	18	Bulgaria	2
Germany	17	Mexico	2
India	17	Pakistan	2
Ukraine	15	Romania	2
China and Taiwan	11	South Africa	2
Sweden	10	Armenia	1
Spain	8	Lithuania	1
Belgium	7	Netherlands	1
Czech Republic	6	Slovenia	1

Source: International Atomic Energy Agency, Nuclear Power Plants Information, "Operational Reactors" (http://www.iaea.org/cgi-bin/db.page.pl/pris.oprconst.htm, accessed June 1, 2009)

Table 1.3   Distribution of operating power reactors by age category (May 2009)

Age	Number of Reactors (Total = 436)
0–2	3
3–5	11
6–8	11
9–11	14
12–14	13
15–17	20
18–20	25
21–23	60
24–26	86
27–29	59
30–35	83
36–40	49
41 +	2

Source: International Atomic Energy Agency, Nuclear Power Plants Information: Operational Reactors by Age (http://www.iaea.org/cgi-bin/db.page.pl/pris.reaopag.htm, last updated May 6, 2009).

and is stored in about a dozen countries. *That is about the same amount [of plutonium] used to produce all of the tens of thousands of devastating weapons in the world's nuclear arsenals today.*[26]

Chemical and Biological Weapons

The technology of designing and building chemical weapons of mass destruction for military use is more complex than the technology of manufacturing toxic chemicals. There must be ways of delivering the chemicals to the target as well as shielding their own forces against them. The delivery system must protect the lethal chemicals against heat, vibration, contamination, shock, and other problems of premature release or deterioration. Some two dozen countries are known or believed to have significant chemical weapons arsenals.[27]

Of course, for terrorist use, deadly chemicals require little if any sophisticated "weaponization." Terrorists can inject toxic chemicals into ventilating systems using something as simple as a pesticide sprayer. They can release nerve gas into subway tunnels with the crudest of devices (as the Aum Shinrikyo cult did in Tokyo in 1995), especially if they are willing to die in the process.

Biological weapons pose similar problems. The knowledge and facilities required to breed highly infectious, lethal microorganisms are widespread and are bound up with the ordinary medical and biological technologies of modern life. But the attacker's problems of weaponization and of protecting its own forces and population are much greater. On a weight basis, they have much more killing power than the most deadly chemicals. Inhaling a hundred millionth of a gram (eight thousand spores) of anthrax bacillus causes illness that is almost 100 percent fatal within five days. The lethal inhaled dose of the highly toxic nerve gas sarin is nearly a hundred thousand times larger (one milligram).[28] Beyond this, when living organisms are released into the environment, there is always the chance that they will mutate or otherwise behave in unpredictable ways, particularly if they are novel, genetically engineered life forms.

Fortunately, the difficulties of controlling deadly biologicals have so far made them less attractive to militaries. But they keep trying. Unfortunately, the difficulties of control would not stop an irrational terrorist group or doomsday religious cult from using them.

Missiles

While neither missiles themselves nor missile technology is classified as dangerous, their proliferation makes effective military delivery of weapons of mass destruction much easier. In 1987, seven nations with advanced missile technology (United States, United Kingdom, Canada, France, Germany, Italy, and Japan) enacted the Missile Technology Control Regime (MTCR) to coordinate and control export of missiles and related technology. By the mid-1990s, more than two-dozen other nations had agreed to abide by MTCR restrictions, including Russia and China—both major missile exporters.[29] The agreement seems to have helped reduce the proliferation of long-range ballistic missile technology, but it was too little and too late to prevent the spread of short-to-medium-range missile technology, making many of the world's nations vulnerable to attack from hostile neighbors.

In 1998, North Korea first launched a missile that over flew Japan, dramatically punctuating that point.[30] In early spring 2009, North Korea tested a longer-range missile, claiming it was launching a communications satellite into orbit. No satellite was orbited, but the Taepodong-2 missile again flew over Japan, landing 1,300 miles from its launch site.[31] Then North Korea launched seven shorter-range missiles less than a week after its second nuclear weapons test on May 25, 2009.[32]

Table 1.4 lists 32 countries that are deploying ballistic missiles that they bought or manufactured, as of September 2007. They include all the countries that currently have nuclear weapons or active nuclear weapons programs, and most that once had an active nuclear weapons program. (Of the 20 countries shown in Table 1.1, 16 are included.) Table 1.4 gives the range of the longest-range missile each country has, along with the sources of all of the missiles in their arsenals. It is worth noting that a missile with a substantial range of up to 1,000 km (620 miles) is still considered "short-range," while a missile that can fly up to 3,000 km (1,860 miles)—more than halfway across the United States—is only considered "medium range".

A potential missile builder can learn the theory of missile guidance and control systems—among the most difficult parts of missile development—from publicly available books, magazines, and university courses. The hardware required is also not that difficult to come by. Furthermore, it is easy to purchase handheld receivers to help guide the missile that access the Global Positioning System (GPS) satellite network. This could tell a missile where it is at any point in time with an accuracy of a few yards.[33]

Beyond using an anti-aircraft missile to bring down a civilian airliner, missile attack is likely to have little appeal for subnational terrorist groups. Without the dedicated resources of a government, it would be difficult to secretly build and successfully launch a missile of any substantial range—and there is no need. Successful terrorist actions to date make it clear that they can deliver any weapon they might construct in simpler, cheaper, and more reliable ways: by boat, truck, airplane—or suitcase.

Toxic Chemicals

Toxic chemical technologies are thoroughly integrated with the operation of the chemicals industry on which so much of modern industrial and agricultural output depends. There is little doubt that industry and agriculture can become less dependent on highly toxic chemicals. However, the knowledge and techniques of chemistry required to make toxic chemicals are so widespread and such a normal part of modern life that the potential for producing dangerous chemicals will always be with us.

Table 1.4 Countries with ballistic missiles or missile programs (September 2007)

Status	Country	Sources of All Missiles of All Ranges in Arsenal
Long Range (With Payload Weights of 1,000 kg–3,200 kg)		
Operational	USA	Domestic Production
Operational	Russia	Domestic Production
Operational	United Kingdom	USA
Operational	France	Domestic Production
Operational	China	Domestic Production/Russia
Operational	Israel	Domestic Production/USA/France
Under Development	India	Domestic Production/U.S.S.R./Russia/France/USA
Intermediate Range (With Payload Weights of 700 kg–1,000 kg)		
Operational	India	Domestic Production/U.S.S.R./Russia/France/USA
Under Development	Pakistan	Domestic Production/North Korea/China
Under Development	North Korea	Domestic Production/U.S.S.R.
Medium Range (With Payload Weights of 700 kg–2,150 kg)		
Operational	Pakistan	Domestic Production/North Korea/China
Operational	North Korea	Domestic Production/U.S.S.R.
Operational	Iran	Domestic Production/China/Libya/North Korea/Russia
Operational	Saudi Arabia	China
Short Range (With Payload Weights of 300 kg–1,000 kg)		
Unknown	Afghanistan	U.S.S.R.
Operational	Armenia	Russia
Operational	Bahrain	USA
Operational	Belarus	U.S.S.R.
Operational	Egypt	Domestic Production/U.S.S.R./North Korea
Operational	Georgia	U.S.S.R.
Operational	Greece	USA
Unknown	Iraq	Domestic Production
Operational	Kazakhstan	U.S.S.R.
Operational	Libya	Domestic Production/U.S.S.R.
Operational	Romania	U.S.S.R.
Operational	Slovakia	U.S.S.R.
Operational	South Korea	Domestic Production/USA

Table 1.4 (Continued)

Status	Country	Sources of All Missiles of All Ranges in Arsenal
Operational	Syria	Domestic Production/U.S.S.R./ North Korea
Operational	Taiwan	Domestic Production/Israel
Operational	Turkey	Domestic Production/USA/China
Operational	Turkmenistan	U.S.S.R.
Operational	Ukraine	U.S.S.R.
Operational	United Arab Emirates	U.S.S.R.
Operational	Vietnam	Domestic Production
Operational	Yemen	U.S.S.R./North Korea

Notes:

1) All ranges and payloads given are for the longer-range missiles in that country's arsenal.
2) Missile ranges are classified as follows: Short (up to 1,000 km /621 miles); Medium (1,000–3,000 km/620–1,860 miles); Intermediate (3,000–5,500 km/1,860–3,415 miles); Long or intercontinental (more than 5,500 km/3,415 miles).
3) All countries (except Saudi Arabia) with medium or longer-range missiles also have shorter-range missiles.
4) The longest-range missiles of the United Kingdom and France are submarine based.

Source: Constructed by author from data provided in Arms Control Association, "Worldwide Ballistic Missile Inventories" (http://www.armscontrol.org/factsheets/missiles; accessed June 5, 2009)

THE PROBLEM AND THE SOLUTION

It was not until the twentieth century that our technological development finally put into our own hands the capacity to do catastrophic damage to ourselves and to the web of life on earth. The weapons of mass destruction, which we so diligently created and still so assiduously work to refine, have spread around the planet, exposing us to grave and perhaps terminal danger. On every major continent, nuclear power threatens us with the chronic problem of hazardous waste, the possibility of encouraging the proliferation of nuclear weapons, and the catastrophic potential of accident. The explosion of chemical technologies has injected a vast array of newly created chemicals into the world, many of them toxic. They came so quickly and on such a scale that the natural environment had no chance to find ways of breaking down and recycling them. Now more and more places all over the globe have become contaminated. Always there is the danger that a terrible accident will convert this chronic threat into an acute disaster. It has happened before. It will happen again.

We may be on the verge of generating a similar problem today through the rapid advancement of biotechnologies. Introducing more and more novel, genetically engineered biological organisms into an unprepared natural world may pose an even greater threat. If we want to survive, let alone prosper, sooner or later we must face up to the limits that human imperfection unavoidably puts on our ability to control the products of our technological brilliance. Sooner would be much better than later.

Much of this book is an attempt to look at the many ways human fallibility and the characteristics of modern technology can combine to cause us to threaten ourselves, to make things go not just wrong, but catastrophically wrong. It is a fate we can most surely avoid. By making the right choices, we can direct technology in ways that will continue to make our material lives better. That too is the subject of this book. But to make intelligent decisions and realize the brighter future to which we all aspire, we must first confront the downside of the technologies we choose—no matter how frightening—and not simply revel in the benefits they appear to provide. Otherwise, our natural tendency to overemphasize the advantages and undercount the costs, to "buy now and pay later," will get us into very deep trouble some day.

Because we humans are the most capable species on earth, we are also the most dangerous. Our confidence that we can indefinitely avoid catastrophe while continuing to develop and use powerful and dangerous technologies makes us more likely to fall into the potentially lethal technology trap. We cannot keep winning a technological game of "chicken" with nature. We are not perfect, and we are not perfectible. We are people, not gods. If we do not always remember the fallibility that is an inherent part of our nature, we will eventually do devastating and perhaps terminal harm to ourselves and to the other species that keep us company on this lovely, blue-green planet.

NOTES

1. The pioneering work of Amartya Sen is of special interest here. See, for example, "Starvation and Exchange Entitlement: A General Approach and Its Application to the Great Bengal Famine," *Cambridge Journal of Economics* (1, 1977); and *Poverty and Famines: An Essay on Entitlement and Deprivation* (Oxford: Clarendon Press, 1981).

2. R. L. Sivard, et al., *World Military and Social Expenditures* (Washington, D.C.: World Priorities, 1993), 21. Overall, the twentieth century saw more than 250 wars that killed more than 100 million people (R. L. Sivard, et al., *World Military and Social Expenditures*, Washington, D.C.: World Priorities, 1996), 7.

3. For many other examples of these so-called technological "revenge effects," see Tenner, Edward, *Why Things Bite Back: Technology and the Revenge of Unintended Consequences* (New York: Vintage Books, 1997).

4. A number of interesting scientific analyses of some of the likely ecological consequences of nuclear war have been published since the mid-1970s. These include National Research Council, *Long-Term Worldwide Effects of Multiple Nuclear-Weapons Detonations* (Washington, D.C.: National Academy of Sciences, 1975); P. R. Ehrlich, C. Sagan, D. Kennedy, and W. O. Roberts, *The Cold and the Dark: The World After Nuclear War* (New York: W. W. Norton, 1984); M. A. Harwell, *Nuclear Winter: The Human and Environmental Consequences of Nuclear War* (New York: Springer-Verlag, 1984); and National Research Council, *The Effects on the Atmosphere of a Major Nuclear Exchange* (Washington, D.C.: National Academy of Sciences, 1985).

5. In 1998, sugar cubes containing anthrax bacilli that were confiscated from a German courier in 1917 (apparently intended for germ warfare use in World War I) were found in a museum in Norway. Stored without any special precautions, the anthrax was revived, still viable and virulent, after 80 years. "Norway's 1918 Lump of Sugar Yields Clues on Anthrax in War," *New York Times* (June 25, 1998).

6. Acquired Immune Deficiency Syndrome (AIDS) is caused by a number of "retroviruses," Whereas most viruses interfere directly with the DNA (deoxyribonucleic acid) of an infected cell and then indirectly affect the cell's RNA (ribonucleic acid), retroviruses affect the RNA directly. The first human retrovirus implicated as a cause of AIDS was HIV-1 (human immuno-deficiency virus-1). Since that time HIV-2 and possibly HIV-3 have been dis-covered. The AIDS viruses have a number of peculiarities that make them particularly resistant to attack by the body's immune system, making the search for a vaccine extremely difficult. See, for example, J. L. Marx, "The AIDS Virus—Well-Known But a Mystery," *Science* (April 24, 1987), 390–392; D. D. Edwards, "New Virus, Growth Factor Found for AIDS," *Science News* (June 6, 1987), 356; and D. M. Barnes, "AIDS: Statistics But Few Answers," *Science* (June 12, 1987), 1,423–1,425.

7. In a sense, biologists are today where physicists were in the early part of this century. Their research having led them to the discovery of the basic forces of the physical universe locked within the nucleus of the atom, physicists col-laborated to use that knowledge to create the most destructive weapon in human history, under the urgings of what was thought to be military necessity. Biologists today have uncovered the basic forces of the biological universe locked within the nucleus of the living cell. It is certain that some will again argue for the military necessity of using that knowledge to create horrendous

weapons. Let us hope the biologists have learned something from the experience of the physicists and that those whose work is centered on the physical understanding of life will not allow themselves to be coerced or convinced to create instead the means of mass destruction.

8. F. Barnaby, "Environmental Warfare," *Bulletin of the Atomic Scientists* (May 1976), 37–43.; and A. H. Westing, *Weapons of Mass Destruction and the Environment* (London: Taylor & Francis and the Stockholm International Peace Research Institute, 1977), 49–63.

9. Op. cit. P. R. Ehrlich, et al., 1984.

10. A. H. Westing, *Ecological Consequences of the Second Indochina War* (Stockholm: Almqvist & Wiksell and the Stockholm International Peace Research Institute, 1976).

11. W. G. Van Dorn, B. LeMehaute, and L. Hwang, *Handbook of Explosion-Generated Water Waves*, Tetra Tech Report No. TC-130 (Pasadena, CA: Tetra Tech Inc., October 1968), prepared for the Office of Naval Research (Contract No. N00014-68-C-0227).

12. National Geographic News, "The Deadliest Tsunami in History?" http://news.nationalgeographic.com/news/2004/12/1227_041226_tsunami .html (accessed June 1, 2009).

13. C. A. Larsen, "Ethnic Weapons," *Military Review* (published monthly by the U.S. Army Command and General Staff College, Fort Leavenworth, KS), November 1970, 3–11. An unexplained, oblique reference to the possibility of this type of weapon appeared decades later in an article in the *New York Times*, with the alarming title, "Iranians, Bioweapons in Mind, Lure Needy Ex-Soviet Scientists" (December 8, 1998, by Judith Miller and William J. Broad). Such developments, we are reassured are "years away."

14. It is not widely appreciated that the majority of high-level nuclear waste generated since the beginning of the nuclear era is from weapons production, not from civilian power plants.

15. S. Diamond, "Union Carbide's Inquiry Indicates Errors Led to India Plant Disaster," *New York Times*, March 21, 1985.

16. W. Sullivan, "Health Crisis Could Last Years, Experts Say," *New York Times*, December 7, 1984.

17. John F. Burns, "India Sets Three Nuclear Blasts, Defying Worldwide Ban; Tests Bring Sharp Outcry," *New York Times*, May 12, 1998, and "Indians Conduct 2 More Atom Tests Despite Sanctions," *New York Times*, May 14, 1998. For more details, see David Albright, "The Shots Heard 'Round the World," *Bulletin of the Atomic Scientists*, July/August 1998.

18. John F. Burns, "Pakistan, Answering India, Carries Out Nuclear Tests; Clinton's Appeal Rejected," *New York Times*, May 29, 1998. See also David Albright, "Pakistan: The Other Shoe Drops," *Bulletin of the Atomic Scientists*, July/August 1998.

19. The 2006 underground test is believed to have produced a yield of less than one kiloton, making it a rather small nuclear device. Federation of

American Scientists, "DPRK Nuclear Weapons Program," http://www.fas .org/nuke/guide/dprk/nuke/index.html (accessed June 1, 2009). The second test, on May 25, 2009, was estimated to have a much higher yield, probably from 10 to 20 kilotons, on the order of the Hiroshima bomb. BBC News, "North Korea Conducts Nuclear Test," May 25, 2009, http://news .bbc.co.uk/2/hi/asia-pacific/8066615.stm (accessed June 1, 2009).

20. Robert S. and Kristensen Norris, "Nuclear Notebook: Worldwide Deployments of Nuclear Weapons, 2009," *Bulletin of the Atomic Scientists*, November/December 2009, 86.

21. GlobalSecurity.org, "Weapons of Mass Destruction (WMD): A. Q. Kahn," http://www.globalsecurity.org/wmd/world/pakistan/khan.htm (accessed June 1, 2009).

22. Patricia A Gilmartin, "U.S. Officials Assess Status of Former Soviet Weapons Programs," *Aviation Week and Space Technology*, January 20, 1992, 27.

23. Judith Miller and William J. Broad, "Iranians, Bioweapons in Mind, Lure Needy Ex-Soviet Scientists," *New York Times*, December 8, 1998.

24. According to the Southern Poverty Law Center's Intelligence Project, which seeks to track such groups, "The number of hate groups active in the United States continued to grow in 2008. . . . [The] annual hate group count identified 926 hate groups—a 54 percent increase since 2000. . . ." Southern Poverty Law Center, SPLC Report, Volume 39, Number 1: Spring 2009.

25. Choe Sang-Hun, "North Korea Threatens Military Strikes on South," *New York Times*, May 28, 2009.

26. Theodore B. Taylor, "Worldwide Nuclear Abolition," in Nuclear Age Peace Foundation, *Waging Peace Bulletin*, Summer 1996, 3.

27. These include Bulgaria, Burma (Myanmar), China, Czech Republic, Egypt, France, India, Iran, Iraq, Israel, Libya, North Korea, Pakistan, Romania, Russia, Saudi Arabia, Slovakia, South Africa, South Korea, Syria, Taiwan, United States, Vietnam, and Yugoslavia (Serbia). E. J. Hogendoorn, "A Chemical Weapons Atlas," *Bulletin of the Atomic Scientists*, September/ October 1997.

28. Jonathan B. Tucker, "The Future of Biological Warfare," in W. Thomas Wander and Eric H. Arnett, *The Proliferation of Advanced Weaponry: Technology, Motivations and Responses*, Washington, D.C.: American Association for the Advancement of Science, 1992, 57.

29. W. Thomas Wander, "The Proliferation of Ballistic Missiles: Motives, Technologies and Threats," in W. Thomas Wander and Eric H. Arnett, *The Proliferation of Advanced Weaponry: Technology, Motivations and Responses*, Washington, D.C.: American Association for the Advancement of Science, 1992, 77; see also Leonard S. Spector, Mark G. McDonough, and Evan S. Medeiros, *Tracking Nuclear Proliferation* (Washington, D.C.: Carnegie Endowment for International Peace, 1995), 185–187.

30. Sheryl WuDunn, "North Korea Fires Missile Over Japanese Territory," *New York Times*, September 1, 1998.

31. William J. Broad, "North Korean Missile Launch Was a Failure, Experts Say," *New York Times*, April 5, 2009; and Blaine Harden, "Defiant N. Korea Launches Missile," *Washington Post*, April 5, 2009.

32. Peter Foster, "North Korea Launches Seventh Missile Off East Coast," *Telegraph.co.uk*, May 29, 2009.

33. Azriel Lorber, "Tactical Missiles: Anyone Can Play," *Bulletin of the Atomic Scientists*, March 1992, 39–40.

2

Dangerous Technologies and the Terrorist Threat

Eighteen minutes after noon on February 26, 1993, a blast shook the twin 110-story towers of New York's World Trade Center with the force of a small earthquake. A van loaded with 1,200 pounds of powerful explosive had blown up in the underground garage. Walls and floors collapsed, fires began to burn, and smoke poured into hallways and stairwells, darkened by the loss of power. Dozens were trapped between floors for hours in many of the Center's 250 elevators. Some 40,000 people in the hundreds of offices and miles of corridors of Manhattan's largest building complex had to find their way out amidst the smoke, darkness, and confusion. It took some of them most of the day. When it was all over, six were dead, more than 1,000 were injured, and property damage was estimated at half a billion dollars.[1]

The terrorists who attacked the World Trade Center that day had apparently hoped to destabilize one of the towers, causing it to fall into the other and bring them both down. That bombing was the worst terrorist incident on American soil to that point. They had caused a lot of damage, injured a lot of people, but when it was all over, they did not succeed. The twin towers still stood tall and strong—at least, as it turned out, for the time being.

On April 19, 1995, a little more than two years after the World Trade Center attack, a rented truck packed with more than 4,000 pounds of explosives made from widely available fertilizers, chemicals, and fuel sat parked by the Alfred P. Murrah Federal building in Oklahoma City. Just after 9:00 AM, when the building's workers were at their jobs and the second-floor day care center was filled with young children, the truck blew up with a deafening roar. Walls, ceilings, and much of the building's north face came down in an avalanche of concrete, steel, and glass. The blast left a crater 20 feet wide and eight feet deep, overturned cars, damaged six nearby buildings, and set dozens of fires. Nearly 170 people were killed, including many of the children in the day care center, and hundreds more were injured. A building that was tiny by comparison to the World Trade Center, in a city a small fraction of the size of New York, had sustained a homegrown domestic terrorist attack that took nearly 30 times as many lives.[2] Terrorism had come to America's heartland.

Then, on the morning of September 11, 2001—eight and a half years after the initial truck bombing of the World Trade Center—19 al Qaeda operatives hijacked four planes, using weapons as primitive as box cutters and knives. In a coordinated, sophisticated terrorist operation, five of the men hijacked American Airlines flight 11, departing Boston at 7:45 AM; five more hijacked United Airlines flight 175, departing Boston a few minutes before 8:00 AM; another four hijacked United Airlines flight 93 out of Newark at 8:01 AM; and the remaining five took over American Airlines flight 77, which left Washington Dulles Airport at 8:10 AM. At about a quarter to nine, the terrorists who had commandeered American 11 deliberately flew it into the North Tower of the World Trade Center, triggering an enormous explosion and fire. About twenty minutes later, with every available news camera in New York already focused on the unfolding drama, the hijackers of United Airlines 175 flew that airliner into the South Tower, triggering another huge explosion and fire. Within 90 minutes, to the shock and horror of a world watching what was already a major disaster, both towers had collapsed into an enormous pile of rubble.[3] More than 2,700 people were dead or dying.[4] The suicide terrorists of al Qaeda had finally succeeded in doing what others had tried and failed to do nearly a decade earlier.

At 9:40 AM, the hijackers of American 77 flew the plane directly into the Pentagon, the symbol of American military power, and 189 more people died. The passengers and crew of the final hijacked aircraft, United 93, fought with the hijackers, and amid the struggle the plane crashed into the Pennsylvania woods.[5] There is still controversy over its intended target, a matter to which we will later return.

In the end, the 9/11 attacks left close to 3,000 people dead, with direct damage to property estimated at $10 billion to $11 billion[6]— 20 times the property damage and 500 times the death toll of the 1993 World Trade Center attack. It was by far the most deadly and the most costly international terrorist attack the world had ever seen.

As terrible as these attacks were, they are dwarfed by the mayhem that could be caused by a successful terrorist assault on a nuclear power plant, toxic chemical manufacturing facility, or radioactive waste storage site—or the magnitude of disaster that terrorists armed with weapons of mass destruction could unleash. Even a crude, inefficient, homemade nuclear weapon would have turned the World Trade Center into rubble so quickly that no one would have had the chance to escape; tens of thousands would have died. A more efficient weapon could have leveled much of Oklahoma City.

THE NATURE OF TERRORISM

Not every form of violent, destructive, and anti-social activity is terrorism. Nor is terrorism defined by the ultimate goals terrorists seek to achieve. Calling groups that use violence "terrorists" when we do not like their objectives and "freedom fighters" when we do is a political game. It will not help us understand what terrorism is, judge how likely terrorists are to use dangerous technologies, or figure out what can be done about it. Instead, we need a working definition that is more than just propaganda or opinion.

Terrorism can be defined by its tactics and strategy: it is violence or the threat of violence carried out against innocent victims with the express purpose of creating fear and alarm in order to influence public opinion or the behavior of the public, the private sector, or the government. When an armed gang shoots bank guards to steal money, that is a violent crime, not an act of terrorism. The violence is perpetrated to prevent the guards from stopping the theft, not to frighten the wider population. When a gang randomly plants bombs on city buses, however, they are not trying to stop the passengers from interfering with them; they are trying to frighten people. Their acts are intended to have effects that reach well beyond the immediate damage they are causing or threatening to cause. Whether their objective is to force the government to release political prisoners or to extort a ransom, they are terrorists because they are trying to terrorize.[7]

Unlike other criminals, terrorists usually try to draw attention to themselves, often claiming "credit" for the acts they have committed. In many ways, terrorism is a perverse form of theater in which terrorists play to an audience whose actions—and perhaps, opinions—they are trying to influence.[8] When they hijack an aircraft or kidnap a group of tourists or business executives, they may be playing to an audience of corporate managers who can assemble a ransom, government officials who can order their imprisoned comrades released, or whoever else has the power to meet their demands. But they are also playing to the public, whose mere presence as well as opinions and actions can put pressure on those in power to do what the terrorists want done. Those actually taken hostage cannot meet the terrorist demands, any more than can those maimed when a pub is bombed or those killed by a murderous spray of gunfire in a hotel lobby. Nor are they in any position to apply pressure to the people who have that power. They just happened to be in the wrong place at the wrong time. Innocent victims, they have become unwitting players caught up in a real-life drama, the cannon fodder of terrorism.

Terrorists are trying to make the public feel vulnerable, unsafe, and helpless. In some cases, choosing victims at random is the best way to accomplish this. If there is no clear pattern as to which particular bus is blown up, which airliner hijacked, which building bombed, there is no obvious way to avoid becoming a victim. That is very frightening. If the terrorist objective is more targeted, on the other hand, choosing victims randomly but within broadly defined categories may be more effective. The mercury poisoning of Israeli oranges in Europe in the late 1970s was targeted randomly but only at consumers of Israeli produce. It was intended to damage Israel's economy by creating fear that their agricultural exports were unsafe.[9] Economic damage was also the goal of the terrorist who poisoned some of the Johnson & Johnson Company's pain-killing Tylenol capsules with lethal cyanide in the early 1980s.[10] The targets of the Oklahoma City bombing were also neither purely random nor very specific, but were chosen to intimidate federal employees and users of federal services to express broad ideological antipathy toward the government.

In sum, acts intended to instill fear in the public, committed against more or less randomly chosen victims not themselves able to meet the attackers' demands, define terrorism and set it apart from many other forms of violence. Bombing the barracks of an occupying military force is an act of war, violent and murderous, but not an act of terrorism. It attacks those who are directly involved in the activity the attackers are trying to oppose, not randomly chosen innocent victims. The act of a habitual sex offender in kidnapping, raping, and murdering a more or less randomly chosen innocent victim is a vicious and brutal crime, but it is also not terrorism. Though it may well instill fear in the public, it is not done for that purpose, and it is not done to influence public opinion or behavior. Suicide bombing a city marketplace to precipitate a change in government policy is an act of terrorism. The more or less randomly chosen victims cannot directly change government behavior, but the indiscriminate slaughter is intended to shock and frighten people into demanding that the government change direction by convincing them that they will be in danger until those policies change. Whether or not the bombing achieves that objective, the act itself is still an act of terrorism.

It is important to emphasize that there is nothing in the definition of terrorism that prejudges the legitimacy or desirability of the terrorists' ultimate goals. Whether a group is trying to overthrow a legitimate democratic government and establish a rigid dictatorship, create a homeland for a long disenfranchised people, trigger a race war, or get more food distributed to malnourished poor people, if the group uses terrorist means, it is a terrorist group.

Terrorism may be despicable, but it is not necessarily irrational. There are a variety of reasons why subnational groups with clearly political goals sometimes choose terrorist tactics to undermine support for the government and/or its policies. They may believe that this is an effective way of convincing the public that the government does not deserve their support because it cannot keep them safe. Or they may believe that provoking widespread and repressive counterterrorist measures will turn the public against the government by exposing just how brutal and overbearing it can be. As paradoxical as it may seem, terrorist groups clearly believe that the end result of their terrible random acts of violence will be an uprising of the public against the government and increased support for the group's political agenda.

Since international terrorists attack only foreigners and their property and usually claim to be the avenging arm of their oppressed brothers and sisters, it is easier to understand why they might believe that their brutal actions will build public support for their cause at home. They may also see international terrorism as the only way to shock the world into paying attention to the plight of their people. For subnational groups, it is certainly true that terrorism is the "weapon of the weak." A powerful and influential group would not need to resort to such desperate and horrible tactics to make itself heard.

There is a tendency to think of terrorists as either small, disconnected groups of half-crazy extremists or expert paramilitary cadres bound tightly together in grand international conspiracies. In fact, the reality most often lies in between. It is true that terrorist groups do cooperate across ideological and political boundaries, sometimes even "subcontracting" with each other or carrying out joint attacks.

But these coalitions are typically loose and transitory. Even al Qaeda is not really the highly disciplined, tightly controlled, unified master terrorist organization it is often made to seem. "Al Qaeda pursues its objectives through a network of cells, associate terrorist and guerilla groups and other affiliated organizations, and shares expertise, transfers resources, discusses strategy and even conducts joint operations with some or all of them. . . . [I]t developed a decentralized, regional structure. . . . [Its] worldwide nodes have no formal structure and no hierarchy. . . . Nor do the regional nodes have a fixed abode."[11] In some ways, that is its strength; its decentralization and amorphousness makes it more difficult to disrupt or destroy.

Governments and Terrorism

Terrorism is not only a tactic of subnational groups. Governments can, and all too often do, carry out terrorist acts. In fact, the term "terrorist" appears to have first been applied to the activities of a government, the French government after the Revolution.[12] The Nazi Gestapo, Iranian Savak, Iraqi Mukhabarat, and many other "secret police" organizations in many other countries have deliberately terrorized the population to suppress opposition and force submission to the edicts of brutal governments. Because of the resources at their command, when governments engage in terrorism, their actions are often far more terrifying than the acts of subnational terrorists.

Governments have sometimes carried out official campaigns of terrorism. The Ethiopian government launched what it called the "red terror" in reaction to a revolutionary group's "white terror" campaign in the late 1970s. Within two months, more than 1,000 people were killed, many of them teenagers. Their dead bodies were displayed in public squares with signs hung on them saying "The red terror must crush the white terror."[13] In 1998, testimony before South Africa's Truth and Reconciliation Commission revealed an apartheid-era campaign of chemical and biological attacks intended to murder political opponents.[14]

State-Sponsored Terrorism

Governments sometimes directly aid subnational groups that stage attacks against the homelands or interests of opposing governments, groups that are "terrorist" by any reasonable definition of the term. Governments have provided safe havens, intelligence, weapons, and even training. On the specious theory that "the enemy of my enemy is my friend," the United States, Saudi Arabia, and others flooded Afghani revolutionary forces with weapons and money after the Soviet military intervened to support the Afghan government in 1979. Much went to groups in Peshawar, the capital of Pakistan's Northwest Province bordering Afghanistan. Long a violent area, the money and guns helped Peshawar descend deeper into lawlessness. According to a high-ranking officer, the Pakistani military was directly involved in training something like 25,000 foreign volunteers to fight with the Afghan guerillas. After the Soviets withdrew from Afghanistan in 1989, a large number of the surviving foreign volunteers stayed in and around Peshawar, working with organizations that have been accused of being fronts for international terrorist groups.[15] Ramzi Ahmed Yousef, convicted of both the 1993 bombing of New York's World Trade Center and a 1995 plot to blow up a dozen American airliners in East Asia, used Peshawar as a base.[16] Several of the eight men the FBI accused in 1993 of plotting to blow up car bombs at U.N. headquarters, the Lincoln and Holland Tunnels, and several other sites in New York City were involved in the Afghan War.[17] So too was Osama bin Laden, the Saudi who was to apply the knowledge of management and organization (and money) gained in his wealthy family's business to deadly purpose as the Emir-General of al Qaeda.[18] In the words of one senior Pakistani official:

> Don't forget, the whole world opened its arms to these people. They were welcomed here as fighters for a noble cause, with no questions asked . . . nobody thought to ask them: when the Afghan Jihad is over, are you going to get involved in terrorism in Pakistan? Are you going to bomb the World Trade Center?[19]

Nuclear deterrence, a mainstay of the official security policy of the nuclear weapons states, is itself a form of international terrorism. After all, nuclear deterrence does not so much threaten to annihilate the leaders of opposing governments—those with the power to make decisions on war or peace—as it holds hostage and threatens to kill the ordinary people of the opposing nation if their government decides to attack. Even in democracies, the general public under threat is not in a position to control the decision of their government to launch a nuclear attack. No referendum has ever been planned for "button-pushing" day. Furthermore, the underlying objective of threatening nuclear attack is precisely to create such widespread fear that the opposing government will feel enormous pressure to avoid behaviors that would result in the threat being carried out. This is terrorism, plain and simple. In fact, during the Cold War the threat of "mutually assured destruction" was officially called a "balance of terror." And a balance of terror is still terror.

Though it does not in any way legitimize or excuse the use of terrorist tactics by subnational groups, the fact is that even democratic governments have provided them something of a model for this horrendous type of behavior.

THE TERRORIST THREAT OF MASS DESTRUCTION

As Bruce Hoffman points out, "The enormity and sheer scale of the simultaneous suicide attacks on September 11 eclipsed anything previously seen in terrorism. . . . [D]uring the entirety of the twentieth century no more than 14 terrorist operations killed more than 100 persons at any one time . . . until the attacks on the World Trade Center and the Pentagon, no single terrorist operation had ever killed more than 500 persons . . . "[20] Yet terrorists have still not committed violence on the scale that could result from a successful attack on a nuclear power plant, toxic chemical manufacturing facility, or hazardous waste dump. They have not yet used a homemade, store-bought, or stolen nuclear weapon. They have not yet poisoned the water supply of a city or the air supply of a major building with deadly chemicals or virulent bacteria.

Why not? If it is because they do not have and cannot develop the capability to use dangerous technologies as a weapon or a target, then we can relegate these frightening scenarios to the realm of science fiction and breathe a collective sigh of relief. But if instead the capability to do such nightmarish damage is within their reach, it is important to know what is holding them back. Is it just a matter of time before this modern-day horror becomes real?

Because their actions seem so immoral, abhorrent, and repulsive, we usually assume that terrorists will do whatever harm they are capable of doing. But though their methods are similar, not all terrorists are alike. Some may actively seek the capability for mayhem that dangerous technologies provide, while others have no desire to do that much damage. It would be very useful to know which is which. At best, that might help us formulate more effective strategies for preventing a terrorist-induced catastrophe. At least, we would know what kinds of groups need to be watched most closely.

Then there is the biggest question of all: whatever the reason terrorists have not yet committed such atrocities, is there any reason to believe this restraint will continue?

CAN TERRORISTS "GO NUCLEAR"?

Despite high priority and lavish government funding, it took years for a collection of the most brilliant scientific minds of the twentieth century to develop the first nuclear weapon. Potent nerve gas weapons emerged from technically advanced laboratories run by teams of highly trained chemists. A great deal of engineering and scientific effort has gone into designing nuclear power plants. They are protected by layers of backup and control systems intended to make catastrophic failure very unlikely, whether by accident or sabotage. Are terrorists really sophisticated enough to get their hands on and successfully use dangerous technologies as weapons or as targets of their attacks?

The image of the terrorist as a demented fanatic who is only capable of stashing a suitcase full of dynamite sticks wired to a crude timing device in some forgotten corner of a building went out the window long before September 11. Such crude forms of terrorism can still be very effective, but decades ago terrorists and criminals using terrorist tactics showed themselves capable of much greater tactical and technological sophistication.

In August 1980, a desk-sized box was delivered to the executive offices of Harvey's Casino in Stateline, Nevada. An extortion note sent to the management warned that it contained a bomb that would explode if any attempt were made to move it. Not knowing whether the threat was real, casino managers called experts from the FBI, the Army bomb disposal team, and the U.S. Department of Energy (the agency in charge of nuclear weapons research and manufacture). X-rays revealed that the box contained 1,100 pounds of explosives. The experts were struck by the highly sophisticated design, but nevertheless believed they could safely disarm it. They were wrong. It exploded, causing $12 million worth of damage to the Casino.[21]

If there were any lingering doubt about the possibility that technologically advanced subnational terrorist groups could arise, it was dispelled in the mid-1990s by the emergence of the Japanese doomsday cult, Aum Shinrikyo. Nearly a dozen of the sect's top leaders were educated in science and engineering at top Japanese universities, as were some other members. When police arrested cult members and accused them of releasing sarin nerve gas in March 1995 on the Tokyo subways, they found hidden laboratories at the cult's compounds for manufacturing the gas. They charged that Aum also had facilities capable of producing biological warfare agents. Furthermore, Japanese police reportedly suspected that the purpose of the 1993 visit of a high cult official to Australia was to obtain uranium to be used in building nuclear weapons.[22] Finally, the level of sophistication shown in planning and coordinating the execution of the September 11 attacks, and the amount of death and destruction those attacks caused, took terrorism to a whole new level.

There have been periodic reports of attempts by al Qaeda to get their hands on much more destructive weapons than the airliners they turned into bombs on that terrible day in 2001. The final report of the 9/11 Commission states, "Al Qaeda has tried to acquire or make nuclear weapons for at least ten years ... and continues to pursue its strategic goal of obtaining a nuclear capability."[23] As long ago as 2002, the CIA reported finding "rudimentary diagrams of nuclear weapons inside a suspected Al Qaeda safe house in Kabul."[24]

Over the past 60 years, thousands of people have been trained in the design and manufacture of nuclear weapons in the United States, the former Soviet Union, Britain, France, China, India, Pakistan, Israel, and now North Korea. These are people of widely differing political, ideological, and religious views, personalities, and life circumstances. Many of Russia's nuclear scientists were living in such economic deprivation during the 1990s that the United States allocated $30 million in 1999 to help create non-military jobs for them in the hope of discouraging them from selling their expertise to rogue nations or terrorists.[25] Can we really be sure that no terrorist gang, political group, or religious cult will ever be able to recruit, coerce, or buy off any of these experienced nuclear weaponeers?

When Libya put an end to its nuclear weapons program and allowed outside investigators into the country in 2003, they found blueprints for a 1960s vintage Chinese nuclear weapon that they had apparently bought from the international smuggling network operated by A. Q. Khan, the Pakistani nuclear scientist who had directed his nation's nuclear program. The following year, investigators in Switzerland were ultimately able to decode encrypted files on computers they seized that reportedly belonged to the Khan network, containing the digitized designs for smaller, even more sophisticated nuclear weapons than that found in Libya.[26] Once the designs are digitized, they can easily be transmitted to anyone, anywhere in the world.

Unfortunately, the degree of technical sophistication required to acquire or use dangerous technologies as weapons is actually much lower than many people think. Poison gas can be made with the chemicals most of us have stored around the house. As long ago as 1977, a British military research laboratory was openly advertising the sale of infectious organisms at bargain-basement prices, including three strains of *Escherichia coli*,[27] the bacterium responsible for a mysterious epidemic of food poisoning that shut down the entire Japanese school system in summer 1996. More than 9,400 people were sickened and 10 died.[28]

In March 1995, four members of the right-wing Minnesota Patriots Council were convicted in federal court of conspiracy to use ricin, a deadly biological toxin, to kill federal agents. They manufactured enough to kill 1,400 people using a mail-order manual.[29] Two months later, a member of the American white supremacist Aryan Nations was arrested for (and subsequently pled guilty to) making another mail order purchase—three vials of frozen bubonic plague bacteria—obtained using false credentials from the food-testing laboratory where he worked.[30] Shortly after the September 11, 2001 attacks, an already traumatized nation was subject to another round of terror. Envelopes containing anthrax spores were received through the U.S. mail. Five died, and 17 more became ill in the worst biological attacks in American history.[31]

Designing a Terrorist Nuclear Weapon

The Khan network's activities aside, the "secret" of designing nuclear weapons has been accessible to anyone moderately well-trained in the physical sciences or engineering for decades. More than 30 years ago, the Public TV's "NOVA" science series recruited a 20-year old M.I.T. chemistry student to design a workable atomic bomb. He was required to work alone, without expert assistance, and use only publicly available information. He began by simply looking up references in the college science library. In the students words:

> the hard data for how big the plutonium core should be and how much TNT I needed to use I got from Los Alamos reference books [purchased from the National Technical Information Service in Washington for

about $5 a piece] and also other reference books I checked out of the library.

I was pretty surprised about how easy it is to design a bomb. When I was working on my design, I kept thinking there's got to be more to it than this, but actually there isn't."[32]

Only five weeks later, the student's fully documented and detailed report was given to a Swedish nuclear weapons expert for evaluation. The verdict: a fair chance that a bomb built to this design would go off, though the explosion would probably be no more powerful than 1,000 tons of TNT, more likely less than 100.

The design was crude and unreliable, the yield unpredictable and small by nuclear standards—it would not be an acceptable military weapon. But none of these deficiencies is much of a problem for terrorists who are accustomed to unreliable bombs with unpredictable yields. An explosion equivalent to even 50 tons of TNT would be gigantic by terrorist standards; that is, 25 times as powerful as the explosives used in the 1995 Oklahoma City and 1993 World Trade Center bombings. Imagine what would have happened had those blasts been 25 times as powerful. Then add the death and destruction that would have been caused by the enormous release of heat and radiation from this crude nuclear weapon—designed by one undergraduate student in less than two months.

One year later, a senior at Princeton University duplicated this design feat, and then some. Working from publicly available sources and taking more time, he designed a 125-pound device about the size of a beach ball, which he estimated would explode with the force of about 5,000 tons of TNT. A specialist in nuclear explosives engineering reviewed the student's 34-page term paper, declaring that the bomb design was "pretty much guaranteed to work."[33]

In April 1979, FBI Director William Webster said sufficient information was available in public libraries to design a nuclear weapon small enough to be carried on a terrorist's back.[34] A few months later, *The Progressive* magazine published an article called "The H-Bomb Secret," after a long legal battle. The court refused to enjoin its publication because all of the information it contained had been available for years in public sources. This included a report giving precise specifications for the hydrogen bomb trigger mechanism and other important design details. It had been on the public shelves in Los Alamos since 1975.[35] The Department of Energy argued that they should never have declassified the report and made it available. Why did they? It was simply human error.[36]

Building a "Homemade" Nuclear Bomb

Skeptics argue that designing a weapon on paper may not be all that difficult, but actually building a nuclear bomb would require large teams of people with advanced skills and access to materials and equipment that is expensive and difficult to come by. Ted Taylor, a noted physicist and Los Alamos nuclear weapons designer credited with designing the most efficient A-bomb ever, disagreed. In his view:

> "Under conceivable circumstances, a few persons, possibly even one person working alone, who possessed about ten kilograms of pluto-nium oxide and a substantial amount of chemical high explosive could, within several weeks, design and build a crude fission bomb ... that would have an excellent chance of exploding, and would probably explode with the power of at least 100 tons of chemical high explosive. *This could be done using materials and equipment that could be purchased at a hardware store and from commercial suppliers of scientific equipment for student laboratories.*
>
> "The key person or persons would have to be reasonably inventive and adept at using laboratory equipment and tools of about the same complexity as those used by students in chemistry and physics labora-tories and machine shops."[37] (emphasis added)

The M.I.T. undergraduate who designed the workable A-bomb for NOVA estimated that if he had the plutonium, he could actually build the bomb from scratch in a year or less, with the help of three to four people and no more than $30,000 for supplies purchased from ordi-nary commercial sources. The finished product would be about as big as a desk and weigh 550 to 1,000 pounds.[38]. In other words, it would be roughly the size and weight of the box terrorists delivered to Harvey's Casino in Nevada five years later.

Are the Necessary Nuclear Materials Available?

About 1,000 metric tons of plutonium are contained in stored spent fuel from nuclear power plants, which is four times as much plutonium as has been used in making all of the world's nuclear weapons.[39] For a long time, the public was told that nuclear weapons cannot be built from this "reactor grade" plutonium without an expensive and techni-cally complex refinement process: it is much too heavily contaminated with plutonium-240 to be usable for a weapon. Not only is that untrue, but it has definitely been known to be untrue for 50 years.

In 1962, the U.S. government assembled a nuclear bomb from the low-grade, contaminated plutonium typically produced by civilian nuclear power plants, and tested it in Nevada. It blew up, producing "a nuclear yield."[40] Fourteen years later, a study done at the Lawrence Livermore nuclear weapons lab came to the conclusion "that the distinction between military and civilian plutonium was essentially false—that *even relatively simple designs using any grade of plutonium* could produce 'effective, highly powerful' weapons with an explosive yield equivalent to between 1,000 and 20,000 tons of TNT."[41]

Natural uranium contains only 0.7 percent U-235, much too low a concentration to sustain a chain reaction in weapons. Although some nuclear power reactors use very low enriched uranium, most use uranium enriched to higher levels of U-235. Uranium enriched to more than 90 percent is an excellent nuclear explosive and is typically military weapons-grade. According to Ted Taylor, however, "It is probable that some kind of fission explosive with a yield equivalent to at least a few tens of tons of high explosive could be made with metallic uranium at *any enrichment level significantly above 10 percent*."[42] Princeton nuclear physicists Frank von Hippel and Alexander Glaser claim, "any uranium with a uranium 235 fraction above 20 percent— must be considered 'direct use' material—that is, usable in nuclear weapons."[43] The small discrepancy between these two estimates of "weapons grade" enrichment is more apparent than real—Glaser and von Hippel do not argue a nuclear weapons cannot be made out of uranium enriched to less than 20 percent, merely that the amount required "becomes too large to fit in a reasonably sized device."[44] What constitutes reasonable size depends on the way the weapon is to be used, by whom, and for what purpose.

Uranium enriched to 20 percent or more, indisputably sufficient for nuclear weapons purposes, is commonly called "highly enriched uranium" (HEU). There are civilian nuclear power reactors fueled by 90 percent enriched uranium, but most use uranium that is less than 10 percent enriched. Military nuclear power reactors may use more enriched uranium. For example, most of the Russian Navy uses uranium fuel enriched to 20 percent to 45 percent.[45] It is also estimated that " More than 50 tons of HEU are in civilian use, dispersed around the globe to support about 140 reactors employed to conduct scientific or industrial research or to produce radioactive isotopes for medical purposes."[46] Thus, the uranium used as fuel in many, but not all nuclear power and research reactors could be used to make bombs without further enrichment. Lower enriched reactor fuel can still be used, but it would have to be processed to raise its concentration of U-235.

In 1996, *Time* magazine reported that 17 scientists at the Los Alamos nuclear weapons laboratory were given the assignment of trying to design and build terrorist-type nuclear weapons using "technology found on the shelves of Radio Shack and the type of nuclear fuel sold on the black market." By early 1996, they had successfully assembled more than a dozen "homemade" nuclear bombs.[47]

If terrorists were willing to settle for a device that dispersed deadly radiation without a nuclear blast, they would have a much wider variety of designs and nuclear materials from which to choose. A so-called "dirty bomb" made by packing radioactive material around a conventional explosive is one possibility. It is unlikely to kill large numbers of people quickly, most likely causing acute illness mainly to those who are directly exposed within at most a few city blocks of the explosion.[48] Radiological attacks in which large numbers of people are caused to inhale a radioactive aerosol released in a restricted space, sprayed with contaminated liquid, or caused to eat or drink radioactive material— say, used to contaminate food at a processing plant—could well prove much more lethal.[49] The skills and equipment needed to build a dirty bomb or other dispersion device are also much simpler and the materials more available than those required for building a nuclear weapon. Biological or chemical weapons dispersal devices would be even simpler. By one estimate, terrorist biological weapons might be developed at a cost of $100,000 or less, "require five biologists, and take just a few weeks, using equipment that is readily available almost anywhere in the world."[50]

Getting Access to Nuclear Explosives

The problems of record keeping and protection of inventories that are detailed in Chapter 3 raise the possibility that it may not be as difficult as one might hope for terrorists to get their hands on "any grade of plutonium" or the sufficiently enriched uranium needed to build a crude nuclear weapon.

There is little doubt that some black market in nuclear materials now exists. In 1996, CIA Director John Deutch put it this way, "The chilling reality is that nuclear materials and technologies are more accessible now than at any other time in history. . . . "[51] In 2005, the International Atomic Energy Agency (IAEA) reported documenting 18 seizures of stolen weapons-grade uranium or plutonium.[52] In April 2006, Russian police arrested a number of "co-conspirators" involved in stealing 49 pounds of low-enriched uranium from the Elektrostal nuclear fuel fabrication plant. Those arrested included a foreman at the facility, which also processes large amount of HEU.[53] In October 2008, IAEA Director Mohamed ElBaradei pointed out that there had been nearly 250 thefts of "nuclear or other radioactive material" over the one-year period from mid-2007 to mid-2008. He went on to say, "Equally troubling is the fact that much of this material is not subsequently recovered".[54]

Once "nuclear materials and technologies" are acquired, they can be smuggled across international borders. In 1985, a West German businessman was convicted of illegally shipping an entire $6 million nuclear processing plant to Pakistan—along with a team of West German engineers to supervise its construction![55]

Prior to 9/11, borders were so porous that it would have been relatively easy to bring nuclear contraband into the United States. In early 1996, *Time* reported, "U.S. intelligence officials admit that a terrorist would have no more difficulty slipping a nuclear device into the U.S. than a drug trafficker has in bringing in bulk loads of cocaine."[56] Decades earlier, special forces teams smuggled simulated nuclear bombs into the United States dozens of times to see if it could be done. They carried dummy weapons across the borders in trucks, small planes, and boats. None of them were ever intercepted.[57]

Border security is much tighter today. Nevertheless, in 2005 undercover investigators from the U.S. General Accountability Office (GAO) successfully smuggled enough radioactive cesium 137 to make two "dirty bombs" across northern and southern borders into the United States. They used counterfeit documents and were allowed through border crossings in Texas and Washington State despite setting off radiation alarms.[58] In 2007, officials from the GAO reported in congressional testimony that their investigators "found various security vulnerabilities along the U.S.-Canadian border that terrorists could exploit to bring an unconventional weapon into this country." The officials testified, "In three of the four locations on the U.S.-Canada border, investigators carried a duffel bag across the border to simulate the cross-border movement of radioactive materials or other contraband."[59]

The *Ottawa Citizen* reported that in one case, officers from U.S. Customs and Border Protection could not even find the investigators after they received reports of "suspicious activity."[60] Earlier that same year,

"GAO auditors found that [radiation] sensors tested by the Homeland Security Department had detection rates as low as 17 percent and never more than about 50 percent," despite the Department of Homeland Security's (DHS) claims that the new detectors, part of their proposed $1.2 billion radiation detector program, "would have a 95 percent detection rate for highly enriched uranium."[61] In a related (not necessarily nuclear) story, during late 2008/early 2009, GAO investigators "successfully smuggled bomb-making materials into ten high-security federal buildings, constructed bombs and walked around the buildings undetected . . ."[62]

According to a *New York Times* report in early 2007, DHS was also "trying to find ways to stop a plot that would use a weapon built within the United States." Reacting to the proposed DHS detection and response system, the newspaper reported that physicist and nuclear terrorism expert Benn Tannenbaum "said the system would never create anything close to an impenetrable barrier, particularly for a nuclear bomb, since the required ingredients have low levels of radioactivity and can easily be shielded."[63]

In light of all this, and the volume of illegal drugs flowing across U.S. borders, it is difficult to be optimistic that the possibilities of smuggling dangerous materials or technologies into the United States are a thing of the past.

Some Incidents

There is a history of nuclear threats and related plots by terrorists and criminals. Examples include the following:

- In 1978, the FBI arrested two men after they tried to recruit an undercover agent to take part in a plot to steal an American nuclear submarine. The men showed him their plan to use a gang of 12 to murder the crew of the nuclear submarine USS Trepang, sail it from its dock in New London, Connecticut to the mid-Atlantic, and turn it over to a buyer they did not identify. The plan included the option of using one of the ship's nuclear missiles to destroy an East Coast city as a diversion to help with the getaway.[64]
- In December 1994, a Czech scientist named Jaroslav Vagner and two colleagues were arrested for nuclear smuggling when Prague police found nearly six pounds of weapons-grade uranium in the back seat of his car. Eighteen months later, interviews with Czech police and newly released documents revealed that allies of the three conspirators had threatened to explode a nuclear weapon at an unspecified Prague hotel unless the prisoners were released. According to the Czech police detective handling

the investigation, "It is possible they have the nuclear material to do it. We found out they were planning to bring out [of Russia] 40 kilos of uranium within several days and . . . one ton within several years. . . . "[65]

- In December 2004, Australian Federal Police arrested three men, charging that the men were part of a plot to attack the Lucas Heights research and medical nuclear reactor complex, with the support of a larger network of Australian terrorists. Nearly a year later, police arrested about 20 others suspected of aiding the plot to attack the nuclear facility and other Australian targets. A 2004 police report indicated that the terror network had bought the same types of chemicals that were later used in the terrorist subway and bus attacks in London in July 2005, and that they had bomb-making information, extremist literature, and instructional videos apparently linked to al Qaeda.[66]

- On the night of November 7, 2007, four intruders breached security, penetrating the perimeter fence at the Pelindaba nuclear plant in South Africa. They entered the control room and shot one of the two workers there. Another group simultaneously breached the plant's fence at a different location, but encountered security forces that drove them off. The plant at Pelindaba holds hundreds of kilograms of HEU removed from South Africa's nuclear weapons when the country decided to dismantle its nuclear arsenal just before the end of apartheid. Both groups of intruders escaped, apparently without taking anything of value. According to Matthew Bunn, "These people cut through a 10,000 volt security fence. They disabled sophisticated electronic intrusion detectors. They went straight to the emergency control center. It does suggest that they had someone inside who was going to help them make sure that the security alarms didn't go off and that security forces didn't respond in time."[67]

In an attempt to deal with the threat of nuclear terrorism, the United States put together the multi-agency Nuclear Emergency Search Team (NEST) in 1975. NEST was set up to evaluate criminal or terrorist nuclear threats and, if necessary, to conduct extensive high-tech bomb searches to find and disable nuclear devices. By early 1996, the team had evaluated 110 threats (including some nuclear emergencies not involving weapons) and mobilized to search or take other action 30 times. All but one of the threats was reportedly a hoax.[68] What are the chances that this specialized team could actually find a real terrorist nuclear device? In 1980, NEST's assistant director put it this way, "If you can cut it down to a few blocks, we have a chance. But if the message is to search Philadelphia, we might as well stay home."[69]

Since 2001, the Department of Energy (DoE) expanded its radiological search capabilities through the Radiological Assistance Program (RAP). In an April 2006 report, however, the DoE's oversight branch was critical of the RAP team's performance during tests. The RAP teams reportedly "could not quickly provide local government with advice on what actions to take to reduce the public's exposure to radiation or whether to evacuate areas."[70] In February 2008, the American Physical Society and the American Association for the Advancement of Science jointly issued a document on U.S. nuclear emergency response capabilities, under the leadership of Michael May, former director of the Lawrence Livermore National [Nuclear Weapons] Laboratory. The document reportedly argued that the United States lacked enough "atomic detectives" to handle a crisis involving a nuclear device.[71]

THREATENING DANGEROUS TECHNOLOGY FACILITIES

Terrorists would not have to steal or build weapons of mass destruction to wreak havoc. They could achieve pretty much the same effect by sabotaging plants that produce or use large quantities of toxic chemicals, attacking nuclear or toxic chemical waste storage areas, or triggering the catastrophic failure of nuclear power plants. Ordinary explosives or incendiary devices, placed in the right places and ignited at the right time, could cause more damage than that caused by the terrible accidents at Bhopal and Chernobyl.

Nuclear Regulatory Commission (NRC) data on "safeguards events" involving nuclear materials, power plants, and other facilities show 2,033 threatening events during the 25 years from 1976 to 2000, as shown in Table 2.1).[72] More than 40 percent (or 833) of these events were bomb-related.

These events follow an uneven pattern over time, as shown in Table 2.2. It is good news that every five-year period of the quarter century from 1976 to 2000 saw substantially fewer bomb-related incidents than the preceding five-year period. Even in the latest five years for which data are readily available (1996 to 2000), however, there were 38 bomb-related incidents, or an average of more than seven incidents a year. There does not seem to be any good reason to believe that credible threats against nuclear facilities will stop anytime soon.

Many of the events in these tables were only threats or hoaxes, but they indicate that nuclear facilities are considered vulnerable enough to make the threats/hoaxes credible. In any case, it is clear that nuclear facilities have been considered attractive targets by terrorists and criminals for some time now—a point punctuated by the Australian and South African nuclear terrorist incidents described earlier.

Table 2.1 Nuclear safeguards incidents by category, 1976–2000

Category	Number of Events (Total = 2,033)	Percent of Total
Bomb-Related	833	41.0%
Firearms-Related	585	28.8%
Tampering/Vandalism	134	6.6%
Intrusion	47	2.3%
Missing or Allegedly Stolen	34	1.7%
Arson	22	1.1%
Transport-Related	12	0.6%
Miscellaneous	366	18.0%

Notes:

1) The NRC's definitions for the categories are:
   Bomb-Related: "Events . . . concerned with explosives or incendiary devices and related threats."
   Firearms-Related: "Event . . . typically describes the discharge, discovery or loss of firearms"
   Tampering/Vandalism: "includes destruction or attempted destruction. . . . which do not directly cause a radioactive release"
   Intrusion: "Incidents of attempted or actual penetration of a facility's barriers or safeguards systems"
   Missing or Allegedly Stolen: "Events in which safeguarded material was stolen, alleged to be stolen, discovered missing or found" [includes missing/stolen during transport]
   Arson: "intentional acts involving incendiary materials resulting in damage to property"
   Transportation-Related: "incidents where safeguarded material was misrouted or involved in an accident"
2) Even though there is some overlap in categories, each event was only included in a single category. There is thus no "double-counting".

Sources: International Safeguards Branch, Division of Safeguards and Transportation, Office of Nuclear Material Safety and Safeguards, *Safeguards Summary Event List (SSEL)* (Washington D.C.: U.S. Nuclear Regulatory Commission, NUREG-0525; Vol. 1, June 1992), pp. vii and pp. A-1through A-15. Also Safety and Safeguards Support Branch, Division of Fuel Safety and Safeguards, Office of Nuclear Material Safety and Safeguards, *Annual Safeguards Summary Event List (SSEL) 2000* (Washington, DC: U.S. Nuclear Regulatory Commission, NUREG 0525 (Vol.4, May 2001), p. A-5: Safeguard Events by Category, 1990–2000.

A Chernobyl by Design?

The same month terrorists used a truck bomb in the 1993 attack on the World Trade Center in Manhattan, a mental patient crashed his station wagon through a door at the infamous Three Mile Island nuclear power plant.[73] Coming so close together, these two events naturally raised the question of what would happen if a disturbed individual or terrorist

Table 2.2   Nuclear safeguards incidents by year, 1976–2000

Year	Total Number of Incidents	Reactor Events Only	Bomb-Related Incidents
1976	72	66	55
1977	34	29	28
1978	47	40	29
1979	118	111	94
1980	109	103	73
1981	74	70	48
1982	82	80	57
1983	56	54	39
1984	58	57	28
1985	67	58	23
1986	96	84	43
1987	95	91	35
1988	137	128	37
1989	148	142	25
1990	103	100	11
1991	152	150	32
1992	94	94	12
1993	117	116	12
1994	119	118	9
1995	60	50	12
1996	32	31	6
1997	25	22	6
1998	12	5	7
1999	21	18	8
2000	16	16	11
TOTAL	**1,942**	**1,833**	**740**

Sources: Operations Branch, Division of Fuel Safety and Safeguards, Office of Nuclear Material Safety and Safeguards, *Safeguards Summary Event List (SSEL)* (Washington D.C.: U.S. Nuclear Regulatory Commission, NUREG-0525, Vol. 2, Rev. 3; July 1995), A–1, A–2, and A–8. Also Safety and Safeguards Support Branch, Division of Fuel Safety and Safeguards, Office of Nuclear Material Safety and Safeguards, *Annual Safeguards Summary Event List (SSEL) 2000* (Washington, D.C.: U.S. Nuclear Regulatory Commission, NUREG 0525 (Vol. 4, May 2001), A-1: Safeguard Events, 1990–2000.

gang attacked a nuclear plant with a truck bomb as powerful as the one that ripped apart six stories of the Trade Center or destroyed the Murrah Federal Building in the Oklahoma City bombing two years later. What about the possibility of terrorists crashing an aircraft into the reactor building, mimicking the attacks that collapsed the Trade Center's twin towers on that terrible day in September 2001?

The reinforced containment structure that covers a nuclear power reactor is the last line of defense against the release of large amounts of radioactive material. Its presence at Three Mile Island and its absence at Chernobyl was one of the most important reasons why the Chernobyl accident did so much more damage. Early in a severe reactor accident, high pressure from the heat of fission products, gas generation, and the like threaten the containment, just as it fills with many radioactive materials in vaporized and aerosol form, easily carried by the wind and easily inhaled. It is the worst possible time for the containment to fail. If a terrorist group were to trigger a major nuclear accident, perhaps by sabotaging the plant with inside help, anything they could do to weaken the containment enough to make it fail would greatly increase the magnitude of the disaster. (It appears likely there was insider involvement in the Pelindaba nuclear incident.) Coordinating sabotage of the plant with a vehicle bomb attack might just do the trick. Is this possible?

In 1984, Sandia National Laboratories completed a study concluding that nuclear facilities in the United States were vulnerable to terrorist truck bombs and that putting a few barricades near the reactor building would not solve the problem.[74] "Unacceptable damage to vital reactor systems" could be done by bombs detonated some distance away, in some cases possibly even off-site.[75]

But of course as we learned on 9/11, trucks are not the only vehicles that terrorists could use to attack a nuclear plant. In the early 1970s, James Schlessinger, Chair of the Atomic Energy Commission (predecessor of the NRC) put it this way:

> If one intends to crash a plane into a facility . . . there is, I suspect, little that can be done about the problem.
>
> "The nuclear plants that we are building today are designed carefully, to take the impact of, I believe a 200,000 pound aircraft arriving at something on the order of 150 miles per hour. It will not take the impact of a larger aircraft."[76]

At least a few dozen nuclear plants still operating in the United States today have containments no stronger than that to which Schlessinger was referring. Even before September 11, 2001, terrorists had hijacked quite a few jet airliners that are considerably heavier and faster than the aircraft whose impact those containments were designed to resist. A Boeing 747, for example, weighs more than 300,000 pounds and can travel at speeds over 500 miles per hour. Sadly, it is no longer a matter of speculation how much damage the crash of a large airliner filled with jet fuel can do to a very substantial building, even at relatively slow speed. A much smaller plane could seriously damage the containment if it were filled with powerful explosives.

The only airliner hijacked on September 11 that did not reach its target was United Airlines flight 93, departing Newark for San Francisco. Similar to the other 9/11 attacks, it is commonly believed that its intended target was an important symbolic building, probably the White House, possibly the Capitol building. But perhaps its hijackers had a different target in mind. The plane flew out of the East Coast headed west and slightly south. After the hijacking, it looped around and headed east again. That Boeing 767 ultimately crashed in Pennsylvania, killing everyone aboard. When the plane crashed it was headed generally toward and only about 120 miles (about 15 minutes flying time) from the Three Mile Island nuclear power plant. That plant, site of one of the two worst nuclear power accidents in history, is also an important symbol, known to people all over the globe. And it is of a vintage similar to that to which James Schlessinger was referring in the quote above.

The NRC has repeatedly made it clear that the containments built around all American commercial nuclear reactors are designed to be strong enough to withstand a plane crash. But the NRC has also admitted that they (especially the older plants) were designed to take the impact of a small airliner flying at say 150 miles per hour—not a Boeing 767 coming in at 500-plus miles per hour. If that fourth hijacked plane had reached and crashed into the containment on the nuclear reactor building at Three Mile Island, we might well have had an American Chernobyl on our hands.

In early 2002, the NRC ordered that a series of upgrades in security measures be made at nuclear plants in the United States. These upgrades included measures to mitigate the effects of the fire and explosion that would accompany the crash of a large commercial aircraft, whether intentional or accidental.[77] But on December 25, 2006, the *New York Times* reported that the nuclear power industry had requested "the government to specify how new nuclear plants should minimize damage from airplane attacks" because several weeks earlier, "the Nuclear Regulatory Commission decided not to institute requirements on building new plants that are tougher than the rules that prevailed decades ago when the old ones were built." The article quoted one NRC commissioner as saying that the NRC's "proposal does not ensure that any new nuclear reactors will be designed to withstand commercial aircraft crashes."[78] Near the end of April 2007, the NRC voted 4 to 1 *against* requiring the design and construction of new nuclear reactors to ensure that they could withstand the impact of large aircraft.[79]

As to sabotaging a nuclear plant, since the late 1970s there have been publicly available documents, written to help nuclear plant operators test their security that would be very useful to a potential saboteur. One included a computer program that could be used to determine the best route for a saboteur to follow.[80]

Then there is the *Barrier Penetration Database*, prepared by Brookhaven National Laboratory under contract to the NRC. It gives detailed information on the types of tools and explosives necessary to break through dozens of barriers, from chain link fences to reinforced concrete walls.[81] A 1993 RAND study (sponsored by DoE) looked at 220 direct attacks by armed bands of guerrillas or terrorists that they found analogous to the kinds of armed assault such groups might launch against American nuclear facilities. The attackers were successful 74 percent of the time.[82]

Of course, sabotage would be easier with the help of insiders working at the facility. (Again, consider the 2007 incident at Pelindaba.) In a 1990 study, *Insider Crime: The Potential Threat to Nuclear Facilities*, RAND found that financial gain was the main motivation in the overwhelming majority of insider crimes they studied.[83] Terrorists might need to do little more than find a sufficiently money-motivated insider and pay him/her off. Guards themselves were responsible for more than 40 percent of the crimes against guarded targets. Their conclusion?

> Beyond a certain point, security considerations in hiring, guarding, controlling, and checking people can become so cumbersome as to actually impede the operation of a facility. This creates a serious dilemma in the case of a nuclear facility . . . total security can never be attained.[84]

It would be much simpler to attack a toxic chemical plant, radioactive, or toxic chemical waste storage dump than a nuclear power plant. Terrorist assaults on other dangerous technology facilities are even more likely to succeed.

WHY HAVE TERRORISTS NOT YET ENGAGED IN ACTS OF MASS DESTRUCTION?

Experts have warned about the dangers of nuclear terrorism and its equivalents at least since the 1960s. Dangerous technologies have already been used to do damage on a scale similar to conventional terrorism (such as in the 1995 Tokyo nerve gas attack). There have also been threats of mass destruction and hoaxes involving nuclear weapons. As yet, however, there has been no publicly reported case of terrorists (or criminals) doing the kind of massive damage that could result from large-scale use of dangerous technologies as either a weapon or target. Why not?

It is clearly not because this kind of attack is beyond their capabilities. It cannot be because of a moral revulsion against taking innocent lives, since that is the terrorist's stock in trade. It is possible that terrorists might be inhibited by a belief that murder and destruction on a scale even more massive than 9/11 would invite ferocious retaliation. But decades of experience show that terrorists are willing to risk ferocious retaliation, and may even be trying to provoke it. Many who seek to retaliate against terrorists are already prepared to do them grievous, even deadly harm. Even in free societies, where the search for terrorists is complicated by constraints against hurting innocent people and forfeiting personal freedoms, terrorists are already pursued with dogged determination and severely punished (witness the death sentence meted out to Timothy McVeigh, convicted of the Oklahoma City bombing).

More likely, most terrorists have simply not found acts of mass destruction even on the scale of 9/11 necessary up to now. If acts of conventional terrorism still provoke enough fear to put pressure on decision makers to do what terrorists want done, there is no particular reason for them to go to the trouble, danger, and expense of acquiring and using means of mass destruction. If acts of conventional terrorism come to be more routine and thus generate less shock and fear, terrorists may someday conclude that they must commit much greater violence to frighten people enough to achieve their objectives. They will find the tactics of mass destruction waiting in the wings.

Another possibility is that so far, with the exception of al Qaeda, terrorists have believed that engaging in mass destruction would interfere with achieving their objectives. The credibility of this explanation depends on what kind of group we are considering. Terrorist groups are not clones of each other. Understanding what makes them different is important to judging how likely it is that any particular group will use dangerous technologies. It is also key to developing more effective countermeasures.

A Taxonomy of Terrorists

The first and perhaps most obvious distinction is between domestic groups and international terrorists. Germany's left-wing Red Army Faction, France's right-wing Federation for National European Action,

America's racist white supremacist The Order, Spain's Basque separatist ETA, and Peru's revolutionary Shining Path are examples of domestic terrorist groups active in the 1980s.[85] On the other hand, the Popular Front for the Liberation of Palestine, the Japanese Red Army, Hamas, the Jewish Defense League, and the Armenian Secret Army are historical examples of internationally focused terrorist groups.[86] Some have been more successful than others, but most of these terrorist groups were still with us in the early twenty-first century.[87] Some have straddled the boundary between domestic and international terrorism. The Irish Republican Army (IRA), for example, carried out most of its attacks inside Northern Ireland, but it was also responsible for terrorist bombings in England.

Secondly, some terrorists have well-defined political goals, while the goals of others are much more vague, general, ideological, and/or anarchic. In years past, the Palestine Liberation Organization (PLO) used terrorist tactics to raise awareness of the plight of disenfranchised Palestinian Arabs and gain support for establishing an independent Palestinian State. The goal of IRA terrorism was also clear, specific, and political: to end British rule in Northern Ireland. By contrast, the Symbionese Liberation Army, made famous by its kidnapping of newspaper heiress Patricia Hearst, had only the most general, ideological, anti-capitalist goals. The long terrorist career of the Unabomber was aimed at promoting vague anti-technology goals. The self-proclaimed goal of Shoko Asahara, the leader of the Japanese terrorist cult Aum Shinrikyo, infamous for releasing poison gas in Tokyo's subways in 1995, was to "help souls on earth achieve 'ultimate freedom, ultimate happiness and ultimate joy.' "[88]

Asahara's statement leads naturally to a third related characteristic that differentiates terrorists from each other: some are rational and some are not. There is a temptation and desire to believe that anyone who would engage in brutal terrorist actions cannot be rational. But that is not true. Rationality is a matter of logic, not morality. It simply means that the tactics used are logically related to the goals pursued. The German Nazis were vicious and their behavior despicable and profoundly immoral, but not illogical. At least in the short term, their tactics led them step-by-step toward their goals of conquest and control.

Nonrational behavior includes both irrationality (in the sense of craziness) and behavior that is driven by something other than logic. Emotion, culture, tradition, and religion are important nonrational drivers of human behavior. To refuse to do business with the lowest-cost supplier because that firm is run by someone who once caused you personal emotional pain may not be rational, but it is easy to understand and certainly not crazy. Similarly, belief in God and adherence to the ceremonies of a particular religion is not a matter of logic; it is a matter of faith. While faith may not be logical, it is not irrational.

Whether or not terrorist tactics actually advanced the political goals of the PLO or the IRA, the decision to use them, though reprehensible, was still rational: terrorism might logically have led them where they wanted to go. It is difficult, however, to imagine by what logic the terrorism of Aum Shinrikyo can be related to their spiritual goal of helping people on Earth to attain "ultimate freedom, ultimate happiness and ultimate joy." Similarly, the brutal murders perpetrated in Los Angeles decades earlier by the "Manson Family" were driven by Manson's psychotic fantasies, not some logical process of choosing tactics that service an achievable goal.

A fourth factor that differentiates terrorist groups is the degree of support for their underlying goals. Whether or not the public approves of their tactics, terrorist groups fighting for a popular cause behave differently from those whose *goals* are considered extremist and out of touch. Everything else being equal, when the *cause* is popular, it will be easier for the group to recruit members, raise funds, and find hiding places for their weapons and for themselves. Those that are rational will cultivate this sponsorship and take care when choosing their targets and tactics to avoid alienating supporters. They will try to avoid catching their potential allies in the web of innocent victims. Groups with little or no public support are less inhibited in choosing victims. They may believe that acts of horrific violence are the only way to shock a complacent or submissive public into action. Despite their lack of support, it is a safe bet that most see themselves as the leading edge of a great movement, and they think that once the public awakens, a mass of supporters will arise and carry them to victory.

Finally, different terrorist groups face off against different kinds of opponents. Some see a very specific enemy, such as executives at a particular company or the political leaders of a specific country. Others see a much larger enemy, such as all nonwhites or all non-Christians, or a vaguely defined group, such as the "international Jewish conspiracy." The nature of their perceived enemy affects the kinds of actions they take and the intensity of the violence they commit. The range of targets and degree of violence will be greater when the enemy is "the federal government" or "the Jews" than when it is the management of Exxon or the ruling party in Britain.

To summarize, the five distinguishing factors are:

- **Geopolitical focus.** Are they domestic or international, and if international are they state-sponsored?
- **Nature, specificity, and achievability of goals.** Are their goals vague and ideological or specific and political?
- **Rationality.** Is their behavior driven by logic?
- **Public support.** How much public support is there for their goals as opposed to their tactics?
- **Size and character of their enemy.** Is their opponent a small and specific group of decision makers or a much larger generalized class of people?

Domestic terrorists with clearly defined and potentially achievable political objectives are most likely to see the terrorism of mass destruction as counterproductive. Because they see their terrorism as resistance and rebellion that will eventually rally their silent, disempowered supporters to the cause, they must always balance the shock effect of the damage they do against the support they will lose if their violence becomes too extreme. Except in situations of complete desperation, groups of this type are almost certain to see in advance that acts of mass destruction would be disastrous blunders.

International terrorist groups with clearly defined and potentially achievable political objectives are somewhat more likely to escalate the level of violence. If they are playing to a domestic audience and if most of the violence they do is outside their home country, they may feel that spectacular acts of destruction will not alienate and may encourage those who support their cause at home. This might be especially true of a group attacking targets inside a nation whose military is occupying the terrorists' home country. If the terrorists are also trying to influence the wider international community to support their cause, however, acts of extraordinary violence are not likely to seem appealing.

Domestic or international terrorist groups whose objectives are more general, ideological, or ill formed and whose attitudes are much more nihilistic—such as doomsday cults, racist or religious ideological extremists, and nuts—are an entirely different story. They are less likely to be deterred by worries about alienating supporters and may find mass destruction an appealing, unparalleled opportunity to exercise their power. Doomsday cults could even see these acts as a way of hastening the salvation they believe will follow the coming cataclysm. Such groups are extremely dangerous.

As an international terrorist network operating under an extremist religious philosophy, al Qaeda is undoubtedly the best-known example of such a group. Al Qaeda has not only proven itself capable of committing acts of massive death and destruction, it has celebrated and sought to enhance them. If we think of al Qaeda in terms of the five factors in our taxonomy, it is no surprise that they would find the tactics of massive destruction attractive. They are clearly an international terrorist group (Factor 1); trying to achieve extremist ideological goals (Factor 2); driven by nonrational, religious beliefs (Factor 3); with even the vast majority of their co-religionists in disagreement with their goals, not to mention their tactics (Factor 4); who appear to see virtually the entire non-Islamic world, as well as non-fundamentalist Muslims, as their enemies (Factor 5).

By 2008, there were some 926 domestic hate groups active throughout the contiguous United States—many (but nowhere near all) of them right wing, violence-oriented, white supremacist militias.[89] Collectively calling themselves the "Patriot" movement, the agendas of some of these groups are extremist by any reasonable definition of the word. Many Americans believe that some federal agencies are oversized, inefficient, and sometimes abusive. But the rhetoric of the Patriots goes much farther. In the words of one analyst who has closely tracked the movement:

> those in the Patriot movement are convinced that the government is evil. It is run by a secret regime ... that seeks to disarm American citizens and subjugate them to a totalitarian world government. Just about anything that can be described with the adjectives 'global,' 'international' or 'multicultural' is a Patriot menace.[90]

These groups were heavily armed, and not without financial resources. In 1995, 11 members of the Patriot group "We the People" were indicted on felony charges related to a scheme in which they allegedly collected almost $2 million from thousands of people they charged $300 each to be part of a phony class-action lawsuit against the federal government. A year before, the host of a popular Patriot radio program and founder of the Patriot group "For the People" took in more than $4 million.[91]

America's right-wing extremist paramilitary militias have many characteristics that make them prime candidates for the terrorist use of dangerous technologies. Some rank high on each of the five factors in the taxonomy of terrorist groups:

- **Factor 1.** While they are domestic rather than international, they see themselves as fighters against a national government that has become the pawn of an international conspiracy. They believe federal authorities constitute

an "occupation government"—some think no level of government higher than the county has any legitimacy. Thus, they see themselves more as fighters against a foreign government than as domestic revolutionaries.
- **Factor 2.** Their goals are ideological, general, and anarchic.
- **Factor 3.** Living in a paranoid world of conspiracies, they are motivated by nonrational beliefs often clothed in the garb of some form of end-time religion. Some have been tied to the violence-oriented, post-millennial Christian Identity movement.
- **Factor 4.** There is very little public support for their nihilistic goals.
- **Factor 5.** Rather than a very specific and limited enemy, to many of the Patriot militias, all non-whites, non-Christians, and many mainstream Christians as well are the enemy.

After studying a wide range of terrorist groups for the Department of Energy in 1986, the RAND Corporation concluded, "Of the terrorist organizations active in this country, the right-wing extremists appear to pose the most serious threat to U.S. nuclear weapons facilities."[92]

To date the lack of unity and coordination among extremist militias has limited their terrorist efforts. There is, however, cross-fertilization among them. They count well-trained American military personnel (including at least some former Green Berets) among their members and are always trying to recruit more. It does not seem that they have yet been able to attract many highly educated and technically skilled people (as did Aum Shinrikyo), but the movement is relatively new and does seem to have significant financial resources, so it is too facile to assume their terrorist activities will stay limited indefinitely. Along with al Qaeda's minions, the most extreme of these homegrown, violence-oriented militias are prospective candidates for becoming nuclear terrorists.

IS THE TERRORIST THREAT GROWING?

There are data available on the frequency and severity of terrorist attacks and credible threats of attack, but they are limited and flawed. Because the "terrorist" label is so politically loaded, governments and other organizations that gather and publish such data often use different definitions of what constitutes terrorism. Beyond this, incidents that involve grave potential danger—such as threats against nuclear power plants, nuclear weapons facilities, or radioactive waste storage areas—tend to be covered up when they can be, on the belief that making them public would spread undue fear and thus give the terrorists a partial victory. Such incidents are also kept secret to create a false sense of security so the public will continue to support particular government institutions or policies.

It is easier to keep attacks quiet when they are carried out by domestic terrorists than when international terrorists are involved. The incentives are also greater: the frequency and severity of domestic terrorist incidents are measures of internal opposition and political turmoil, something no government likes to publicize. As a result, while all data on terrorism must be treated with caution, data on international terrorism are likely to be more reliable (see Table 2.3).

According to the U.S. Department of State, there were nearly 14,813 separate incidents of international terrorism from 1968 to 2003—an average of more than 410 per year for close to four decades. The average number of terrorist incidents per year rose from 160 in the late 1960s to more than 400 during the 1970s, peaking at more than 540 in the 1980s. With the end of Cold War rivalries and the beginning of serious ongoing peace negotiations in the Middle East, the average number of terrorist incidents per year dropped to less than 400 during the 1990s, and then dropped farther to a little less than 300 in the first few years of the 2000s. Still, there were 84 percent more incidents per year on average from 2000 to 2003 than there had been in 1968 and 1969.

International terrorism has also become a more deadly business over the years. The average number of people killed each year skyrocketed from 15 in 1968 and 1969 to 150 during the 1970s, then grew rapidly to more than 500 in the 1980s. Fatalities dropped even more sharply than incidents in 1990 to 1999, falling to about 250 per year. Even so, despite the Middle East peace process and the end of the Cold War, the average number of people killed each year by international terrorists was 66 percent higher in 1990 to 1999 as it had been in the 1970s, and more than 16 times as high as in the late 1960s. In the first four years of the 2000s, the average death toll from terrorism per year jumped up to more than 1,300, largely, but not solely, due to the events of September 11.

These data are far from perfect, but they would have to be very far off the mark to overturn the basic picture they paint. International terrorism is, sad to say, still going strong—and becoming more lethal. It shows no signs of fading away any time soon.

In the United States we have become accustomed to thinking of international rather than domestic terrorism as the gravest threat to life and limb. Yet fewer people were killed worldwide in 1995 in all 440 international terrorist incidents than died that year in the single domestic terrorist bombing in Oklahoma City.

Table 2.3 Incidents of international terrorism, 1968–2003[*]

Year	Number of Incidents[**]	Fatalities[†]	Injuries	Total Casualties[‡]
1968	120	20		
1969	200	10		
1970	300	80		
1971	280	20		
1972	550	140		
1973	350	100		
1974	400	250		
1975	345	190		
1976	457	200		
1977	419	150		
1978	530	250		
1979	434	120		
1980	499	150		
1981	489	380		
1982	487	221	840	1,061
1983	497	720	963	1,683
1984	565			1,100
1985	635	825	1,217	2,042
1986	612	576	1,708	2,284
1987	665	633	2,272	2,905
1988	605	658	1,131	1,789
1989	375	407	427	834
1990	437	200	677	877
1991	565	102	242	344
1992	363	93	636	729
1993	431	109	1,393	1,502

Year				
1994	321	315	663	977
1995	440	165	6291	6,456
1996	296	314	2,912	3,226
1997	304	221	693	914
1998	274	741	5,952	6,693
1999	392	233	706	939
2000	423	405	791	1,196
2001	346	3,547	1,080	4,627
2002	199	725	3,295	4,020
2003	208	625	3,646	4,271
TOTAL	**14,813**	**13,895**	**37,535**	**50,469**
	(over 36 yrs)	*(over 35 yrs)*	*(over 21 yrs)*	*(over 22 yrs)*
AVERAGE	**411/yr**	**397/yr**	**1,787/yr**	**2,338/yr**

*The U.S. State Department excluded intra-Palestinian violence beginning with 1984, apparently because it was considered to be domestic terrorism. Such terrorism had previously been included because the "statelessness" of the Palestinian people made the violence international.

**Data for 1968 to 1974 were extracted from an unnumbered bar graph; they are approximations, rounded to the nearest ten. The underlying data are unpublished.

†Data for 1968 to 1981 were extracted from an unnumbered line graph; they are approximations, rounded to the nearest ten. The underlying data are unpublished.

‡Includes only years for which data on *the sum of both* fatalities and injuries were available. Therefore, the total of this column is less than the sum of the totals of the separate columns for fatalities and injuries.

Sources: Data on the number of incidents for 1975 to 1994 from the Office of the Coordinator for Counter-Terrorism, U.S. Department of State, Patterns of Global Terrorism, 1994, p. 65, 1995 data from the 1995 edition, p. 4, 1996 and 1997 data from the 1997 edition; data on number of incidents for 1968 to 1974 from Office of Counterterrorism, U.S. Department of State (unpublished) as cited in Kegley, Charles W. Jr, *International Terrorism: Characteristics, Causes, Controls* (New York: St. Martin's Press, 1990), p. 15.

Data on fatalities for 1968 to 1983 and injuries for 1982 to 1983 from Cordes, Bonnie, et al., *Trends in International Terrorism, 1982 and 1983* (Santa Monica, CA: Rand Corporation, August 1984), pp. 6–7; these data were apparently based U.S. State Department data. Data for fatalities and injuries for 1985 to 1997 from *Patterns of Global Terrorism* annual editions for 1985 to 1997; approximate total casualties data for 1984 were from the 1992 edition, p. 60. Data for 1998 to 2003 from U.S. Department of State, Country Reports on Terrorism (various years): *Patterns of Global Terrorism, The Year in Review,* for years 1998 to 2003 (accessed at http://www.state.gov/; April 2009).

Terrorists have not yet used dangerous technologies to do catastrophic damage, as weapons or as targets. But there is nothing inherent in the nature of terrorism that makes it self-limiting. Those who are ready, even eager, to die for their cause, who stand willing to abandon every constraint of civilized behavior and moral decency against the slaughter of innocents, cannot be expected to permanently observe some artificial restriction on the amount of havoc they wreak.

Many terrorists are political rebels, in the classic sense. They have chosen reprehensible methods to fight for rational, limited, clearly defined objectives. But others are striking out against a world of enemies, out of touch with reality and trying to punish, even to destroy a world which they cannot live in and do not understand.

NOTES

1. Robert D. McFadden, "Blast Hits Trade Center, Bomb Suspected, 7 Killed, Thousands Flee Smoke in Towers: Many Are Trapped for Hours in Darkness and Confusion," *New York Times* February 27, 1993; and Richard Bernstein, "Trade Center Bombers Get Prison Terms of 240 Years," *New York Times*, May 25, 1994.

2. John Kifner, "At Least 21 Are Dead, Scores Are Missing After Car Bomb Attack in Oklahoma City Wrecks 9-Story Federal Office Building: 17 Victims Were Children in 2d-Floor Day Care Center," *New York Times*, April 20, 1995; and Joseph B. Treaster, "The Tools of a Terrorist: Everywhere for Anyone," *New York Times*, April 20, 1995.

3. U.S. State Department, *Patterns of Global Terrorism 2001* (Washington, DC: May 2002), 1; Eric Lipton, "Towers' Wind-Force Design Questioned," *New York Times*, June 19, 2004.

4. CBS News, "Official 9/11 Death Toll Climbs by One," *CBSNews.com*, July 10, 2008: http://www.cbsnews.com/stories/2008/07/10/national/main4250100.shtml (accessed June 8, 2009).

5. Op. cit., United States Department of State, 1.

6. Op. cit., CBSNews. See also Insurance Information Institute, "9/11 and Insurance, One Year Later—Terror Attacks Most Complex Disaster in History," New York: September 5, 2002; http://www.iii.org/media/updates/press.635680/ (accessed June 8, 2009); Institute for the Analysis of Global Security, "How Much Did the September 11 Terrorist Attack Cost,"http://www.iags.org/costof911.html (accessed June 8, 2009).

7. A similar, simple definition of terrorism can be found in the Rand Corporation's terrorism reports. See, for example, Cordes, Bonnie, Hoffman, Bruce, Jenkins, Brian M., Kellen, Konrad, Moran, Sue and Sater, and William,

Trends in International Terrorism, 1982 and 1983 (Santa Monica, CA: Rand Corporation, August 1984), 1.

8. One reason the 9/11 attacks drew such a strong reaction was that, for all their horror, they were a spectacular piece of theater. Before they realized what they were really seeing, some watching early television coverage of the attacks reported they first thought they were seeing a clip from a gruesome new Hollywood movie.

9. "Some Poisoned Israeli Oranges Discovered in Europe," *New York Times*, February 2, 1978. In a letter sent to the West German Ministry of Health, a group calling itself "Arab Liberation Army—Palestine Commando" claimed responsibility.

10. "Medical Sciences," *1984 Yearbook of Science and the Future* (Chicago: Encyclopaedia Britannica, Inc., 1984), 353. It is unclear whether the Tylenol poisonings were the work of someone with a grudge against the company or someone intending to extort money.

11. Rohan Gunaratna, *Inside Al Qaeda: Global Network of Terror* (New York: Columbia University Press, 2002), 95.

12. The so-called "Reign of Terror" was carried out by the infamous Maximilien Robespierre's Jacobin government, between the summer of 1793 and the summer of 1794.

13. John Darnton, "Ethiopia Uses Terror to Control Capital," *New York Times* (February 10, 1978).

14. According to the testimony, the campaign included cigarettes infected with anthrax, candies spiked with botulism, sugar containing salmonella, and whiskey laced with the herbicide paraquat See Suzanne Daley, "In Support of Apartheid: Poison Whisky and Sterilization,"*New York Times*, June 11, 1998.

15. John F. Burns, "A Network of Islamic Terrorism Traced to a Pakistan University," *New York Times*, March 20, 1995.

16. Ibid. See also Christopher S. Wren, "U.S. Jury Convicts 3 in a Conspiracy to Bomb Airliners," *New York Times*, September 6, 1996, and Benjamin Weisler, "Mastermind Gets Life for Bombing of Trade Center," *New York Times*, January 9, 1998.

17. Robert D. McFadden, "FBI Seizes 8, Citing a Plot to Bomb New York Targets and Kill Political Figures," *New York Times*, June 25, 1993.

18. For much more detail on bin Laden, see Gunaratna, Rohan, op. cit., 16–53.

19. Op. cit., John F. Burns.

20. Bruce Hoffman, "Rethinking Terrorism and Counterterrorism Since 9/11" in Gus Martin, ed., *The New Era of Terrorism* (Thousand Oaks, CA: Sage Publications, 2004), 2.

21. Bruce Hoffman, Peter deLeon, Bonnie Cordes, Sue Ellen Moran, and Thomas C. Tompkins, *A Reassessment of Potential Adversaries to [sic] U.S. Nuclear Programs* (Santa Monica, CA: Rand Corporation, March 1986), 5.

22. "Aum Paid $4 Billion to Foreign Companies," *Japan Times*, July 2, 1995. Members of Aum Shinrikyo reportedly visited the Russian nuclear research institute at Obninsk in 1994 seeking nuclear materials. Obninsk is the same institute that U.S. investigators believed to be the source of the small amount of weapons-grade nuclear material smuggled into Germany in 1994; see Gregory Katz, "Uranium Smuggling Case Confirms Security Fears," *Dallas Morning News*, May 26, 1996.

23. As quoted by Graham Allison in "Nuclear 9/11," *Bulletin of the Atomic Scientists* (September/October 2006), 37.

24. In the early twenty-first century, the CIA's reputation for accuracy with respect to intelligence about covert nuclear weapons programs was damaged by its public support of the Bush Administration's contention that Iraq still had such a program (which it did not) in justifying the invasion and occupation of Iraq. Central Intelligence Agency, "Unclassified Report to Congress on the Acquisition of Technology Relating to Weapons of Mass Destruction and Advanced Conventional Munitions, 1 January to 30 June 2001," January 30, 2002, as quoted by Allison, Graham in "Nuclear 9/11," *Bulletin of the Atomic Scientists* (September/October 2006), 37.

25. This money was to be spread across all 10 "closed cities" in the Russian nuclear weapons complex to help start commercial ventures. Michael R. Gordon, "Russia and U.S. Plan to Guard Atom Secrets," *New York Times* (September 23, 1998).

26. David E. Sanger, "Nuclear Ring Reportedly Had Advanced Design," *New York Times* (June 14, 2008).

27. Peter T. Kilborn, "British Group's Sale of Bacteria Stirs Fear of Use by Terrorists," *New York Times* (February 11, 1977).

28. Sherly WuDunn, "Elusive Germ Sows Panic in Japan," *New York Times*, August 23, 1996.

29. Klanwatch Project, *False Patriots: The Threat of Anti-Government Extremists* (Montgomery, AL: Southern Poverty Law Center, 1996), 31.

30. Ibid., 25. See also Todd S. Purdum, "Two Charged with Possessing Toxin Thought to Be Anthrax," *New York Times*, February 20, 1998.

31. U.S. Federal Bureau of Investigation, "Amerithrax Investigation," http://www.fbi.gov/anthrax/amerithraxlinks.htm (accessed June 9, 2009).

32. Michael J. Ambrosino, Executive Producer, "The Plutonium Connection" (Boston: WGBH-TV, NOVA #214, first transmission PBS March 9, 1975), 8. See also David Burnham, "Bill Asks Curb on Plutonium Use to Prevent Building of Homemade Bomb," *New York Times*, February 27, 1975.

33. *Newsday* (the news daily of Long Island, New York), October 10, 1976.

34. Ronald J. Ostrow, "Backpack A-Bomb Possible, FBI Says," *Los Angeles Times*, April 26, 1979.

35. Howard Morland, "The H-Bomb Secret: To Know How Is to Ask Why," *The Progressive*, November 1979, 14–23.

36. Dierdre Carmody, "U.S. Hunting for Copies of Classified Data on Bomb," *New York Times*, May 18, 1979.

37. Mason Willrich and Theodore B. Taylor, *Nuclear Theft: Risks and Safeguards* (Cambridge, MA: Ballinger Publishing Company, 1974), 20–21.

38. David Burnham, "Bill Asks Curb on Plutonium Use to Prevent Building of Homemade Bomb," *New York Times* (February 27, 1975).

39. Theodore B. Taylor, "Nuclear Power and Nuclear Weapons," Global Security Study No.22 (Santa Barbara, CA: Nuclear Age Peace Foundation, July 1996), 2.

40. Robert Gillette, "Impure Plutonium Used in '62 A-Test," *Los Angeles Times*, September 16, 1977.

41. Robert Gillette, "Proves Civilian Plants Can Be Atomic Arsenals," *Los Angeles Times*, September 14, 1977.

42. "The critical mass of metallic uranium at 10% enrichment, with a good neutron reflector [like beryllium] is about 1000 kilograms ... a sphere ... about a foot and a half in diameter. At 20% enrichment, the critical mass drops to about 250 kilograms ... and at 50% enrichment, it is fifty kilograms," op. cit., Willrich and Taylor, 17.

43. Alexander Glaser and Frank von Hippel, "Thwarting Nuclear Terrorism," *Scientific American*, February 2006, 59.

44. Ibid. They point out that 22 kilograms would be enough for a critical mass if 93 percent enriched uranium were used (covered with a beryllium neutron reflector), while about 400 kilograms would be needed if the uranium were only enriched to 20 percent.

45. In 1993, 4.5 kilograms of enriched reactor fuel was stolen from a Russian Navy base at Kola Bay. See Oleg Bukharin and William Potter, " 'Potatoes Were Guarded Better,' " *Bulletin of the Atomic Scientists*, May/June 1995, 46.

46. Op. cit., Alexander Glaser and Frank von Hippel, 58. On the same page Glaser and von Hippel also reported, "As of the end of 2005, some ten metric tons of exported bomb-grade HEU still resided in countries that do not possess nuclear weapons—enough to make 150–200 gun-type [nuclear] explosive devices."

47. Douglas Waller, "Nuclear Ninjas: A New Kind of SWAT Team Hunts Atomic Terrorists," *Time*, January 8, 1996, 39–41.

48. Chris Schneidmiller, "Dirty Bombs Present Quick Threat, Extended Challenge," *Global Security Newswire*, February 20, 2007: http://www.nti.org/d_newswire/ (accessed February 22, 2007).

49. "Radiological Poisoning Attacks Possible, Experts Say," *Global Security Newswire*, August 1, 2007: http://www.nti.org/d_newswire/ (accessed August 1, 2007).

50. "Germ Warfare: New Threat from Terrorists," *Science News*, May 18, 1996, 311.

51. Ibid.

52. Peter D. Zimmerman and Jeffrey G. Lewis, "The Bomb in the Backyard," *Foreign Policy*, November/December 2006, 38.

53. Op. cit., Graham Allison, 38.

54. "ElBaradei Warns of Nuclear Thefts," *Global Security Newswire*, October 28, 2008: http://www.nti.org/d_newswire/ (accessed October 29, 2008).

55. Leonard S. Spector, "Clandestine Nuclear Trade and the Threat of Nuclear Terrorism," in Paul Leventhal and Yonah Alexander, ed., *Preventing Nuclear Terrorism* (Lexington, MA: Lexington Books, 1987), 80.

56. Op. cit., Waller, Douglas, 39–41.

57. Andrew Cockburn, "Can We Stop A Bomb Smuggler?," *Parade Magazine*, November 3, 1985, 12.

58. "GAO Gets Radioactive Material Across U.S. borders," *Global Security Newswire*, March 28, 2006: http://www.nti.org/d_newswire/ (accessed March 28, 2006).

59. "GAO Describes Porous U.S.-Canada Border Security," *Global Security Newswire*, September 28, 2007: http://www.nti.org/d_newswire/ (accessed October 3, 2007).

60. Ibid.

61. "DHS Oversold Radiation Sensors, Report Says," *Global Security Newswire*, July 20, 2007: http://www.nti.org/d_newswire/ (accessed August 1, 2007).

62. Ed O'Keefe, "GAO: Major Security Flaws at Federal Buildings," in "Federal Eye: Keeping Tabs on the Government," *The Washington Post*, July 7, 2009.

63. Eric Lipton, "New York to Test Ways to Guard Against Nuclear Terror," *New York Times*, February 9, 2007.

64. Associated Press, "FBI Arrests Two in Plot to Steal Nuclear Sub," *Washington Post*, October 5, 1978.

65. Gregory Katz, "Uranium Smuggling Case Confirms Security Fears," *Dallas Morning News*, May 26, 1996.

66. Michael Perry, "Sydney Nuclear Reactor Terror Plot Target," *Reuters*, November 14, 2005; "Australia Nuclear Reactor Believed Target," *Adnkronosinternational*, November 14, 2005; David King, "Nuclear Attack in Jihad Plot," *The Australian*, November 15, 2005.

67. "Nuclear Terror Risks Demand Global Attention, Report Says," *Global Security Newswire*, November 18, 2008: http://www.nti.org/siteservices/ (accessed November 19, 2008; and "South African Nuclear Plant Raiders May Have Had Inside Help," *Global Security Newswire*, November 24, 2008: http://www.nti.org/siteservices / (accessed November 24, 2008).

68. Mahlon, E. Gates, "The Nuclear Emergency Search Team," in op. cit., Paul Levanthal and Yonah Alexander, 397–402; also, op. cit, Douglas Waller, 39–42.

69. John R. Emshwiller, "In Atom Bomb Scare, Federal NEST Team Flies to the Rescue," *Wall Street Journal*, October 21, 1980.

70. Jon Fox, "U.S. Radiological Response Vulnerable to Attack, GAO Says," *Global Security Newswire*, September 22, 2006: http://www.nti.org/d_newswire/ (accessed October 19, 2006).

71. Chris Schneidmiller, "More Nuclear Detectives Needed in Crisis, Groups Say," *Global Security Newswire*, February 20, 2008: http://www.nti.org/d_newswire/ (accessed February 25, 2008).

72. Operations Branch, Division of Fuel Safety and Safeguards, Office of Nuclear Material Safety and Safeguards, *Safeguards Summary Event List (SSEL)* (Washington, D.C.: U.S. Nuclear Regulatory Commission, NUREG-0525, Vol. 2, Rev. 3; July 1995).

73. Matthew L. Wald, "Doubts Raised on New Rules to Protect Nuclear Reactors," *New York Times*, February 11, 1994.

74. "Weekly Information Report to the NRC Commissioners," April 20, 1984, enclosure E, p.3, as cited in Daniel Hirsch, "The Truck Bomb and Insider Threats to Nuclear Facilities," in op. cit., Paul Leventhal and Yonah Alexander, 209.

75. Daniel Hirsch, "The Truck Bomb and Insider Threats to Nuclear Facilities," in op. cit., Paul Leventhal and Yonah Alexander, 210.

76. James R. Schlessinger, quoted from transcript of "Meet the Press" in a *Senator Mike Gravel Newsletter*, October 31, 1973, by David Krieger, "Terrorists and Nuclear Technology," *Bulletin of the Atomic Scientists*, June 1975, 32.

77. "U.S. Nuclear Plant Anti-Terrorist Measures Consolidated," *World Nuclear News*, July 3, 2007: http://www.world-nuclear-news.org/regulation Safety/ (accessed July 12, 2007).

78. Matthew Wald, "Nuclear Firms Request Rules to Combat Aerial Attacks," *New York Times*, December 25, 2006. Commissioner Jaczko was in favor of strengthening the NRC's proposal to include tougher rules for protecting reactors against aircraft crashes.

79. "New U.S. Nuclear Reactors Not Required to Withstand Strike by Large Aircraft, NRC Rules," *Global Security Newswire*, April 27, 2007: http://www.nti.org/d_newswire/ (accessed July 12, 2007).

80. Associated Press, "Information on How to Sabotage Nuclear Plant Available to Public," *Atlanta Journal and Constitution*, November 25, 1979.

81. A. Fainberg, and A. M. Bieber, Brookhaven National Laboratory *Barrier Penetration Database*, prepared under Contract No. E-1-76-C-02-0016 to the Division of Operating Reactors, Office of Nuclear Reactor Regulation, U.S. Nuclear Regulatory Commission (NUREG/CR-0181, published July 1978).

82. Christina Meyer, Jennifer Duncan, and Bruce Hoffman, *Force-on-Force Attacks: Their Implications for the Defense of U.S. Nuclear Facilities* (Santa Monica, CA: Rand Corporation, 1993).

83. Bruce Hoffman, Christina Meyer, Benjamin Schwarz, and Jennifer Duncan, *Insider Crime: The Threat to Nuclear Facilities and Programs* (Santa Monica, CA: Rand Corporation, February 1990), 33.

84. Ibid., 41.

85. Konrad Kellen, "The Potential for Nuclear Terrorism: a Discussion," in Paul Leventhal and Yonah Alexander, ed., *Preventing Nuclear Terrorism* (Lexington, MA: Lexington Books, 1987), 109. Note that some of these groups continue to be active, while the activities of others have waned.

86. Ibid.

87. Intelligence Resource Program of the Federation of American Scientists, "Liberation Movements, Terrorist Organizations, Substance Cartels, and Other Para-State Entities," http://www.fas.org/irp/world/para/index .html (accessed March 19, 2007).

88. Nicholas D. Kristof, "At Trial in Tokyo, Guru Says Aim Was to Give 'Ultimate Joy,'" *New York Times*, April 25, 1996.

89. Southern Poverty Law Center, "Active U.S. Hate Groups," http:// www.splcenter.org/intel/map/hate.jsp (accessed June 13, 2009).

90. Klanwatch Project, *False Patriots: The Threat of Anti-Government Extremists* (Montgomery, AL: Southern Poverty Law Center, 1996), 8. See also Intelligence Project, Southern Poverty Law Center, *Intelligence Report* (Spring 1998), 6–28.

91. Ibid., 17.

92. Bruce Hoffman, *Terrorism in the United States* (Santa Monica, CA: Rand Corporation, January 1986), 51.

Part II

What Could Go Wrong?

3

Losing Control
(of Dangerous Materials)

The proliferation of dangerous technologies has spread vast stockpiles of nuclear weapons, radioactive materials, toxic chemicals, and the like all over the globe. At the same time, there has been an enormous growth in the organization and sophistication of organized criminals and terrorists, the two groups that most find these inventories a tempting target. It is therefore urgent that strict control be exerted over dangerous inventories at all times and in all places. They must be protected against theft or damage by outsiders and safeguarded against damage or unauthorized use by insiders. The fences, walls, vaults, locks, and guards that impede the entry of outsiders will not stop insiders who know protection systems well enough to circumvent them. There are always some insiders authorized to get past all the protective barriers. Preventing unauthorized use by authorized personnel requires record-keeping "detection" systems designed to track the location and condition of everything in the inventory and warn of anomalies quickly enough to be useful.

"Protection" and "detection" are problems common to all inventories, whether they contain candy bars, shoes, diamonds, gold, toxic waste, nerve gas, or nuclear weapons. The systems designed to solve these problems also have a lot in common, including their inherent limitations. Because there are such strong incentives to track and protect stockpiles of fissile material and weapons of mass destruction as completely as possible, looking at systems designed to control them should help us understand the theoretical and practical limits of our ability to control inventories of dangerous material in general.

KEEPING TRACK OF DANGEROUS MATERIALS

It is impossible to know the exact location and condition of every item in any inventory of size constantly and with certainty. There is always some degree of error, reflected in all inventory control systems by a special "margin of error" category. The U.S. Nuclear Materials Management and Safeguard System is the official government system for keeping track of fissionable materials such as plutonium and highly enriched uranium. It too has such a category. Before 1978, it was called "material unaccounted for" (MUF). After 1978, it was renamed "inventory difference" (ID). MUF or ID, it is defined as "the difference between the quantity of nuclear material held according to the accounting books and the quantity measured by a physical inventory."[1]

If it is known that some plutonium was destroyed in a fire, that missing material is considered "accounted for" and is not included in MUF/ID. Plutonium believed lost as a result of machining operations (such as cutting or grinding the metal) or trapped somewhere in the miles of piping in a facility is also not included in MUF/ID. Even if plutonium has been lost but permission has been granted to write it off for whatever reason, it is still not included in the MUF/ID. In other words, the MUF/ID reflects the difference between what the record-keeping system shows should be in the inventory and what a direct physical check of the inventory shows is actually there. That much material is out of control.

Since some discrepancy between what the books say and what is actually in the physical inventory is inevitable, a decision must be made as to how large a difference is too large. Decades ago, the NRC set standards for this "acceptable" level of inaccuracy, calling it the "limit of error on inventory differences," or LEID. Concluding that the NRC's statistical tests on shipments of nuclear fuels had become so muddled that they were meaningless, staff statisticians decided to create a new LEID index in 1980. It was defined so that there would statistically be 95 percent confidence that no thefts or losses had occurred as long as the absolute value of the actual inventory difference (ID) was less than the LEID. They applied this test to ID and LEID data on 803 nuclear materials inventories between 1974 and 1978. Assuming no loss or diversion, the absolute value of the actual ID should have been greater than LEID in only about 40 cases (5% of 803). In fact, ID exceeded LEID for 375 inventories, nearly half the cases. The staff's understated conclusion: "Something is wrong!"[2]

It is important to understand that the mere fact there is an error in the inventory does not mean that anything has been stolen or otherwise "diverted" to unauthorized use. But it does mean that it *could have been* stolen or diverted without the record-keeping system noticing. In the words of the Department of Energy (DoE) itself, "It is not prudent to discount the fact that a small inventory difference could possibly be due to theft. . . . "[3]

How Good Can Detection Be?

In the mid-1970s, the Atomic Energy Commission (AEC) decided to put together a special study group to look into the problem of safeguarding weapons-grade nuclear materials. The group included a former assistant director of the FBI, an MIT mathematician, a consultant on terrorism, and one representative each of the nuclear weapons laboratories and nuclear industry. They addressed the question of just how good a record-keeping system could be. With respect to both plutonium and uranium, the study group concluded:

> Because of the finite errors in the methods of analysis; because these errors cannot be reduced to a size which is much smaller than has been experienced in the past; . . . because of the human factors and the statistics of the system; there does not appear to be any way in which the measurement of the total inventory of an operating plant, or even a large segment of that inventory, can ever be known to better than one-tenth of one percent. At the present time, the real possibilities are much closer to one percent error. . . .[4]

A decade and a half later, in June 1988, the European Parliament inquired into the inventory safeguards system of the International Atomic Energy Agency (IAEA), the agency charged with detecting the diversion of nuclear materials. Like the AEC study group, the IAEA concluded that a 1 percent uncertainty in measuring plutonium inventories was to be expected.

IAEA's standards require that it be able to detect the diversion of any "*significant* quantity" of nuclear material, defined as enough to make one crude bomb or several sophisticated weapons. For example, 8 kilograms (17.6 pounds) of plutonium is considered a "significant quantity." Yet in 1986, a leaked confidential IAEA safeguards report indicated that the agency's real goal was only to detect the diversion of one to six significant quantities of nuclear material, a disturbingly large margin of error. Even if IAEA performance were up to its own standards, there would still be a 5 percent chance that a significant diversion of plutonium would not be detected.[5]

In February 1996, the Department of Energy (DoE) issued a landmark report, *Plutonium: The First 50 Years*, focused on America's production, acquisition, and use of plutonium from the dawn of the nuclear age (1944) through 1994. It gives 50-year MUF/IDs for each of the seven major DoE sites around the United States as well as the combined total for the inventories of the Departments of Energy and Defense.[6] The MUF/ID averaged 2.5 percent of the inventory—much higher than any of the preceding analyses predicted.[7]

The DoE report claims that only 32 percent of the inventory difference occurred since the late 1960s, indicating that tracking accuracy improved. That is heartening news. But even if the system had been that accurate throughout the entire 50-year period, the MUF/ID would have still been close to one percent.

If the IAEA detects anything unusual at the facilities it oversees (or if the host government is interfering with its inspections, as in Iraq before the 2003 U.S.-led invasion), its governing board can only notify its member nations, the UN Security Council and the UN General Assembly. It has no direct enforcement powers. What happens if the MUF/ID is too high at a nuclear facility within the United States? According to the DoE, " ... both DoE and contractors operating DoE facilities have always investigated and analyzed each and every inventory difference to assure that a theft or diversion has not occurred."[8] However, a former NRC safeguards official painted a different picture:

> The NRC shows the flag and sends a taskforce to investigate, they have one or two reinventories, they find or lose 3 or 4 kilograms and arrive at a final figure. They then look at the whole operation and say the material unaccounted for could have disappeared into the river or hung up in the pipes or whatever, but say they find no direct evidence of theft or diversion. It's just a ritual ... they are playing a game and nobody really believes in the accounting system.[9]

That may have been the case in the era before 9/11, but surely strong efforts have been made to improve the accuracy of accounting and other detection systems since 2001. Precisely how successful these efforts have been I cannot say. But it is hard to be optimistic, considering the kinds of roadblocks (including actions by the NRC itself) that have interfered with the need, even more obvious since 9/11, to strengthen nuclear plants to withstand large airplane crashes (discussed in Chapter 2) for more than eight years now since 9/11.

Of course, the MUF/ID for fissile materials is not just a problem in the United States. As of January 2009, there were nine nations armed with at least some nuclear weapons and 436 nuclear power reactors operating in 30 countries on five continents. The seriousness of these inventory-control problems in Russia led to the establishment of the Material Protection Control and Accounting Program between the U.S. DoE's national labs and their Russian counterparts in 1993.[10] The Soviet system of record keeping was so poor that at least as late as 1996 Russia still did not have accurate records of the quantity, distribution, and status of nuclear materials at many of the 40 to 50 nuclear locations and 1,500 to 2,000 specific nuclear areas throughout the former Soviet Union.[11] Though security practices in Russia have improved since then, too much dangerous material and technology may have already slipped out of control.

In the United Kingdom, the 1999 MUF for the Sellafield nuclear complex alone included 24.9 kg of plutonium (about 4.0 to 5.0 kg is adequate for a well-designed nuclear weapon); another 5.6 kg was included in the MUF for the complex in 2001, and 19.1 kg more was included in 2003.[12] A spokesperson for the operating company said that plutonium MUF was "normal:" "It is impossible to measure absolutely exactly that amount of material going into the plant and the amount coming out because of the changes material undergoes in the process."[13] In 2004, 29.6 more kilograms of plutonium was MUF at Sellafield. The U.K. Atomic Energy Authority said, "The MUF ... figures for 2003/04 were all within international standards of expected measurement accuracies. ... "[14]

There are limits to how well even the best and most sophisticated detection systems can operate when fallible human beings are involved—and fallible human beings are always involved. MUF/ID (or its equivalent) will never be zero for any large-size inventory over any extended period of time.

In 2005, Frank von Hippel, former Assistant Director for National Security in the White House Office of Science and Technology Policy, presented estimates of world weapons-grade inventories at 155 to 205 metric tons of plutonium and 850 to 1,550 metric tons of uranium to the U.N. Secretary Generals' Advisory Board on Disarmament Affairs. Considering that 25 kg of uranium or 4 kg of plutonium are sufficient to make a nuclear weapon, he estimated these inventories were the equivalent of 75,000 to 115,000 nuclear weapons.[15] Even achieving the lowest MUF/ID considered theoretically possible by the AEC study (0.1%) would leave enough fissile material in the "uncontrolled fringe" for 75 to 115 city-destroying nuclear weapons or 1 million lethal doses if dispersed as an aerosol terrorist weapon.[16]

Tracking Weapons and Related Products

There is no publicly available information on the structure, capabilities, or performance of the systems used to keep track of weapons of mass destruction. We could use the MUF/ID from the plutonium accounts as an estimate of the weapons MUF/ID, but tracking the flow of materials passing through processing plants is different from keeping track of finished products like canisters of nerve gas or nuclear weapons. There is some spotty and largely anecdotal information publicly available on the control of military inventories of conventional weapons, ammunition, and other supplies, which might be an alternative guide.

According to the Army Inspector General, record keeping was so bad in the mid-1980s that the amount of ammunition and explosives lost by the Army each year *could not even be determined*.[17] The GAO concluded, "Controls are inadequate to detect diversion."[18] Overall inventory adjustments for the Air Force supply system *doubled* from 1982 to 1985, from 2.6 percent of inventory value to 5.2 percent.[19] At the same time, the Navy's Judge Advocate General was reporting that deficiencies in inventory control were systemic to the Navy's "afloat supply system."[20]

The Pentagon and Congressional officials admit that some of the lost arms and explosives wind up in the hands of criminals and terrorists.[21] According to the Bureau of Alcohol, Tobacco and Firearms, military explosives were used in 445 bombings in the United States from 1976 to 1985.[22] Overall, "inventory adjustments" were nearly 5 percent of total inventory value for the American military as a whole in 1985, much higher than the MUF/ID rate for plutonium.[23]

In 1994, GAO looked into military record keeping for portable, handheld, non-nuclear, highly explosive, and "extremely lethal" missiles like Stinger, Dragon, and Javelin that could easily be used to destroy an airliner. They reported that "the services did not know how many handheld missiles they had in their possession because they did not have systems to track by serial numbers the missiles produced, fired, destroyed, sold and transferred ... we could not determine the extent to which any missiles were missing from inventory." Three years later, GAO found that "Discrepancies still exist between records of the number of missiles and our physical count . . . missiles may be vulnerable to insider theft."[24]

Apparently we even have trouble keeping track of the track-keeping records. In late 1997, the *New York Times* reported, "The Energy Department has disclosed in private correspondence that it cannot locate the records proving that it dismantled and destroyed as many as 30,000 nuclear bombs at weapons plants across the nation between 1945 and 1975."[25] The day before September 11, 2001, Secretary of Defense Donald Rumsfeld gave a talk at the Pentagon in which he made this extraordinary comment on the U.S. military's inability to keep accurate records: "According to some estimates, we cannot track $2.3 trillion in transactions. We cannot share information from floor to floor in this building because it's stored on dozens of technological systems that are inaccessible or incompatible."[26]

It would be comforting to believe that all of this was a problem only in the past, and tempting to assume that it has been taken care of since 9/11. Unfortunately, that does not seem to be the case. For example, a 2009 GAO auditing report found that the U.S. military was not properly keeping track of some "87,000 rifles, pistols, mortars and other weapons that it had shipped to Afghan security forces," about a third of the light arms it shipped them between December 2004 and June 2008—and had lost track of about another 190,000 light weapons it shipped to Iraq's security forces in 2004 and 2005. The auditors also concluded that "until June 2008, the military did not even take the elementary step of recording the serial numbers of some 46,000 weapons the United States had provided to the Afghans, making it impossible to track or identify any that might be in the wrong hands." The U.S. military apparently had no idea where another 41,000 weapons were, even though those serials numbers on those weapons had been recorded.[27]

In 2005, a New Jersey biodefense laboratory lost track of live mice infected with the plague, and in late 2008 the same lab was again unable to account for two plague-infected mice that had been frozen.[28] A U.S. Army investigation begun in 2008 found that the U.S. Army Medical Research Institute of Infectious Diseases (USARMIID) at Fort Detrick, Maryland, the largest U.S. military facility for work on deadly pathogens, could not find three vials of potentially lethal Venezuelan equine encephalitis virus. One Army official was quoted as saying, "We'll probably never know exactly what happened."[29] A spot inspection at USARMIID in early 2009 found four vials of that same virus that were not in its records. This is the same lab where the FBI's chief suspect in the deadly anthrax mailings of 2001 worked—the lab that suspended all research in February 2009 because of concern about whether its inventory of deadly pathogens was accurate.[30]

A report by the U.S. Army Audit Agency issued in February 2009 indicated that records for lethal nerve gas agents stored at bases around the United States did not match records for the amounts of material disposed of. According to the report, "They did not have effective procedures in place to insure amounts destroyed were accurately recorded in the recording system. Consequently CMA [the Army Chemical Materials Agency] didn't have complete assurance that amounts recorded in the system were accurate."[31] Even small amounts of nerve gas are extremely deadly. That same month, the Inspector General of the DoE released a report saying, "the department cannot properly account for and effectively manage its nuclear materials maintained by domestic licensees and may be unable to detect lost or stolen material." Auditors found that the DoE did not have an accurate accounting of the amounts and location of nuclear materials in 37 percent (15 of 40) of the facilities audited.[32]

Comparing records from five DoE reports that were publicly available, the Institute for Energy and Environmental Research found that roughly 300 kg of plutonium were unaccounted for at the Los Alamos National Laboratory (LANL), one of the nation's three nuclear weapons research labs, between 1996 and 2004.[33] In 2004, LANL was completely shut down for seven months after multiple incidents involving lack of proper security procedures in handling classified material;[34] in 2006, close to 1,600 pages of classified information—"including nuclear weapons material, were removed from a vault by a contractor who downloaded information onto a thumb-drive and removed it from the lab."[35] As a result, in early 2007 the head of the National Nuclear Security Administration (a key agency in overseeing the U.S. nuclear weapons program) resigned under pressure.[36] Although these lapses did not bear directly on the plutonium ID problem, they were indicative of less-than-ideal attention to security matters at the lab. As the chair and ranking member of the House Committee on Energy and Commerce put it in 2008, "Most frustrating was a culture that treated America's nuclear secrets like leftover napkins."[37]

Record-keeping problems will continue across many types of dangerous materials, despite best efforts. MUF/ID or its equivalent is an inherent feature of inventory record keeping; it will never be eliminated. But it is disconcerting to note that our efforts are all too often less than the best. Reflecting on the problem of controlling inventories of the nation's nuclear weapons in a November 2008 interview with the *Los Angeles Times*, Air Force Chief of Staff General Norton Schwartz said, "It seems to me there is a more modern way to maintain inventories of weapons and nuclear materials." By "modern" he had in mind using such innovations as barcodes or Global Positioning System trackers—both technologies that had long since passed into common civilian use.[38] If the general was right, it implies that at least as of late 2008 the average supermarket used a more modern approach to inventory tracking than the U.S. military was using to keep track of the nation's nuclear arsenal.

It is more conservative to use the much lower margin of error for plutonium to estimate the MUF/ID for nuclear weapons stockpiles. The lowest estimate of plutonium MUF/ID *capability* is 0.1 percent. As of the beginning of April 2009, there were an estimated 23,335 nuclear warheads in the global stockpile.[39] Therefore, even if the records of *all* the nuclear nations were 99.9 percent accurate (MUF/ID = 0.1%), there would still be more than 23 nuclear weapons unaccounted for at any point in time. That is enough to do catastrophic environmental damage and take tens of millions to hundreds of millions of lives.

It is important to emphasize that, by itself, none of this means that even a single nuclear weapon or microgram of plutonium has actually been stolen or diverted. It also does not mean that the amount of weaponry or plutonium in the MUF/ID is lying around in vacant lots or unprotected basements. What it does mean is that—as long as there are large inventories—even the best record-keeping systems cannot track products as dangerous as nuclear weapons or materials as dangerous as HEU and plutonium well enough to *assure* us that nothing has been stolen or otherwise found its way into unauthorized hands.

Why Is Accurate Detection Critical?

Despite the guards, locks, fences, and alarms, imperfect record keeping matters because we must still protect against the threat of internal theft or blackmail. Internal theft is theft by those authorized to have access to the inventory, including the guards assigned to protect it. Accurate record keeping is the first line of defense against the internal thief for whom the critical problem is avoiding detection, not getting in.

Flaws in detection give internal thieves the room they need, especially if what they are trying to steal can be removed a little at a time. One of the most notorious losses of nuclear materials was the disappearance of nearly 100 kilograms (about 220 pounds) of HEU over six to eight years from the Nuclear Materials and Equipment Corporation (NUMEC) plant in Apollo, Pennsylvania in the 1960s.[40] Four Hiroshima-power bombs could be made with that much uranium, yet over six years it could have been stolen by removing as little as 1.5 ounces a day (about the weight of a candy bar). Since uranium is heavier than lead, 1.5 ounces is a very small amount.

Internal theft and malevolence is an age-old, worldwide problem. In 1986, two government studies found that government employees committed most crimes against federal computer systems. According to one report, officials responsible for security "are nearly unanimous in their view that the more significant security problem is abuse of . . . systems by those authorized to use them."[41] Half a year earlier, the Chinese government revealed that top officials on Hainan Island took advantage of economic reform policies to embezzle $1.5 billion, earned by importing and reselling foreign consumer goods over 14 months. The amount of goods involved was staggering. It included 89,000 motor vehicles and 122,000 motorcycles, roughly equal to one-third of China's total annual domestic production.[42]

Inaccurate detection may also be enough to allow a blackmail threat to succeed. If an otherwise credible blackmailer threatens to attack a city with a stolen nerve gas bomb, extremely toxic chemicals, lethal biologicals, or a nuclear weapon that he/she has built from stolen materials—the threat must be taken seriously unless the authorities can be sure it is bogus. Decades ago, city officials in Orlando, Florida got a real-life lesson in exactly what it was like to face such a threat.

On October 27, 1970, a letter arrived at City Hall threatening to blow up Orlando with a nuclear weapon unless a ransom of $1 million was paid and safe conduct out of the country guaranteed. Two days later, city officials received another note, including a diagram of the bomb that the blackmailer claimed to have built, adding that the nuclear material required had been stolen from AEC shipments. City police immediately took the diagram to nearby McCoy Air Force Base and asked a weapons specialist there whether or not the bomb would work. It was not the best of designs, they were told, but yes, it would prob-ably explode. The authorities then contacted the FBI, which asked the AEC if any nuclear material was missing. They were told that there was nothing known to be missing, but then again there was this MUF. . . . So city officials put the ransom together and were about to pay it when the police caught the blackmailer. He was a 14-year-old high school honors student.[43]

Detection systems also play a critical role in highlighting unexplained losses of dangerous materials. With a sharp and timely warning, an intensive search can be launched to find out what happened and quickly recover any materials that have been diverted or simply misplaced. If there is no warning, there is no search. In January 1987, for example, a Mexican lumber dealer bought 274 wooden U.S. Army surplus ammunition crates from Fort Bliss in El Paso, Texas. When he unloaded the first truckload at his lumberyard in Mexico, he found *24, 4.5 foot long fully-armed rockets with high explosive warheads*, the type fired by helicopter gun ships against ground targets. Fortunately, the dealer immediately returned the weapons. According to procedure, the crates should have been inspected twice before they were sold, with Army inspectors signing "several documents" to certify that they were empty. Not only had the inspectors done an inexcusably sloppy job, but nothing in the Army's detection system had indicated that anything was wrong. Fort Bliss officials apparently had no idea that the rockets were missing until the lumber dealer told them.[44]

PROTECTION: SAFEGUARDING DANGEROUS INVENTORIES

There is a basic conflict between protecting inventories from theft or loss and keeping their contents readily accessible. More fences, walls, gates, guards, and checkpoints make it more difficult to divert materials, but these same controls also make it more difficult to get to the inventory for legitimate purposes. This is an especially vexing problem when rapid access is critical. In crisis or high alert, for example, militaries consider rapid access to battlefield weapons (even weapons of mass destruction) vital to their usefulness. Such "readiness" pressures guarantee that there will be less-than-perfect protection.

An ideal protection system would provide quick and ready entry for authorized uses and permit no entry otherwise. It is not enough to be able to withstand a violent and determined assault. It is also necessary to rapidly distinguish between authorized and unauthorized personnel and between legitimate and illegitimate uses by authorized personnel. Accomplishing this every time without error is simply not possible in the real world of imperfect human beings and the less-than-perfect systems they design and operate.

The full range of human and technical fallibility comes into play here. Guard duty is not a very pleasant job. Most of the time it is excruciatingly boring because most of the time nothing happens. Because nothing happens, it does not really matter if guards are alert or drowsy, clear-headed, and ready to act or half asleep—most of the time. At that crucial moment when thieves try to climb the wall and break into the warehouse or when terrorist commandos attack, if the guards are not fully alert, clear-headed, quick acting, and forceful, everything can be lost. The problem is, those crucial moments can come at completely unpredictable times. Therefore, guarding dangerous inventories requires constant vigilance. Yet constant vigilance is less and less likely as hours turn into days, weeks, months, and years in which nothing unusual ever happens.

Boredom dulls the senses. Extreme boredom is stressful, even maddening. It can lead to sluggishness, unpredictable mood swings, poor judgment, and hallucinations (see Chapter 7). People will do almost anything to distract themselves when they are trapped in deeply monotonous routines—from reckless and ultimately self-destructive things such as abusing drugs and alcohol (see Chapter 6) to things that are just plain stupid. Consider, for example, the disturbing, but perhaps not-so-strange case of Tooele Army Depot.

Guarding Nerve Gas at Tooele Army Depot

Investigative reporter Dale Van Atta of the Salt Lake City *Deseret News* did a series of articles about security conditions at Tooele Army Depot (TAD) in 1978.[45] Located in the desert of western Utah, at the time Tooele was the premier nerve gas storage area in the United States. An estimated 10 million gallons of deadly GB and VX nerve gases was stored there. Fatal within minutes of skin contact, even 1 percent that much nerve gas is enough to kill every person on earth.

The largely civilian guard force at Tooele does not appear to have been the world's most vigilant. According to Van Atta's sources, the "biggest problem in the force is constant sleeping on the job." One guard reportedly fell asleep on the transmitter button of his radio and snored undisturbed over the airwaves for an hour and a half. When a relatively new guard argued with his partner about pulling their vehicle over and going to sleep, "He pulled a gun on me and held it for ten minutes, ordering me to sleep," an incident that shows a certain lack of emotional stability, clear thinking, and good judgment apart from what it says about security practices. Not all that many people could go to sleep under orders at gunpoint. The supervisor to whom this incident was reported took no action, even though the guard who had threatened his partner admitted what he had done.[46]

When they were awake, the guards sometimes distracted themselves from the boring routine by drag racing on isolated roads within the Depot. "There was at least one every night, and it usually averaged about three races every night," one source said. According to one of the guards, "some supervisors will 'hot rod' their vehicles and try to elude guards chasing them. 'It's just ... something for them to do.' "[47]

Another thing for them to do was play cards. Marathon card and cribbage games among the guards were not rare. Arsonists burned down an old railroad station at TAD while guards on the night shift played poker. Alcohol abuse was also a problem. According to one guard, "If they were to give a breathalyzer when we all line up for duty, a couple dozen men would be sent home throughout the shifts." An Army captain, a security shift supervisor, reportedly came to work so drunk one night that one of his subordinates sent him home. "He knows he's an alcoholic, and all the men know it."[48]

When not sleeping, drinking, drag racing, or playing cards, some of the poorly paid guards were stealing. But one reported incident of malfeasance that came to light was far more serious than repeated petty theft. TAD used munitions workers and guards to count the nerve gas weapons, not independent inventory employees. In the summer of 1976, workers taking physical inventory at TAD discovered that there were 24 fewer 105mm nerve gas shells than the records showed in storage. Instead of reporting the loss, the men created a dummy pallet of 24 empty shells painted to look like live shells that contained deadly GB nerve gas, and then reported that the inventory matched the records. According to Van Atta's sources, at least eight men—including a lieutenant colonel—were involved in the cover-up.[49]

When the incident later came to light, the Army claimed that no nerve gas shells were actually missing. The problem, they said, had been an error in TAD's records, which showed 24 more shells received than the manufacturer had made.[50] Maybe so. Still, the willingness of TAD workers to cover up the apparent loss of two dozen deadly nerve gas shells is a horrifying breakdown of the inventory safeguard system. The workers who built the dummy shells had no way of knowing that the real shells were not in the hands of criminals or terrorists. There was enough GB gas in these shells to kill thousands of people. Yet they took deliberate action, which if undiscovered would have stopped anyone from trying to track down the missing nerve gas.

Suppose it had been taken by criminals or terrorists and used as a blackmail threat. Skeptical officials might have failed to take the threat seriously because the records showed nothing missing. Would the blackmailers, now angry and frustrated, have hesitated to carry out their deadly threat?

Unfortunately, this is not the end of the story. A few months later, TAD workers restacking pallets of nerve gas canisters found that one of the 24-shell pallets had been broken open, and one of the shells removed. The Pentagon reported five months later that its investigation of this incident was completed and "no conclusive accounting for the discrepancy was determined."[51] In plain English, they had no idea what happened to the missing nerve gas.

The most compelling test of any protection system comes when someone attempts to penetrate it. If guards can respond quickly and effectively at critical moments, it is not so important that they are not all that impressive the rest of the time. Could the troubled guard force at Tooele pull itself together and repel any outside intruders that might have attacked? According to Van Atta's sources, Tooele's high security areas were successfully infiltrated more than 30 times in the decade preceding his testimony. Fortunately, most were "friendly" assaults; that is, they were tests of the security system by special units of the Army and the Utah Army National Guard. During test intrusions, one group of men entered the most sensitive chemical area at the Depot and painted the date and time of their mission on top of two buildings they infiltrated; another group commandeered three vehicles and drove around a secured area for hours without being caught or even challenged. In another test, four prostitutes created a diversion that distracted six guards, while intruders entered a high security area and left a message behind. Other test intruders let the air out of the tires of vehicles in which guards were sleeping and left dummy dynamite sticks on the door of a chemical storage igloo.[52]

Then there were the real infiltrations. Several times guards discovered the tracks of motorcycles that appeared to have been shoved under outer fences and ridden in unauthorized areas. Unidentified individuals systematically broke the windows and slashed the tires of so many military vehicles that it was estimated it would have taken more than four hours for several men to do that much damage. A number of times, unauthorized armed men discovered inside the Depot told guards that they were rabbit hunters who had strayed— in every case the "rabbit hunter" was let go and the incident was never reported. In the words of an official letter of complaint signed by one frustrated TAD guard, "It is my opinion that the Tooele Army Depot Guard Force could not effectively counter an offensive by a troop of Girl Scouts. . . ."[53]

All this would have an air of comedy about it if it were not for the fact that these people were at the core of the protection system at a facility which stored 100 times enough nerve gas to kill every man, woman, and child on earth.

It was probably true that the majority of the guards at Tooele were conscientious, hardworking, and competent people. But it is certainly true that the protection system was a disaster waiting to happen. Why is all of this—which after all happened decades ago at a single facility—important to the broader issue of protecting dangerous inventories today?

If there were any real grounds for believing that highly peculiar circumstances had conspired to create a situation at Tooele unique in time and place, it would still be frightening to know that it could happen. What makes it important is that the most unusual thing about the situation at Tooele is that it became public knowledge. The conditions of work, human failings, attitudes, and organizational problems exposed by Van Atta's reporting are much more common than most of us would like to believe. That is not to say that all, or even most, dangerous inventories face protection as disastrously flawed as at Tooele in the 1970s. Yet there is plenty of evidence that the kinds of problems seen at Tooele are far from unique in time or place.

In 1996, the Vice President in charge of Westinghouse Hanford's tank program was quoted as describing the tank farm at DoE's Hanford nuclear reservation as "a Siberia for a lot of derelicts on the site."[54] These "derelicts" were tending 177 huge underground tanks (many the size of the U.S. Capitol dome) filled with enormous quantities of extremely radioactive, high-level nuclear waste, and 54 of these tanks of lethal radioactive material are known to occasionally build up flammable gases that could lead to a chemical explosion.[55]

Protecting American Nuclear Weapons

Beginning in the mid-1970s, there was a great deal of Congressional concern about the security of American nuclear weapons sites around the world. In September 1974, Rep. Clarence Long reported, "At . . . [one] overseas location . . . over 200 nuclear weapons are in storage . . . .

Less than 250 feet from the facility is a host nation slum which has harbored dissidents for years.... [T]he United States deploys nuclear weapons at some foreign locations hundreds of miles from the nearest American installation."[56] Long also revealed a Defense Department consultant's report that found: "[nuclear] weapons stored in the open, and [on] aircraft visible from public roads in the United States;... vehicles and forklifts in restricted areas which could be used by an attacking force to capture or carry away nuclear weapons;... superficial checks of restricted areas;... [and] inspections which have failed to report deficiencies.... "[57]

In September 1974, General Michael Davison, Commander of U.S. forces in Europe had told a West German audience that his troops would have difficulty protecting nuclear weapons against a determined terrorist attack.[58] Additional funding was provided to the Department of Defense for improving protection at American nuclear weapons facilities. Years later, a Cox Newspapers reporter named Joe Albright decided to investigate whether the deficiencies had been corrected. In February 1978, he testified before Congress:

> "... posing as a fencing contractor I talked my way past the security guards at two SAC [Strategic Air Command] nuclear weapons depots and was given a tour of the weak links in their defenses against terrorist attacks... I also purchased government blueprints [by mail] showing the exact layout of the two weapons compounds and the nearby alert areas where [nuclear-armed] B-52s are ready to take off in case of war.... [They] disclose a method of knocking out the alarm circuits... [and]. two unguarded gates through the innermost security fence....
>
> "As an imposter at that SAC base... I came within a stones throw of four metal tubes... that the Air Force later acknowledged were hydrogen bombs. At that moment, I was riding about 5 MPH in an Air Force pickup truck that was being driven by my only armed escort, an Air Force lieutenant. Neither the lieutenant nor anyone else had searched me or inspected my bulky briefcase, which was on my lap."[59]

Albright wrote a series of newspaper articles about his exploits. *After* they were published:

> An envelope containing a set of revised blueprints, disclosing... the wiring diagram for the solenoid locking system for the B-52 alert area, was mailed to me by the Corps of Engineers several days after Brig. Gen. William E. Brown, chief of Air Force security police, issued a worldwide directive to all Air Force major commands re-emphasizing vigilance against intruders... [60]

Protecting American Inventories: The Problems Continue

Neither the public revelations of the 1970s nor giving the Pentagon more money to beef up protection has ended these problems. The record of dangerous inventories protection problems is long and disturbing. A small sample follows:

- In the summer of 1982, a team of trained commandos hired to play "mock terrorists" walked into the center of the plutonium manufacturing complex at Savannah River, South Carolina. They also "visited" the nuclear bomb parts manufacturing operation at Rocky Flats in Colorado and the final assembly point for all American nuclear weapons at the Pantex plant in Texas.[61]
- In 1984, "mock" terrorists succeeded in three simulated infiltrations of the S site at Los Alamos National Laboratories. In two, they would have walked away with weapons-grade plutonium. In the third, they would have stolen an unlocked nuclear device constructed for the Nevada test site that could have been exploded within hours.[62]
- In October 1985, working with DoE investigators, an employee of the Pantex nuclear weapons plant smuggled a pistol, a silencer, and explosives into the top-secret facility. A few days later, those same weapons were used to steal nuclear bomb components, including plutonium from the plant. The DoE also discovered that some of the guards at Pantex had found yet another way to alleviate the boredom of guard duty—they were bringing friends on site and having sex with them in the guard towers. Although DoE officials said the security problems had been corrected, Texas Congressman John Bryant commented, "[W]e had such lax procedures that an amateur could get out of there with the most dangerous weapons in the world." Rep. John Dingel, Chair of the House oversight committee, added, "We found safeguards and security to be in a shambles at many of the nuclear sites."[63]
- During another exercise in 1985, mock terrorists escaped with plutonium after successfully "assaulting" the Savannah River Plant in South Carolina. The attackers succeeded even though the guards knew in advance that a test was coming—they had been given special guns that did not fire real bullets to be used during the test. The guards in a helicopter dispatched to chase the escaping "terrorists" could not shoot at them because they had forgotten their weapons. Twenty minutes after the attackers escaped, some guards at the facility were still shooting—at each other. Yet that same year the DoE gave the security company at the plant a nearly $800,000 bonus for "excellent performance."[64]
- In 1986, it was revealed that the Navy was using at least two ordinary cargo ships with *civilian* crews to ferry long-range, submarine-launched nuclear missiles across the Atlantic.[65]

- In August 1987, the U.S. Army in Europe finally managed to locate 24 Stinger missiles that Army Missile Command had asked them to find *nearly a year earlier.* Some Stingers were not stored securely, kept in lightweight metal sheds with "Stinger" stenciled on the outside.[66] Yet these surface-to-air missiles can destroy aircraft flying three miles high.[67]
- In October 1996, seven men, two of whom were civilian employees at Fort McCoy Army Base, were charged with stealing $13 million of military equipment, including 17 mobile TOW missile launchers and a Sheridan battle tank! All the equipment stolen was in operating condition.[68]
- During the six years up to September 2001, it was Rich Levernier's job to test the protection systems at 10 U.S. nuclear weapons facilities. Once a year he would direct military commandos posing as mock terrorists in attacks. The facility security forces always knew what day the attacks would occur months before. In spite of this, Levernier reported, "Some of the facilities would fail year after year. . . . In more than 50 percent of our tests of the Los Alamos facility, we got in, captured the plutonium, got out again, and in some cases . . . we didn't encounter any guards."[69] He said that security at nuclear weapons facilities as "all smoke and mirrors. On paper it looks good, but in reality, it's not . . . there are gaping holes in the system, the sensors don't work, the cameras don't work, and it just ain't as impressive as it appears."[70]
- In October 2004, the interim Iraqi government warned the United States that "nearly 380 tons of powerful conventional explosives—used to demolish buildings, make missile warheads and detonate nuclear weapons" were missing from the huge Al Qaqaa military installation that was supposed to be under U.S. military control. U.S. government officials and the American military acknowledged that the explosives disappeared sometime after the March 2003 invasion of Iraq.[71]
- In 2008, a U.S. Air Force investigation reported that most European sites used for deploying American nuclear weapons *did not meet the U.S. Defense Department's minimum security requirements.* In peacetime, the weapons at each of these sites are under U.S. Air Force control, but in time of war they would be transferred to the Air Force of the nation in which the site is located (upon authorization by the U.S. National Command Authority).[72]

A 1999 study by the President's Foreign Intelligence Advisory Board, "Science at Its Best, Security at it Worst," reportedly called the DoE "a large organization saturated with cynicism, an arrogant disregard for authority, and a staggering pattern of denial . . . [which has] been the subject of a nearly unbroken history of dire warnings and attempted but aborted reforms." It was simply "incapable of reforming itself."[73] Despite decades of investigations, hearings, newspaper accounts, special funding for improved security, and protestations by the Departments of Energy and Defense that decisive action has been taken,

American military stockpiles of weapons, explosives. and other dangerous materiel are still less than completely protected.

Both the private sector and local/state government have also experienced some of the very same problems. The U.S. Nuclear Regulatory Commission (NRC) tested the ability of armed terrorists to penetrate commercial nuclear power facilities. Serious security problems were discovered at almost half the nation's nuclear plants. In at least one mock attack, "terrorists" were able to sabotage enough equipment to cause a core meltdown. Yet the security testing program was eliminated in 1998 in the interests of cutting costs.[74]

Safeguarding Inventories in the Former Soviet Union

The military was not immune to the rapid rise in crime rates in the former Soviet Union during the late 1980s and early 1990s. The biggest problem was theft of small weapons, sold in a growing black market that helped arm the notorious organized crime gangs plaguing Russian society. But not all the weapons stolen were small.

In the spring of 1990, the Internal Affairs Ministry of the Republic of Armenia announced that 21 battle tanks had been stolen by two groups of armed men. The weapons were ultimately recovered. In two other separate incidents in Armenia, 40 armed men attacked one military base and as many as 300 armed men attacked another. The heavily armed Soviet state was breaking down. Fighting in January between Armenian and Azerbaijani groups near Baku posed a serious enough threat to major nuclear storage sites outside the city to cause elite Russian airborne units to be dispatched to the area.[75]

After the collapse of the Soviet Union, officials at the Russian nuclear regulatory agency (Gosatomnadzor) revealed that there were serious problems with the physical protection systems at both military and civilian nuclear facilities in Russia. They lacked operational devices to monitor doors and windows and to detect unauthorized entry. Guards were poorly paid, not all that well qualified, and not well protected. There were no vehicle barriers around the facilities, and communications between guards on-site and authorities off-site were primitive.[76]

Combined with the increasingly depressed state of the Russian economy, these protection problems predictably led to thefts, some of which have become public. In 1993, 4.5 kilograms of enriched uranium were stolen from one of the Russian Navy's main nuclear fuel storage facilities, the Sevmorput shipyard near Murmansk. Six months later, three men were arrested and the stolen fuel recovered: the man charged with climbing through a hole in the fence, breaking a padlock on the door, and stealing the uranium was the deputy chief engineer at Sevmorput; the man accused of hiding the stolen uranium was a former naval officer; and the alleged mastermind of the operation was the manager in charge of the refueling division at the shipyard![77]

Apparently, stealing uranium from Sevmorput was not that big a challenge. According to Mikhail Kulik, the chief investigator:

> On the side [of the shipyard] facing Kola Bay, there is no fence at all. You could take a dinghy, sail right in—especially at night—and do whatever you wanted. On the side facing the Murmansk industrial zone there are . . . holes in the fences everywhere . . . where there aren't holes, any child could knock over the half rotten wooden fence boards.[78]

If the thief had not made the amateurish mistake of leaving the door of the burglarized shed open, in Kulik's view the theft "could have been concealed for ten years or longer."[79]

At about the same time, a colleague of mine working for the House Armed Services Committee traveled to the former Soviet Union as part of an official group invited to visit military facilities there. Among other places, the group was taken to a nerve gas storage facility, a fairly simple wooden structure with windows in the woods. It had no apparent barriers or extensive guard force surrounding it, let alone any sophisticated electronic detectors. It did have what he described as a "Walt Disney padlock" on the front door. When the group asked how many active nerve gas shells were stored in this building, they were told that no accurate records had been kept.

In 1994, while he was a White House official, physicist Frank von Hippel visited the Kurchatov (nuclear research) Institute in Moscow, accompanied by American nuclear materials security experts. He reported that "There, in an unguarded building . . . [we] were shown 70 kilograms of almost pure weapons-grade uranium disks stored in what looked like a high school locker."[80]

Visits like these were part of growing concern in the United States about the security of weapons and dangerous materials in the former Soviet Union, leading to more U.S.-Russian cooperation on improving security during the 1990s. In 2006, Von Hippel and Glaser wrote, "This effort was spurred by theft of fresh, unburned HEU fuel in Russia and other countries of the former Soviet Union . . . No one outside Russia—and perhaps no one inside—knows how much may have been stolen."[81] U.S.-Russian cooperation has helped improve control of these inventories. But there is still work to be done. Glaser and von Hippel reported that Russia's Institute of Physics and Power Engineering (IPPE) in Obninsk had a large inventory of highly enriched uranium: "8.7 tons, mostly in tens of thousands of thin . . . disks about two inches in diameter. . . . " They go on to say, "Ensuring that no one walks out with any disks constitutes a security nightmare."[82]

In one of the more bizarre incidents of the post-Soviet Russia, a local state electric power utility, annoyed that Russia's Northern Fleet was $4.5 million behind in paying its electric bills, cut off power to the fleet's nuclear submarine bases on the Kola Peninsula in April 1995. A fleet spokesperson said "switching off the power for even a few minutes can cause an emergency," though he refused to elaborate.[83] From 1992 to 1995, there were 16 cases of electricity being deliberately cut off to Russian military bases because of nonpayment—including a cutoff of power to the central command of Russia's strategic land-based nuclear missile forces.[84] Aside from complicating inventory protection, such stoppages could cause dangerous disruptions in military command and control. The turmoil extended into the civilian sector. In December 1996, more than a dozen employees took over the control room of the nuclear power plant supplying most of the electricity to St. Petersburg and threatened to cut off power unless they received months of back pay. The next day 400 coworkers joined the protest.[85]

Without doubt, the most frightening news to emerge from Russia in recent times came from General Alexander Lebed. While national security advisor to President Boris Yeltsin, Lebed became concerned about the security of Russia's stockpile of some 250 one-kiloton nuclear "suitcase" bombs—nuclear weapons built to look like suitcases and carried by hand. Designed to be detonated by one person with 20 to 30 minutes preparation, no secret arming codes were required. Lebed ordered an inventory to assure that they were all safe. In an American television interview in September 1997, when he was asked if there were a significant number that are missing and unaccounted for, he replied, "Yes. More than 100." He went on to say the suitcase bombs were "not under the control of the armed forces of Russia. I don't know their location. I don't know whether they have been destroyed or whether they are stored. I don't know whether they have been sold or stolen." Next he was asked, "Is it possible that everybody knows exactly where all these weapons are, and they just didn't want to tell you?" He replied, "No." He was then asked, "You think they don't know where they are?" His reply: "Yes indeed."[86]

U.S. concerns about Russian weapons-grade material being stolen by or sold to terrorists, criminals, or hostile nations led to an agreement that the United States would pay Russia $12 billion for 500 metric tons of weapons-grade uranium extracted from the nuclear weapons it was scrapping. The uranium would be diluted to 4 percent U-235 by blending with natural uranium and used as reactor fuel. Unfortunately, implementation of the landmark agreement was slowed by a variety of problems.[87] By early 1998, only 36 of the 500 metric tons of highly enriched uranium (7%) contracted had actually been delivered.[88] Fortunately, the "Megatons to Megawatts" program accelerated during the next decade. By December 31, 2008, USEC had converted a total of 352 metric tons of HEU to reactor fuel—the equivalent of more than 14,000 warheads.[89]

Fabricating diluted weapons HEU into low-enriched uranium reactor fuel and the parallel option of turning plutonium into mixed oxide reactor fuel are safer than leaving them as weapons-grade metals.[90] But nuclear power is also a dangerous technology. "Burning" uranium and mixed oxide fuel in civilian nuclear power reactors does not eliminate the protection problem. It also generates more nuclear waste, which must then be securely stored. There are other ways of handling the plutonium and HEU extracted from weapons, but to date there is no "safe" option. This is one legacy of the nuclear arms race that we are going to have to live with, one way or another, for a long time to come.

SMUGGLING DANGEROUS MATERIALS

The inescapable limits imposed by both human fallibility and the inevitability of tradeoffs in technical systems prevent us from ever doing well enough at either record keeping or protection to completely safeguard inventories of materials that pose a potentially catastrophic threat to human life. But these limits do not inherently mean that catastrophic potential will be realized. We know we have lost track of frighteningly large amounts of dangerous materials. We also know that at least some of that material has been stolen. But is there any real evidence that the material has made its way to the black market or been put to unauthorized use?

More than once, bomb quantity shipments of nuclear materials were apparently diverted unintentionally—aboard planes hijacked to Cuba—during the 1960s and early 1970s. Fortunately, apparently neither the hijackers nor Cuban airport officials knew about the nuclear cargos, and the planes were returned with the nuclear materials still aboard.[91] The circumstances surrounding the loss of 220 pounds of highly enriched uranium (enough for 4 to 10 nuclear weapons) from the NUMEC plant in Pennsylvania in the mid-1960s were very suspicious. NUMEC had been repeatedly cited by federal inspectors for violating security regulations. More than a decade later, two declassified documents revealed that American intelligence agencies suspected Israel might have obtained the missing uranium by clandestine means and used it to manufacture nuclear weapons.[92]

According to a British television documentary aired in the late 1980s, an illicit trade in nuclear materials centered in Khartoum, Sudan began in the 1960s and persisted for decades. First Israel and later Argentina, Pakistan, and South Africa allegedly bought nuclear materials secretly on Khartoum's black market. When Sudanese authorities seized four kilograms of enriched uranium smuggled into the country in 1987, Sudan's Prime Minister reportedly acknowledged there was a black market in Khartoum. A few months later he reversed himself, denying that Khartoum was a center of the black market.[93] A former Sudanese state security officer claimed that the Israeli's were still buying HEU on Khartoum's black market during the 1980s, while lower-quality nuclear materials had probably been sold to Iraq, Iran, and Libya. Former CIA Director Stansfield Turner reportedly acknowledged that there was a black market in Khartoum for uranium that he believed was usually stolen from nuclear industry in Western Europe.[94]

In the 1990s, a series of reports of smuggling nuclear and other dangerous materials out of the former Soviet Union emerged:

- On June 30, 1992, NBC-TV reported on the trial of four Russians arrested by German police three months earlier for trying to sell nearly 1.5 kilograms of enriched uranium allegedly smuggled out of the former Soviet Union. NBC's Moscow correspondent claimed that a clandestine tape shot in Moscow showed an agent of the Russian military posing as a representative of a Polish firm. The Russian allegedly offered to sell 30 Soviet fighter planes, 10 military cargo planes, 600 battle tanks, and a million gas masks."[95]
- On May 10, 1994, German police accidentally found 5.6 grams of nearly pure plutonium 239 in the garage of a businessman named Adolf Jäkle while searching for other illicit material. Jäkle claimed he had gotten the sample of plutonium through a Swiss contact, and that as much as 150 kilograms of plutonium (enough for nearly 20 nuclear weapons) might have already been successfully moved out of Russia to storage in Switzerland.[96]

- On December 14, 1994, Czech police arrested three men and confiscated nearly three kilograms of HEU (87.7% uranium-235). Two were former nuclear workers (one from Belarus, the other from Ukraine), and the third was a Czech nuclear physicist.[97]
- In January 1996, seven kilograms of highly enriched uranium were reportedly stolen from the Sovietskaya Gavan base of Russia's Pacific Fleet. A third of it later appeared at a metals trading firm in Kaliningrad, 5,000 miles away.[98]
- In February 1998, 14 members of the Italian Mafia were arrested by Italian police and charged with attempting to sell 190 grams of enriched uranium and what they said were eight Russian missiles. This was the first documented case of trade in nuclear materials involving organized crime.[99]
- In September 1998, Turkish agents arrested eight men and seized 5.5 kilograms of U-235 and 7 grams of plutonium powder the men had allegedly offered to sell them for $1 million. One of the suspects was a colonel in the army of Kazakstan.[100]

In early 1995, a Western European intelligence report, which excluded "fake" smuggling by con artists, claimed that there had been 233 nuclear smuggling cases between 1992 and 1994.[101] International Atomic Energy Agency records listed 130 confirmed cases from 1993 to 1996.[102] Variations in numbers notwithstanding, there is no doubt that nuclear smuggling is no longer a theoretical possibility. In the 1990s, it was a growth industry.

The United States has had, and continues to have, problems of its own in trying to forestall smuggling of nuclear and other dangerous materials into the country. The *Washington Post* reported that between September 11, 2001 and April 2006, the United States had spent more than $5 billion on homeland security systems designed to detect nuclear or radiological material that smugglers might be trying to take into the country. Not all of the equipment performed up to specifications. For example, one $300 million system could not tell the difference between uranium and cat litter; another $1.2 billion set of baggage screening systems proved no better than previous equipment.[103] In 2006, GAO investigators successfully smuggled enough radioactive cesium 137 across the Canadian and Mexican borders into the United States to make two "dirty bombs." They set off radiation alarms, but they were still allowed to cross into Washington state and Texas, respectively.[104] In 2007, GAO's assistant director for forensic audits and special investigations, John Cooney, put it this way, "Security on ... [the Canadian] border has not really increased too much since the French and Indian War. ... " Senator Charles Grassley said, "They're simply wide open, waiting to be crossed by anyone carrying anything—even a dirty bomb or a suitcase-type nuclear device."[105]

Of course, these problems are not confined to the United States and the countries of the former Soviet Union. There have been more than three decades of media reports in India and elsewhere concerning alleged thefts of yellowcake (partly processed uranium) from India's Jaduguda mine complex.[106] Yellowcake is a raw material that can be used for producing nuclear fuel or nuclear weapons, but it is not highly radioactive and it does requires substantially more processing to be converted to a form that is directly useful for these purposes. On August 23, 2007, the state-run media in China "reported that four men from China's Hunan province . . . were standing trial for attempting to illegally sell eight kilograms of refined uranium." It too was probably yellowcake.[107] In addition, the problems posed by nuclear smuggling are not restricted to weapons and nuclear materials per se. In June 2008, the *New York Times* reported a chilling story: "American and international investigators say that they have found the electronic blueprints for an advanced nuclear weapon on computers that belong to the nuclear smuggling network run by Abdul Qadeer Khan, the rogue Pakistani nuclear scientist." The design was found on computers in a number of countries, including Switzerland and Thailand.[108] Since they were digitized and easy to copy, investigators had no way of knowing how many copies were in circulation.

There is no question that the economic and political turbulence of the early post-Soviet years made adequate control of dangerous inventories more of a problem for the countries of the former Soviet Union. Yet, as decades of experience with dangerous U.S. inventories has shown, inadequate control is a problem even in the absence of such turbulence. The real lesson of this long, troubled history has nothing to do with the particulars of American, Soviet/Russian, or other national societies. What we are seeing is not the result of some odd set of inexplicable flukes. It is instead a painfully clear demonstration that there are inherent, unavoidable limits to securing dangerous inventories anywhere.

In every specific case, there are particular personnel deficiencies, technical flaws, or organizational problems to cite as the reason for the failure of control. Because there is always a proximal cause, there is always the temptation to think that some clever organizational or technical fix will eliminate the problem by eliminating that particular source of trouble. It will not. There are two basic problems involved:

the inherent, unavoidable limits to what technology can accomplish and the inability of fallible human beings to do anything perfectly all the time.

We can and certainly should do better than we have done. The technology, the design and organization of detection and protection systems, and the training of personnel can all be improved. No matter how much we improve them, however, there is no question that from time to time, in one form or another, there will still be losses of control.

NOTES

1. U.S. Department of Energy, *Plutonium: the First 50 Years* (Washington, D.C.: February 1996), 52.

2. Eliot Marshall, "Nuclear Fuel Accounts Books in Bad Shape," *Science* (January 9, 1981), 147–148.

3. Op. cit., U.S. Department of Energy, 52.

4. Special AEC Safeguards Study Group report as excerpted in U.S. Senate, *The Congressional Record: Proceedings and Debates of the 93rd Congress, Second Session* (Washington, D.C.: April 30. 1974), S 6625.

5. Richard Bolt, "Plutonium for All: Leaks in Global Safeguards," *Bulletin of the Atomic Scientists* (December 1988), 14–17.

6. Total cumulative acquisitions of plutonium from all sources amounted to 111,400 kilograms. Total cumulative inventory difference (ID) was given as 2,800 kilograms. Op. cit., U.S. Department of Energy, 22 and 54.

7. Ibid., 22 and 52–54.

8. Ibid., 52.

9. Constance Holden, "NRC Shuts Down Submarine Fuel Plant," *Science*, October 5, 1979, 30.

10. Leslie G. Fishbone, "Halting the Theft of Nuclear Materials," *Scientific American*, February 2006, 62.

11. Michael R. Gordon, "Russia Struggles in Long Race to Prevent an Atomic Theft," *New York Times*, April 20, 1996.

12. Liam McDougall, "Enough Plutonium for Five Bombs 'Missing' at Sellafield," *The Sunday Herald*, December 28, 2003.

13. Ibid.

14. Alan Jones, "Accounts Blamed for Missing Plutonium," *The Independent*, February 17, 2005.

15. Frank Von Hippel, "The Nuclear Fuel Cycle: Fissile Material Control," Presentation before the UN Secretary–General's Advisory Board on Disarmament Affairs, New York: February 24, 2005, unpublished typescript.

16. In W. H. Langham, "Physiology and Toxicology of Plutonium–239 and Its Industrial Medical Control," *Health Physics* 2 (1959), 179, it is assumed that

humans would respond to plutonium as experimental animals do. On a weight-adjusted basis, this implies that a dose of 20mg in systemic circulation would be lethal to half the exposed humans within 30 days. Using Langham's 10 percent lung absorption rate, 200mg would have to be inhaled.

17. U.S. General Accounting Office, "Inventory Management: Problems in Accountability and Security of DOD Supply Inventories," Washington, D.C.: May 1986, 16–18.

18. Ibid.

19. Ibid., 30–32.

20. Ibid., 43.

21. Richard Halloran, "Defense Department Fights to Curb Big Losses in Arms and Supplies," February 12, 1987.

22. Op. cit., U.S. General Accounting Office, May 1986, 16.

23. U.S. General Accounting Office, "Inventory Management: Vulnerability of Sensitive Defense Material to Theft " (September 1997), 59. These "inventory adjustments" involve a wide variety of ordinary goods and materials, not just weapons and explosives. The weapons and explosives they do include are conventional, mostly small arms and ammunition.

24. Ibid., 1, 2, and 5.

25. Ralph Vartabedian, "Nuclear Bomb Records Are Lost, U.S. Says," *New York Times*, October 24, 1997.

26. Donald H. Rumsfeld, "DOD Acquisition and Logistics Excellence Week Kickoff—Bureaucracy to Battlefield," Remarks as Delivered by Secretary of Defense Donald H. Rumsfeld, The Pentagon, September 10, 2001: http://www.defenselink.mil/speeches/2001/s20010910–secdef.html (accessed October 12, 2004).

27. Eric Schmitt, "Afghan Arms Are at Risk, Report Says," *New York Times*, February 12, 2009.

28. "Dead Plague-Infected Mice Lost at New Jersey Biodefense Lab," *Global Security Newswire* (9, 2009: http://gsn.nti.org/siteservices/print_friendly.php?ID=nw_20090209_6075 (accessed June 17, 2009).

29. "Army Investigating Disappearance of Lethal Pathogen at Fort Detrick," *Global Security Newswire*, April 23, 2009: http://gsn.nti.org/siteservices/print_friendly.php?ID=nw_200904239776 (accessed June 17, 2009).

30. Yudhijit Bhattacharjee, "Army Halts Work at Lab After Finding Untracked Material," *Science*, February 13, 2009. Also Marcus Stern, "Top Military Biolab Suspends Research After Pentagon Finds Trouble with Tracking Pathogens," February 10, 2009: www.propublica.org (accessed June 27, 2009).

31. "Discrepancies Found in U.S. Nerve Agent Storage, Destruction Numbers," *Global Security Newswire*, February 9, 2009: http://gsn.nti.org/siteservices/print_friendly.php?ID=nw 20090209_8393 (accessed June 18, 2009).

32. Katherine Peters, "U.S. Energy Department Cannot Account for Nuclear Materials at 15 Locations," *Global Security Newswire*, February 9,

2009: http://gsn.nti.org/siteservices/print_friendly.php?ID=nw_20090224
_7895 (accessed June 18, 2009).

33. Keay Davidson, "Plutonium Could Be Missing from Lab," *San Francisco Chronicle*, November 30, 2005.

34. Kenneth Chang, "Los Alamos Stops Work in Crisis Over Lost Data," *New York Times*, July 17, 2004.

35. Derek Kravitz, "GAO: Security Lapses Continue at Los Alamos," Washington Watchdogs, *Washington Post Investigations*, July 2008.

36. David Morgan, "Nuclear Weapons Agency Chief Quits over Lapses," *Reuters*, January 4, 2007: http://www.reuters.com/article/topNews/idUSN 0420745220070105 (accessed June 15, 2009).

37. Op. cit., Derek Kravitz.

38. "U.S. Seeking to Modernize Nuclear Oversight, Air Force Chief Says," *Global Security Newswire*, November 26, 2008: http://gsn.nti.org/siteservices/ print_friendly.php?ID=nw_20081126_8719 (accessed June 18, 2009).

39. Federation of American Scientists, "Status of the World's Nuclear Forces 2009," http://www.fas.org/programs/ssp/nukes/nuclearweapons/nukestatus .html (accessed June 17, 2009).

40. David Martin, "Mystery of Israel's Bomb," *Newsweek*, January 9, 1978, 26–27.

41. David Burnham, "Computer Safety of U.S. Is Faulted," *New York Times*, February 26, 1986.

42. John F. Burns, "China Reveals Major Scandal Among Top Island Officials," *New York Times*, August 1, 1985.

43. Timothy H. Ingram, "Nuclear Hijacking: Now Within the Grasp of Any Bright Lunatic," *Washington Monthly*, January 1973, 22.

44. Michael Sawicki, "Fort Bliss Sends Rockets to Mexico in Procedural Error," *Dallas Times Herald*, January 30, 1987.

45. The following discussion of security conditions at Tooele Army Depot is based on the written and oral Congressional testimony of Dale Van Atta, including a series of articles he did for the *Deseret News* of Salt Lake City, which were reprinted along with his testimony. The reprinted articles by Dale Van Atta in the *Deseret News* included "World's Death Stored at TAD, and Security is Substandard" (1978); "Charges Loom in Nerve Gas Inventory Quirk" (January 16, 1978); " 'Don't Talk to Deseret News'—— Keep Mum TAD Workers Told" (February 6, 1978); "Anyone Free to Fly Over TAD Cache" (February 14, 1978); "Security at TAD Broken 30 Times in Decade" (February 17, 1978); and "FAA Restricts Flights Over TAD" (February 20, 1978). See *Military Construction Appropriations for 1979*, Hearings before the Subcommittee on Military Construction Appropriations of the Committee on Appropriations, House of Representatives, 95th Congress, Second Session (Washington, D.C.: U.S. Government Printing Office, 1978), 179–209.

46. Ibid., 177.

47. Ibid., 177 and 188.

48. Ibid., 176 and 187–188.

49. Ibid., 177–178 and 182.

50. Ibid., 177–178 and 182–184.

51. Ibid., 182.

52. Ibid., 176.

53. Ibid., 176–177 and 189.

54. Glenn Zorpette, "Hanford's Nuclear Wasteland," *Scientific American*, May 1996, 91.

55. Ibid.

56. Long, Representative Clarence, "Views of the Honorable Clarence D. Long, Submitted to Accompany Fiscal 1975 Military Construction Appropriations," September 24, 1974, 2–3.

57. Ibid., 4.

58. United Press International, September 27, 1974.

59. Joseph Albright, "Prepared Statement," *Military Construction Appropriations for 1979*, Hearings Before the Subcommittee on Military Construction Appropriations of the Committee on Appropriations, U.S. House of Representatives, 95th Congress, Second Session (Washington, D.C.: U.S. Government Printing Office, 1978), 139–140.

60. Ibid., 142.

61. Cox News Service, "Mock Raiders Break Security at Bomb Plant," *The West Palm Beach Post*, September 15, 1982.

62. Daniel Hirsch, "The Truck Bomb and Insider Threats to Nuclear Facilities," in Paul Leventhal and Yonah Alexander, ed., *Preventing Nuclear Terrorism*, Lexington Books, 1987, 218.

63. Bob Drummond, "Weapons Plant Fails 'Terrorist Test,' " *Dallas Times Herald*, February 12, 1987.

64. Ibid.

65. Associated Press, "Civilian Crews Transport N–Missiles Across Atlantic," *Dallas Morning News*, September 22, 1986.

66. U.S. General Accounting Office, *Army Inventory Management: Inventory and Physical Security Problems Continue* (Washington, D.C.: October 1987), 36–37.

67. Andy Lightbody and Joe Poyer, *The Complete Book of U.S. Fighting Power* (Lincolnwood, IL: Beekman House, 1990), 290.

68. Reuters News Service, LTD, "Seven Indicted for Thefts from Army Base," October 3, 1996, Yahoo! Reuters Headlines, http://www.yahoo.com/headlines/961003/news/stories/military_1.html.

69. Mark Hertsgaard, "Nuclear Insecurity," *Vanity Fair* (November 2003), 181.

70. Ibid., 183–184.

71. James Glanz, William Broad, and David Sanger, "Huge Cache of Explosives Vanished from Site in Iraq," *New York Times*, October 25, 2004.

72. DW–WORLD.DE, "European Nuclear Weapons Sites Lack Security, Says US Report," *Deutsche Welle*, June 21, 2008: http://www.dw–world.de/dw/article/0,2144,3428362,00.html (accessed June 18, 2009).

73. Op. cit. Mark Hertsgaard, 194.

74. Frank Clifford, "U.S. Drops Anti-Terrorist Tests at Nuclear Plants: Shrinking Budget Is Cited," *Los Angeles Times*, November 3, 1998.

75. John Fialka, "Internal Threat: The Soviets Begin Moving Nuclear Warheads Out of Volatile Republics," *Wall Street Journal*, June 22, 1990.

76. Oleg Bukharin and William Potter, " 'Potatoes Were Guarded Better,' " *Bulletin of the Atomic Scientists*, May/June 1995, 49.

77. Ibid., 46.

78. Ibid., 48.

79. Ibid.

80. Alexander Glaser and Frank von Hippel, "Thwarting Nuclear Terrorism," *Scientific American*, February 2006, 61–62.

81. Ibid., 59.

82. Ibid., 62.

83. Associated Press, "A Power Struggle in Russia's Military," *Dallas Morning News*, September 23, 1995.

84. Ibid.

85. Michael Specter, "Occupation of a Nuclear Plant Signals Russian Labor's Anger," *New York Times*, December 7, 1996.

86. Andrew and Leslie Cockburn, producers, CBS Television News, "60 Minutes," Steve Croft reporting (air date September 7, 1997).

87. Matthew L. Wald, "U.S. Privatization Move Threatens Agreement to Buy Enriched Unranium from Russia," *New York Times*, August 5, 1998. See also Peter Passell, "Profit Motive Clouding Effort to Buy Up A-Bomb Material," *New York Times*, August 28, 1996.

88. Op. cit., Matthew L. Wald.

89. U.S. Enrichment Corporation, "Megatons to Megawatts Program," http://www.usec.com/megatonstomegawatts.htm (accessed June 19, 2009).

90. Wolfgang K. H. Panofsky, "No Quick Fix for Plutonium Threat," *Bulletin of the Atomic Scientists*, January/February 1996, 3; and op. cit., Frans Berkhout et al., 29.

91. Op. cit., Timothy Ingram, 24. Though he would not give any details, the former chief of safeguards at the AEC at the time confirmed the general character of these reports.

92. "Mystery of Israel's Bomb," *Newsweek*, January 9, 1978, 26; David Burnham, "Nuclear Plant Got U.S. Contracts Despite Many Security Violations," *New York Times*, July 4, 1977; and David Burnham, "U.S. Documents Support Belief Israel Got Missing Uranium for Arms," *New York Times*, November 6, 1977.

93. "The Uranium Trade: All Roads Lead to Khartoum's Black Market," *Africa Report*, January–February 1988, 6. (The title of the British television documentary cited in the *Africa Report* article is "Dispatches: The Plutonium Black Market.")

94. Ibid.

95. Jim Maceda, NBC–TV, *NBC Nightly News*, June 30, 1992, transcript 6.

96. Mark Hibbs, "Plutonium, Politics and Panic," *Bulletin of the Atomic Scientists*, November/December 1994, 25–26; and op. cit., Phil Williams and Paul Woessner, 43.

97. Jane Perlez, "Tracing a Nuclear Risk: Stolen Enriched Uranium," *New York Times*, February 15, 1995; Michael R. Gordon, "Czech Cache of Nuclear Material Being Tested for Bomb Potential," *New York Times*, December 21, 1994; and op. cit., Phil Williams and Paul Woessner, 43.

98. Rensselaer Lee, "Smuggling Update," *Bulletin of the Atomic Scientists*, May/June 1997, 55.

99. Steve Goldstein, "A Nightmare Scenario Catches a Second Wind," *Philadelphia Inquirer*, January 10, 1999.

100. Ibid.

101. The report further claimed that Russian civilian nuclear research institutes did not even do physical checks of their inventories of nuclear materials, they only kept book records. Craig R. Whitney, "Smuggling of Radioactive Material Said to Double in Year," *New York Times*, February 18, 1995.

102. Op. cit. Lee, Rensselaer, 52.

103. "U.S. Readies Billions for Nuclear Detection," *Global Security Newswire*, April 17, 2006, http://www.nti.org/d_newswire/ (accessed April 18, 2006).

104. "GAO Gets Radioactive Material Across U.S. Borders," *Global Security Newswire*, March 28, 2006, http://www.nti.org/d_newswire/ (accessed March 28, 2006).

105. "GAO Describes Porous U.S.-Canada Border Security," *Global Security Newswire*, September 28, 2007, http://www.nti.org/d_newswire/ (accessed October 3, 2007).

106. "Smuggling of Uranium from India: Stories Persist," *WMD Insights*, a project of the Advanced Systems and Concepts Office at the Defense Threat Reduction Agency, June 2006, http://www.wmdinsights.com/I6/I6_SA2 _SmugglingOfUranium.htm (accessed June 20, 2009).

107. Stephanie Lieggi, "Uranium Smuggling Case in China Raises Concern," *Outside Publications by CNS Staff*, Center for Nonproliferation Studies, Monterrey Institute of International Studies, October 2007.

108. David E. Sanger, "Nuclear Ring Reportedly Had Advanced Design," *New York Times*, June 14, 2008.

4

Accidents Do Happen

Because all systems that human beings design, build, and operate are flawed and subject to error, accidents are not bizarre aberrations but rather a normal part of system life.[1] We call them accidents because they are not intentional, we do not want them to happen, and we do not know how, where, or when they will happen. But they are normal because, despite our best efforts to prevent them, they happen anyway.

Accidents and failures of dangerous technological systems differ from accidents and failures of other technological systems only in their consequences, not in their essence. Car radiators corrode and leak, and so do toxic waste tanks. Textile factories explode and burn, and so do nuclear power plants. Civilian airliners crash and burst into flame, and so do nuclear-armed bombers. The reasons can be remarkably similar; the potential consequences could hardly be more different.

Although all accidents involving dangerous technologies are cause for concern, accidents involving weapons of mass destruction are the most worrying. The capacity of these weapons to wreak havoc is the very reason they were created. Their frightful destructiveness makes them attractive to aggressors determined to terrorize opponents into submission as well as to those intent on defensive deterrence. We always knew just having these weapons was dangerous, but we reasoned that if our military forces were powerful enough, no nation—no matter how aggressive or antagonistic—would dare to attack us. We would be safe at home, and safe to pursue our national interest abroad without forceful interference. It seemed worth the risk.

In a world of nations that still have a lot to learn about getting along with each other, there is something to be said for having the capability to use force. Yet there are limits to what the threat or use of military power can achieve. The benefits do not continue to increase indefinitely as arsenals grow, but the economic costs do.[2] The likelihood of catastrophic human and technical failure also grows as they become larger, more sensitive, and more complex (see Chapter 9). With limited benefits and growing costs and risks, it is inevitable that ever-larger arsenals will eventually become a liability rather than an asset.

ACCIDENTS WITH WEAPONS OF MASS DESTRUCTION: POTENTIAL CONSEQUENCES

The accidental nuclear explosion of a nuclear weapon or the accidental dispersal of chemical and biological warfare agents is potentially as devastating as the intentional use of the weapons. The explosive power, radiation, and heat released by the detonation of a modern strategic nuclear weapon would dwarf that of the bombs that destroyed Hiroshima and Nagasaki in 1945. In a matter of seconds, those primitive bombs flattened what were thriving, bustling cities. Within minutes, much of what was left caught fire, engulfing many of those who survived the initial blast. Within weeks, thousands more died a sickened and lingering death as a result of the lethal doses of radiation they had received. The long-term effects of lower doses of radiation claimed still more victims—10, 20, 30, even 40 years after the explosion.

Nuclear weapons use conventional (chemical) explosives to trigger the nuclear explosion that gives rise to the weapon's devastating blast, radiation, and heat effects. Very precise conditions must be met for the conventional explosion to trigger the runaway fission/fusion reactions that produce a nuclear explosion.[3] As a result, if the weapon is not armed, there is very little chance that an accident could cause a full-scale nuclear detonation. However, the fissile material contained in the core of every nuclear weapon is highly radioactive. An accident in which the core was scattered by the conventional explosive, burned up, or otherwise released to the biosphere could do a great deal of damage, even in the absence of a nuclear explosion.

The nuclear core is made of either highly enriched uranium (HEU) or more commonly, plutonium (Pu).[4] From the standpoint of radiological damage to biological organisms, plutonium is much more dangerous. According to the respected sourcebook, *Dangerous Properties of Industrial Materials*, "The permissible levels of plutonium are the lowest for any of the radioactive elements. . . . Any disaster which could cause quantities of Pu or Pu compounds to be scattered about the environment can cause great ecological stress and render areas of the land unfit for public occupancy."[5]

A tiny amount of plutonium is extremely harmful. It is 30 times as potent as radium in producing tumors in animals.[6] It can be conservatively estimated from animal studies that inhaling no more than 200 milligrams (0.007 oz.) of plutonium will kill half the humans exposed within 30 days;[7] an inhaled dose of as little as 1 to 12 milligrams (0.00004–0.0004 oz.) will kill most humans (from pulmonary fibrosis) within one to two years.[8] Even a single microgram (0.000001 gram or 0.000000035 oz.) can cause lethal cancer of the liver, bone, lungs, and so forth after a latency period of years to decades. And a microgram of plutonium is a barely visible speck.[9]

It is reasonable to estimate that a "typical" plutonium-based nuclear weapon in the arsenal of one of today's nuclear armed countries would contain about 9 to 11 pounds (4 to 5 kilograms).[10] Taking the smaller figure to be conservative, a single "typical" nuclear weapon would contain a theoretical maximum of 20,000 doses lethal enough to kill within 30 days; 330,000 to 3,300,000 doses lethal enough to kill at least half the people exposed within one to two years; or 4 billion doses capable of causing cancer over the longer term. If an accident involving one such weapon resulted in people inhaling even 1 percent of its nuclear material, that would still be enough to kill more than 1,600 to 16,000 people within two years or to give up to 40 million people large enough doses to potentially cause them to eventually develop life-threatening cancer.

Chemical and Biological Weapons

Chemical and biological toxin weapons sicken and kill by disrupting the normal biochemical processes of the body. Probably the best-known class of chemical weapons of mass destruction are nerve gases, such as the World War II vintage gas sarin, used in the 1995 terrorist attack in the Tokyo subways. They are neurotoxins that kill by interfering with the enzyme cholinesterase that breaks down the chemical acetylcholine, which triggers muscular contractions. Symptoms of nerve gas poisoning include abnormally high salivation, blurred vision, and convulsions, usually leading to death by suffocation.[11]

In the years following the 1991 Persian Gulf War, thousands of American soldiers began to complain of a mysterious set of debilitating ailments including memory loss, chronic fatigue, joint pain, and digestive problems that came to be called the Gulf War Syndrome. It was not until June 1996 that the military finally admitted that American troops might have been exposed to Iraqi nerve gas. Over the next half year, it slowly emerged that Czech and French troops had detected chemical weapons seven times in the first eight days of the war, that U.S. jets had repeatedly bombed an Iraqi arms depot storing chemical weapons during the war, and that Army engineers had blown up a captured Iraqi storage bunker at Kamisiyah only days after the war ended that the Pentagon knew (but inadvertently failed to tell them) contained nerve gas.[12] The air strikes could have caused chemical agents to drift over thousands of American troops.[13] The Pentagon first estimated that only 400 American troops might have been exposed to poison gas when Army engineers destroyed the Kamisiyah bunker. They later reluctantly upped their estimate to 5,000, then to 20,000, then to 100,000.[14]

Live biological pathogens can also be used as weapons of mass destruction. Rather than poisoning the body directly, these organisms infect their victims, causing sickness and death through the same biological pathways they would follow if the infection had occurred naturally. Unfortunately, biological warfare has been a reality for a very long time. One of the worst epidemics in history began with an outbreak of the Black Death (a form of the plague) in Constantinople in 1334. In 1345 and 1346, the Mongol army laid siege to Kaffa, an outpost of Genoa on the Crimean peninsula. During the siege, the plague swept through the Mongol forces. The Mongols used a powerful catapult to fling the diseased corpses of their own troops into the city.[15] This early use of biological warfare is thought to have caused an outbreak of Black Death not only in Kaffa, but in much of Europe as well, when Genoese merchants later carried the plague to other Mediterranean ports.[16] Within 20 years the Black Death killed between two-thirds and three-quarters of the population of Europe and Asia.[17]

In the mid-1990s, the world began to learn about the horror of Japanese germ warfare activities during World War II. Japan had conducted a large-scale research program, experimenting with anthrax, cholera, and the plague, along with a number of other highly infectious pathogens. The Japanese Army carried out tests on captive human subjects in China who were forced to participate. At one testing area, victims were tied down while airplanes bombed or sprayed the area with deadly bacterial cultures or plague-infested fleas to see how lethal these biological weapons would be.[18] They also dropped infected fleas over cities in eastern and north central China, and succeeded in starting outbreaks of plague, typhoid, dysentery, and cholera. In the late 2000s,

China was still trying to get Japan to dispose of the remnants of up to 2 million chemical bombs (many corroded and leaking) still in Manchuria, where the Japanese had manufactured both those weapons and the deadly germ warfare agents 60 years earlier. It took more than four decades for news of Japan's World War II germ warfare activities to surface because American officials had made a deal with the Japanese not to reveal these activities or prosecute the perpetrators as war criminals in order to gain access to the data that the Japanese experimenters had gathered.[19]

During the Persian Gulf War of the early 1990s, the specter of biological attack arose again—U.N. inspectors searching Iraq after the war discovered that some Iraqi Scud missile warheads had secretly been loaded with anthrax, a deadly and extremely persistent infectious agent.[20]

In recent decades, remarkable advances in genetics have opened the possibility of engineering still more deadly organisms that do not exist in nature. There are those who claim that the recent appearance of previously unknown or rare "emerging viruses" may be the result of accidental releases from biological warfare research labs.[21] Whatever the truth may be, some emerging viruses are good prospects for bioweapons.

For example, the Ebola virus was first identified in Sudan and Zaire in 1976, nine years after its less dangerous but still deadly hemorrhagic cousin, the Marburg virus.[22] Ebola acts quickly, causing fever, headache, muscular pain, sore throat, diarrhea, vomiting, sudden debilitation, and heavy internal bleeding. It kills 50 percent to 90 percent of those infected within a matter of days. People of all ages are vulnerable, the incubation period is only 2 to 21 days, and there is no known cure. However, to date the virus has only been shown to be transmitted by direct contact with body fluids (such as blood or semen) and secretions, although there has been some debate as to whether it can also be transmitted less directly.[23] If some genetically modified version of Ebola and/or Marburg viruses proved to be transmittable through air or water, they would be nearly ideal biowarfare agents (provided, of course, that the side using it had some means of protecting its own people).

ACCIDENTS WITH WEAPONS OF MASS DESTRUCTION: LIKELIHOOD

The best way to assess the likelihood of any system's failure is to carefully analyze its structure, operating characteristics, and history. Because of the secrecy that surrounds the design and operation of systems involving weapons of mass destruction, it is very difficult to assess the likelihood of serious accidents this way. Furthermore, the historical record of accidents with these weapons tends to be incomplete and largely inaccessible. Until 1980, the U.S. military's official list of major nuclear weapons accidents, called "Broken Arrows," included a total of 13 accidents. Late that year, Reuters News Service made public a Defense Department document listing 27 Broken Arrows.[24] *After* Reuters' disclosure, the Pentagon officially admitted that even that list was incomplete—their records actually included 32 major U.S. nuclear weapons accidents since 1950, 31 of which occurred between 1950 and 1968. Where did these 18 additional Broken Arrows (31 minus 13) come from? Why were they not part of the Pentagon's official list in the late 1960s? When the *Associated Press* asked the Pentagon this, "officials said they are unable to explain why the Pentagon at that time [the late 1960s] did not report the larger figure of 31 accidents occurring prior to 1968."[25] Then, as 1990 approached, a newly declassified report from the nuclear weapons laboratories revealed that there had actually been 272 nuclear weapons involved in serious accidents between 1950 and 1968—accidents where there was an impact strong enough to possibly detonate the weapons' conventional high explosives.[26]

In June 1995, Alan Diehl, chief civilian safety official at the Air Force from 1987 to late 1994, publicly accused the military of covering up and playing down investigations of dozens of aircraft accidents. He documented 30 plane crashes (not necessarily involving nuclear weapons) that killed 184 people and destroyed billions of dollars worth of aircraft. Diehl charged that they had been covered up to protect the careers of senior military officers. The *New York Times* reported, "Crashes of military planes are commonplace, Pentagon records show. The Air Force has experienced . . . about one every 10 days in recent months. Five of its 59 Stealth 117-A fighters . . . have crashed in recent years."[27] According to Diehl, that is normal. About 10 percent to 20 percent of Air Force planes crash. A safety record like that would bankrupt any commercial airline.

The Public Record of Nuclear Weapons-Related Accidents

Table 4.1 of the Appendix shows the date, weapons system, location, and description of serious, publicly reported accidents involving U.S. nuclear weapons and related systems. Appendix Table 4.2 gives the same data for nuclear weapons-related accidents of other nuclear-armed countries.[28] Both tables include some accidents that involve major nuclear weapons-capable delivery vehicles (e.g., missiles, aircraft) where the presence of nuclear weapons is either unclear or was specifically denied. There are three reasons for including such accidents.

First, the track record shows that the involvement of weapons of mass destruction in an accident is downplayed or denied whenever possible. In fact, it is the U.S. military's longstanding official policy to neither confirm nor deny the presence of American nuclear weapons anywhere. Second, there is evidence that the distinction between major nuclear-capable delivery systems and those which actually carry nuclear weapons has often been more apparent than real. Testifying before Congress, retired Admiral Gene La Rocque stated, "My experience . . . has been that any ship that is capable of carrying nuclear weapons, carries nuclear weapons . . . they normally keep them aboard ship at all times except when the ship is in overhaul or in for major repairs."[29] But even if nuclear weapons were only present where their presence was specifically confirmed, accidents of this type would still be relevant. The absence of nuclear weapons at the time of an accident involving a system designed to carry them is fortunate, but their presence would obviously not have prevented the accident.

There is no clear evidence in the public record of the accidental nuclear explosion of a nuclear weapon. The closest we have come was a major disaster in the southern Ural Mountains of the USSR in 1957 and 1958, in which nuclear waste from the production of nuclear warheads exploded, apparently because of the buildup of heat and gas. The underground waste storage area erupted "like a volcano." Hundreds of people were killed outright, and many thousands more were exposed to radiation and forced to relocate. A 375-square-mile area of Russia was heavily contaminated with radioactivity.[30] There was also an incident in early February 1970, in which an explosion rocked the main Soviet nuclear submarine yard at Gorki, killing an unspecified number of people. Exactly what exploded was never made public, but it is noted in reports that radioactive material was subsequently found contaminating the Volga River downstream of the shipyard.

On January 24, 1961, we came very close to an accidental nuclear explosion of an American weapon when a B-52, mainstay of the strategic nuclear bomber force, crashed near Goldsboro, North Carolina. The plane carried two 24-megaton nuclear weapons. There was no nuclear explosion. One bomb fell into a field, where it was found intact. According to Ralph Lapp, former head of the nuclear physics branch of the Office of Naval Research, five of the six interlocking safety mechanisms on the recovered bomb had been triggered by the fall. A single switch prevented the accidental explosion over North Carolina of a nuclear weapon more than 1,000 times as powerful as the bomb that leveled Hiroshima.[31] Detonated at 11,000 feet, such a weapon would destroy all standard housing within a circle 25 miles and ignite everything burnable within a circle 70 miles in diameter.

On January 17, 1966 near Palomares, Spain, a B-52 bomber, carrying four 20 to 25 megaton hydrogen bombs, was being refueled in the air by a KC-135 tanker plane. Suddenly the planes lurched, and the fueling boom tore a hole in the side of the B-52. All four of the bombs fell out of the plane as it went down. One landed undamaged; the chemical explosives in two others detonated, scattering plutonium over a wide area of farmland.[32] The fourth bomb fell into the Mediterranean, triggering the largest underwater search undertaken up to that time. After three months, the bomb was found dented but intact, lying in the soft mud.

We have not been as lucky with all of the nuclear weapons accidentally lost at sea. On January 21, 1968, an American B-52 bomber with four megaton-class H-bombs on board crashed and burned in Greenland, melting through the seven-foot thick ice covering Thule Bay. The conventional explosives in all four of the bombs went off, scattering plutonium. What was left sank and remained below the waters of the bay. In October 1986, a fire and explosion caused a Soviet nuclear ballistic missile submarine to sink in water three miles deep, 600 miles northeast of Bermuda. The sub went down with 16 ballistic missiles and two nuclear torpedoes aboard—a total of 34 nuclear warheads. Eight years later, Russian scientists informed American experts that the sub had broken up, that "the missiles and warheads were 'badly damaged and scattered on the sea floor,' " and that it was "certain that the warheads are badly corroded and leaking plutonium and uranium."[33] Strong ocean currents in the area raised the likelihood of wide environmental contamination.

According to a report by William Arkin and Joshua Handler issued in June 1989 and updated by Handler in 1994, there were more than 230 accidents involving nuclear-powered surface ships and submarines between 1954 and 1994. Although some of these were not very serious, many were. Arkin and Handler report that as of the late 1980s, there were "approximately forty-eight nuclear warheads and seven nuclear power reactors on the bottom of the oceans as a result of various accidents."[34]

Then, on August 12, 2000, a torpedo (possibly two) exploded inside the Kursk, one of the newest and most powerful nuclear attack submarines in the Russian Navy's fleet—officially listed as "unsinkable," according to the *Moscow Times*.[35] The "unsinkable" Kursk, with its two nuclear reactors, sank to the bottom of the Barents Sea. Although officials denied it at the time, it is very likely that there were a number of nuclear weapons on board as well. All of the sub's 118 crewmembers died.[36] The Kola Peninsula, off the coast of which the Kursk sank, had already become a junkyard for nuclear-powered subs from the Soviet era. By 2008, there were 150 decommissioned subs in the junkyard, and a German company, EWN, had been working with the Russians for more than two years to extract the nuclear reactors from the hulks of the ships and to build and transfer the reactors to better storage facilities. This cooperative effort was slowly defusing a potential ecological "time bomb" or nuclear accident waiting to happen.[37]

Over the 60-year period from 1950 to 2009, 65 publicly reported nuclear weapons-related accidents occurred involving American nuclear forces, and 35 more involving the forces of other nuclear armed countries: 29 Soviet/Russian, 5 French, and 2 British, with one accident involving both French and British nuclear forces (see Appendix Tables 4.1 and 4.2). I do not claim that this is even a comprehensive list of publicly reported accidents, let alone a complete list of all the accidents that have occurred.

There is a curious pattern to these data. There are many fewer accidents reported publicly for the United States in the second half of this period than in the first (18%), but there are many more reported for the Soviets/Russians in the second half than in the first (63%). The American pattern could be the result of changes in policy. After 1968, for example, the Air Force stopped insisting that part of the strategic nuclear bomber force be constantly airborne. The increase in reported Russian accidents might be due in part to greater openness, with 12 of the 29 Soviet/Russian accidents occurring during or after 1985, when "openness" became official policy. In any case, it is reasonable to assume that there are more than twice as many U.S. as Soviet/Russian accidents listed because almost every kind of information has been more available in American society.

The Chinese are notable by their complete absence from Appendix Tables 4.1 and 4.2. It is highly unlikely that it is because they somehow managed to avoid all accidents involving nuclear weapons-related systems.[38] China has always been very quiet about any accidents involving their military forces. But things may be changing: China's military itself actually publicly announced a terrible accident involving a diesel submarine in the spring of 2003 that killed 70 crew members. "That it was disclosed at all surprised many defense experts here and abroad because military accidents in the past have been kept secret," one American reporter wrote from Beijing.[39]

Even assuming that every nuclear weapons-related accident that ever occurred was publicly reported, and is included in Appendix Tables 4.1 and 4.2, leaves us with a total of 100 nuclear weapons-related accidents from 1950 to 2009, *an average of close to one major nuclear weapons-related accident every seven months for 60 years*. Fewer accidents (25) are listed between 1985 and 2009, but that is still an average of one major nuclear weapons-related accident a year for a quarter of a century.

Other Accidents Related to Weapons of Mass Destruction

There have been a variety of other potentially dangerous accidents involving systems related to nuclear and other actual or potential weapons of mass destruction, associated equipment, and facilities. Some of these are listed in Appendix Table 4.3, along with a number of cases of errant military missiles. Two of the most serious occurred years ago in the former Soviet Union, and they are still shrouded in mystery and controversy.

In an accident that mimicked a biological warfare attack, a cloud of deadly anthrax spores was released into the air in April 1979 when a filter burst at Military Compound 19, a Soviet germ warfare facility in the Ural Mountains. As the population downwind inhaled the airborne spores, an epidemic of pulmonary anthrax broke out. Somewhere between 200 and 1,000 people died.[40] The second incident occurred on September 12, 1990, when an explosion and fire at a nuclear fuel plant in Ust-Kamenogorsk in eastern Kazakhstan released a "poisonous cloud" of "extremely harmful" gas over most of the city's population of 307,000. It is unclear from reports whether the cloud was radioactive or some gaseous compound of the highly toxic element beryllium. There was no indication of casualties, but Kazakh officials did ask the central Soviet government to declare the area around the city an ecological disaster zone.[41]

Table 4.3 also lists two examples of the military tendency to cover up questionable activities for a long time. It was not until 1975 that the U.S.

Army revealed that 10 to 20 years earlier, three employees at their biological warfare research facility in Ft. Detrick, Maryland had died of rare diseases being studied at the Lab.[42] In 1975 and 1976, the U.S. Army also admitted that it had conducted 239 secret, "open air" germ warfare tests from 1945 through the 1960s, in or near major American cities. In one test, live germs were released into the New York City subway tunnels.[43]

Design Flaws and Manufacturing Defects

Military systems—even those involving weapons of mass destruction—are not immune to such endemic problems of technical systems as flaws in design and manufacture (see Chapter 9). On May 23, 1990, nuclear arms experts from the Department of Energy and Congress announced that more than 300 American tactical nuclear warheads deployed in Europe had been brought back to the United States and secretly repaired over the preceding year and a half. The recall was to correct design defects that could have led to accidental detonations of their conventional explosives powerful enough to disintegrate the core of the weapons and scatter plutonium into the environment.[44] Unfortunately, this does not seem to have been all that rare an event.

A few days later, it was reported that more than a dozen types of American nuclear warheads with different designs had been temporarily withdrawn for repairs since 1961.[45] On June 8, 1990, then Secretary of Defense Cheney told the Air Force to withdraw hundreds of SRAM-A short-range nuclear-armed attack missiles from the bomber force until safety studies on their W-69 warheads were completed. Cheney finally did so two weeks after the directors of all three of the nation's nuclear weapons design laboratories (Los Alamos, Livermore, and Sandia) urged that the missiles be removed from bombers kept on 24-hour alert. The directors told the Senate Armed Services Committee about their concern that the warheads might explode, dispersing plutonium and other deadly radioactive materials, if they were subject to the extreme stress of a fire or crash.[46]

Space-Based Military Nuclear Power Accidents

Nuclear-powered military satellites and spacecraft also pose a real danger. On April 21, 1964, an American SNAP-9A military navigation satellite carrying a nuclear generator fueled by plutonium-238 failed to attain orbit. It burned up in the atmosphere.[47]

A launch malfunction on April 25, 1973 caused a Soviet nuclear-powered satellite to fall into the Pacific Ocean north of Japan.[48] Then, four years later came the most spectacular and widely reported failure of a nuclear-powered satellite to date. In January 1978, after months of orbital decay, the crippled Soviet nuclear-powered military spacecraft Cosmos 954 re-entered the atmosphere, scattering some 220 pounds of radioactive debris over an area of Canada the size of Austria. The satellite's reactor contained more than 100 pounds of uranium-235. Thousands of radioactive fragments were recovered after a painstaking, months-long search. Fortunately, the area over which the debris fell was only sparsely populated.[49] Five years later, another nuclear-powered Soviet military satellite, Cosmos 1402, fell into the Indian Ocean.[50] In late 1996, an automated Russia Mars probe carrying 200 grams of plutonium failed to escape the earth, re-entered the atmosphere, and plunged into the Pacific.[51]

The tragic explosion of the space shuttle Challenger during launch on January 28, 1986 could easily have been much, much worse. The very next Challenger launch was scheduled to contain a liquid-fueled Centaur rocket carrying an unmanned spacecraft powered by 46.7 pounds of plutonium. Had Challenger exploded on that mission and triggered an explosion of the Centaur, the satellite's plutonium would have been dispersed into the air. According to John Goffman, former Associate Director of the Lawrence Livermore nuclear weapons laboratory, if that had happened, "the amount of radioactivity released would be more than the combined plutonium radioactivity returned to earth in the fallout from all of the nuclear weapons testing of the United States, Soviet Union, and the United Kingdom—which I have calculated has caused 950,000 lung cancer fatalities. If it gets dispersed over Florida, kiss Florida good-bye."[52]

Many of the 70-plus nuclear-powered systems in space are in near-earth orbit. Nine nuclear-powered American and Soviet spacecraft—more than 10 percent of the number in space—have failed to achieve orbit or otherwise re-entered the atmosphere.[53]

There is also the danger posed by "space junk." Nicholas Johnson, a scientist and an expert on space debris at the Johnson Space Center in Houston estimated that by 2009 there were about 17,000 objects, from used up satellites to lost tools, circling the earth.[54] If a piece of this space debris smashed into an orbiting reactor, it could shatter it into "a million tiny pieces." Twenty years earlier, when Johnson was at Teledyne Brown Engineering, he estimated that some 1.6 tons of radioactive reactor fuel had already been put into orbit by the late 1980s. As to the chances that this radioactive material would be involved in a space collision, Johnson said in 1988 that if the number of satellites and space debris continued its pattern of increase, it was "a virtual certainty."[55]

He turned out to be right. On February 10, 2009, an operational Iridium 33 commercial satellite smashed into a no longer operational Russian Kosmos 2251 military satellite 500 miles over Siberia. The satellites had been in orbits perpendicular to each other. Even a very small piece of debris traveling at orbital speed can cause catastrophic damage to a spacecraft. But these were not small objects; the Iridium 33 weighed about 1,230 pounds, and the Kosmos 2251 weighed almost 2,100 pounds. The collision obliterated both satellites, creating a debris cloud of some 600 objects spreading out in altitude from the point of collision to add to the burden of already existing space junk. Johnson pointed out that the debris added to the chances of a slow motion orbital chain reaction that could eventually threaten other satellites.[56] Many of these are powered by devices that contain radioactive material.

Accidents with Nuclear Waste

On July 23, 1990, more than 30 years after the spectacular explosion of Soviet nuclear weapons waste in the Ural Mountains (described earlier), the report of a U.S. government advisory panel warned that it could happen here. The panel was headed by John Ahearne, a physicist who was formerly a high official in the Departments of Defense and Energy and Chair of the U.S. Nuclear Regulatory Commission. According to the authoritative study, there are millions of gallons of highly radioactive waste, accumulated during four decades of producing plutonium for nuclear weapons and stored in 177 tanks at the Department of Energy's Hanford nuclear reservation in Washington State. Heat generated inside the tanks or a shock from the outside could cause one or more of them to explode. Although the explosion would not itself be nuclear, it would throw a huge amount of radioactivity around. According to the panel, "Although the risk analyses are crude, each successive review of the Hanford tanks indicates that the situation is a little worse."[57]

In 1992, the DoE released another report on Hanford, listing four decades of accidents at that nuclear weapons facility. It was the most thorough overview of human error, management mistakes, and sloppiness at Hanford ever undertaken. It looked at fires, accidental explosions, failures of safety systems, incidents of fuel melting, and other events that exposed the workers to radioactivity and toxic chemicals. The site was found to be extensively contaminated, and it was judged that the waste tanks were still in danger of exploding or otherwise dispersing radioactive material.[58]

On March 11, 1997, the plutonium recovery facility at Tokai was the site of the worst nuclear accident in Japan's history. A fire started in the remote-controlled chamber in which liquid waste from the plutonium extraction is mixed with asphalt. Ten hours later, an explosion blew out most of the windows and some of the doors in the four-story concrete building. Containment systems failed, causing 37 workers to be exposed to radiation and allowing plutonium and other radioactive materials to escape into the air. Radiation was subsequently detected up to 23 miles away. Plant managers and the government said (as usual) that neither the workers nor the public had been exposed to harmful levels of radiation.[59]

ACCIDENTS WITH OTHER DANGEROUS TECHNOLOGIES

There have been many serious accidents involving other dangerous technologies—toxic chemicals and nuclear power among them. There is already an extensive and accessible literature on nuclear power that includes considerable discussion of the likelihood and consequences of accidents. And accidents at nuclear power plants and facilities handling dangerous chemicals also fit more naturally into, and illustrate more strikingly, our subsequent discussions of technical failure and human error (especially Chapter 9).

Accidents during shipment of hazardous chemicals are quite common. On February 29, 1987, or example, a 21-car train with nine tank cars filled with butane derailed in Akron, Ohio. The butane in two of the tankers escaped, starting a fire that spread to a nearby chemical plant with large storage tanks on-site. It would not have taken much more to turn what was a routine accident into a calamity.[60] In January 2002, five tank cars carrying liquefied ammonia gas ruptured when a Canadian Pacific Railway freight train derailed outside Minot, North Dakota. The toxic fumes killed one person and injured more than 300. In the summer of 2004, three people died from chlorine poisoning when a steel tank car filled with the gas broke open after a train derailment in Texas. On January 6, 2005, two freight trains collided in Graniteville, South Carolina. A ruptured car leaked chlorine gas that killed nine people, hospitalized 58, and sent hundreds more for treatment. The same train was also carrying other hazardous chemicals.[61]

Although all accidents with dangerous technologies are cause for concern, those involving weapons of mass destruction have among the greatest catastrophic potential. As long as massive arsenals of these weapons continue to exist, accidents will continue to happen. The more of these weapons there are, the more actively they are moved around, and the more nations possess them, the greater the probability of devastating accident. It is ironic. We have built these weapons to make ourselves secure. But especially in a post-Cold War world, where overwhelming nuclear deterrence adds little to the nation's security, the threat of accidental disaster these weapons pose has made them a source of insecurity instead.

The problem of individual weapons accidents is serious enough, but there is even a greater potential for disaster in the possibility that these weapons will again be used in war someday. As we shall see in the next chapter, that too can happen by accident.

NOTES

1. For an interesting sociological analysis of this phenomenon, see Charles Perrow, *Normal Accidents: Living with High Risk Technologies* (New York: Basic Books, 1984).

2. For a theoretical macroeconomic analysis of the long-term costs of using economic resources for military purposes, see Lloyd J. Dumas, *The Overburdened Economy: Uncovering the Causes of Chronic Unemployment, Inflation and National Decline* (Berkeley, CA: University of California Press, 1986).

3. The most common design for a nuclear weapon, the so-called "implosion design," wraps wedges of conventional explosive around a spherical nuclear core in which the atoms are not densely enough packed at normal pressure to sustain a runaway chain reaction. By directing the force of the conventional explosives inward, the implosion design causes a strong enough shock wave to pass through the nuclear core to temporarily overcome the weak electron forces that normally determine the spacing of the atoms. The nuclei are forced much closer together, raising the probability that the neutrons flying out of a splitting nucleus will strike and split another nucleus. A runaway nuclear fission chain reaction occurs, producing a devastating nuclear explosion. If the shock wave is not precisely uniform from every direction or the nuclear core is not spherical, the device will not work.

4. Samuel Glasstone and Philip J. Dolan, ed., *The Effects of Nuclear Weapons* (Washington, D.C.: U.S. Department of Defense and U.S. Department of Energy, 1977), 5.

5. N. Irving Sax, et.al., *Dangerous Properties of Industrial Materials* (New York: Van Nostrand Reinhold Company, 1984), 2247.

6. See W. H. Langham, "Physiology and Toxicology of Plutonium-239 and Its Industrial Medical Control," *Health Physics* 2 (1959): 179.

7. In W. H. Langham, "Physiology and Toxicology of Plutonium-239 and Its Industrial Medical Control," *Health Physics* 2 (1959): 179, it is assumed that humans would respond to plutonium as experimental animals do. On a weight-adjusted basis, this implies that a dose of as little as 20 mg in systemic circulation would be lethal to half the exposed humans within 30 days. Using the lung absorption rate of plutonium of 10 percent given by Langham, this would mean that a total dose of 200 mg would have to be inhaled in order for 20 mg to get into systemic circulation.

8. The following calculation is based on the data presented in John T. Edsal, "Toxicity of Plutonium and Some Other Actinides,"*Bulletin of the Atomic Scientists* (September 1976), 28–30. In experiments carried out on dogs, at doses of as little as 100–1,000 nanocuries per gram of lung tissue, most died from pulmonary fibrosis within one to two years (1 nanocurie=37 radioactive disintegrations per second; for plutonium, this means that 37 alpha particles are shot off per second). Given an average weight of a human lung of 300–380 grams (*source:* Dallas Medical Examiner's Office), there is about 760 grams of lung tissue in an average person. Assuming that every gram of lung tissue is dosed at 100–100 nanocuries, a total dosage of 76,000–760,000 nanocuries should certainly be lethal. For plutonium-239, one nanocurie equals 16.3 nanograms (billionths of a gram). Therefore, 76,000 nanocuries equal 1.2 milligrams (76,000 nc × 16.3 ng/nc = 1,238,800 ng = 1.2 mg); and 760,000 nanocuries equal 12.4 milligrams (760,000 nc × 16.3 ng/nc = 12,388,000 ng = 12.4 mg).

9. Paul Craig and John A. Jungerman, *Nuclear Arms Race: Technology and Society* (New York: McGraw-Hill, 1986), 304.

10. Depending on the particular type and design of a nuclear weapon, there will be varying amounts of plutonium (or highly enriched uranium) in its nuclear core. The more sophisticated the design, the smaller the amount of nuclear material required to produce a blast of a given size. Less sophisticated weapons and/or weapons designed to produce larger explosions will contain more nuclear material.

11. Robert Harris and Jeremy Paxman, *A Higher Form of Killing: The Secret Story of Chemical and Biological Warfare* (New York: Hill and Wang, 1982), 54; See also F. Bodin and F. Cheinisse, *Poisons* (New York: McGraw-Hill, 1970), 129–130; and William H. Harris and Judith S. Levey, ed., *The New Columbia Encyclopedia* (New York and London: Columbia University Press, 1975), 1910 and 2426.

12. Philip Shenon, "Gulf War Illness May Be Related to Gas Exposure," *New York Times,* June 22, 1996; "Czechs Say They Warned U.S. of Chemical Weapons in Gulf," October 19,1996; "U.S. Jets Pounded Iraqi Arms Depot Storing Nerve Gas," October 3, 1996; and "Pentagon Says It Knew of Chemical Arms Risk," February 26, 1997.

13. Philip Shenon, "U.S. Jets Pounded Iraqi Arms Depot Storing Nerve Gas," *New York Times*, October 3, 1996; and "Pentagon Says It Knew of Chemical Arms Risk," February 26, 1997.

14. Philip Shenon, "New Study Raises Estimate of Troops Exposed to Gas," *New York Times*, July 24, 1997.

15. The catapult they used, called a "trebuchet," was an even older technology. The Chinese invented it more than 1,500 years earlier.

16. Paul E. Chevedden, Les Eigenbrod, Vernard Foley, and Werner Sodel, "The Trebuchet," *Scientific American*, July 1995, 68.

17. William H. Harris and Judith S. Levey, ed., *The New Columbia Encyclopedia* (New York and London: Columbia University Press, 1975), 2161.

18. Nicholas D. Kristof, "Japan Confronting Gruesome War Atrocity: Unmasking Horror," *New York Times*, March 17, 1995.

19. Patrick E. Tyler, "China Villagers Recall Horrors of Germ Attack," *New York Times*, February 4, 1997. In one of the more bizarre episodes of World War II, the Japanese began to launch huge balloons containing conventional explosives at the American heartland in 1944. Carried by prevailing winds, 200 or so landed in the Western states. Some Japanese generals wanted to load the balloons with plague or anthrax-carrying germ warfare agents. Ironically, that plan was vetoed by the infamous general Hideki Tojo, whom the United States later executed as a war criminal.

20. Ibid. See also Leonard A. Cole, "The Spectre of Biological Weapons," *Scientific American*, December 1996, 62.

21. " 'Emerging Viruses' in Films and Best Sellers," *New York Times*, May 12, 1995; Jon Cohen, "AIDS Virus Traced to Chimp Subspecies," *Science*, February 5, 1999, 772–773. See also Bernard Le Guenno, "Emerging Viruses," *Scientific American*, October 1995, 58–59.

22. R. Preston, *The Hot Zone* (New York: Random House, 1994); and *Outbreak* (Los Angeles: Warner Brothers Studios, 1995); see also "Marburg Virus," *BBC News*, April 4, 2005, http://news.bbc.co.uk/2/hi/health/medical_notes/4408289.stm (accessed June 20, 2009).

23. Lawrence K. Altman, "Deadly Virus Is Identified in the Outbreak in Zaire," and contiguous boxed story "A Closer Look: The Ebola Virus," *New York Times*, May 11, 1995.

24. Reuters News Service, "List Reveals Pentagon Hid 14 N-Accidents," *Dallas Morning News*, December 22, 1980.

25. Associated Press, "U.S.: 32 Nuclear Weapons Accidents Since 1952," *The West Palm Beach Post*, December 23, 1980.

26. Chuck Hansen, "1,000 More Accidents Declassified," *Bulletin of the Atomic Scientists*, June 1990, 9.

27. Tim Weiner, "Safety Officer Says Military Hid Crashes," *New York Times*, June 24, 1995.

28. The sources of those accidents cited in the discussion that follows, which are not specifically noted here, can be found in Tables 4.1 and 4.2.

29. Admiral Gene La Rocque, Statement on "Proliferation of Nuclear Weapons," Hearings, Subcommittee on Military Applications, Joint Committee on Atomic Energy, 1974.

30. More than 40 years after this terrible accident, new details continue to emerge. See, for example, Richard Stone, "Retracing Mayak's Radioactive Cloud," *Science*, January 9, 1999.

31. Ralph Lapp, *Kill and Overkill* (New York, Basic Books, 1962), 127; and J. Larus, *Nuclear Weapons Safety and the Common Defense* (Columbus, OH: Ohio State University Press, 1967), 93–99.

32. Decontamination required removal of 1,750 tons of radioactive soil and vegetation. By some reports, the material was returned to the United States and buried at sea off the coast of the Carolinas.

33. William J. Broad, "Soviet Sub Sunk in '86 Leaks Radioactivity Into the Atlantic," *New York Times*, February 8, 1994.

34. William M. Arkin and Joshua Handler, *Neptune Papers No.3: Naval Accidents 1945–1988* (Washington, D.C.: Greenpeace/Institute for Policy Studies, June 1989), 3; Table 5 "Nuclear Reactors Lost in the Oceans," 98; and Table 6 "Accidents Involving Nuclear Powered Ships and Submarines," 99. See also Viking O. Eriksen, *Sunken Nuclear Submarines* (Oslo: Norwegian University Press, 1990).

35. Pavel Felgenhauer, "Sunk Sub Raises Questions," *Moscow Times*, September 4, 2003.

36. Lloyd J. Dumas, "The Nuclear Graveyard Below," *Los Angeles Times*, September 3, 2000, reprinted as "A Sunken Nuclear Graveyard Poses Threats to Human Life" in the *International Herald Tribune*, September 7, 2000; see also "The Kursk Disaster: Day by Day," *BBC News Online*, August 24, 2000: http://news.bbc.co.uk/2/hi/europe/894638.stm (accessed June 20, 2009).

37. DW-WORLD.DE, "Working Together to Defuse Russia's Nuclear Timebomb," *Deutsche Welle*, June 1, 2008: http://www.dw-world.de/dw/article/0,,3036461,00.html (accessed June 20, 2009). See also Lloyd J. Dumas, "The Nuclear Graveyard Below," *Los Angeles Times*, September 3, 2000.

38. In August 1989, the Chinese government finally admitted that nuclear accidents in China between 1980 and 1985 had killed 20 people and injured 1,200 more. They did not specify the number of accidents and gave little information about the type, claiming only that they resulted from the careless handling of nuclear waste and other radioactive materials. However, since there were no nuclear power plants operating in China at that time, the main source of nuclear waste must have been the military production of their nuclear weapons. See Associated Press, "China Admits Nuclear Accident Deaths," *Dallas Morning News*, August 6, 1989.

39. Erik Eckholm, "China Reports 70 Sailors Die in Submarine Accident in Yellow Sea" and "Two Top Chinese Officials Comfort Families of Submarine Victims," *New York Times*, May 3, and May 6, 2003, respectively.

40. The Soviets did not admit that the epidemic was the result of an accident until 1991. Questions remain as to whether the biological warfare program that gave rise to this accident remains in operation despite Boris Yeltsin's order to abandon it in 1992. Judith Miller and William J. Broad, "Germ Weapon's in the Soviet Past or in the New Russia's Future," *New York Times*, December 28, 1998. Jonathan B. Tucker, "The Future of Biological Warfare," in *The Proliferation of Advanced Weaponry*, ed. W. T. Wander and E. H. Arnett (Washington, D.C., American Association for the Advancement of Science, 1992), p 59–60. See also Milton Leitenberg, "Anthrax in Sverdlovsk: New Pieces to the Puzzle," *Arms Control Today*, April 1992.

41. Francis X. Clines, "Disaster Zone Urged in Soviet Nuclear Explosion," *New York Times* (September 29, 1990); see also Associated Press, "Soviet Region Seeks Kremlin Aid after Nuclear Fuel Blast," *Dallas Morning News* (September 29, 1990).

42. Treaster, Joseph B., "Army Discloses Three New Cover-Ups," *New York Times* (September 20, 1975).

43. These tests typically involved bacteria such as Bacillus subtilis and Serratia marcescens. See Leonard A. Cole, *Clouds of Secrecy: The Army's Germ Warfare Tests over Populated Areas* (Totowa, NJ: Rowman and Littlefield, 1988). See also Harold M. Schmeck, "Army Tells of U.S. Germ War Tests; Safety of Simulated Agents at Issue," *New York Times*, March 9, 1977; Lawrence K. Altman, "Type of Germ Army Used in Test Caused Infection Four Days Later," *New York Times*, March 13, 1977; Nicholas M. Horrock, "Senators Are Told of Test of a Gas Attack in Subway," *New York Times*, September 19, 1975.

44. Keith Schneider, "U.S. Secretly Fixes Dangerous Defects in Atomic Warheads," *New York Times*, May 24, 1990.

45. Keith Schneider, "Interest Rises in Studies of Atomic Shells," *New York Times*, May 28, 1990.

46. Keith Schneider, "Nuclear Missiles on Some Bombers To Be Withdrawn," *New York Times*, June 9, 1990; see also Keith Schneider, "Interest Rises in Studies of Atomic Shells," *New York Times*, May 28, 1990.

47. "Plutonium Fallout Theory Revealed Here," *Japan Times*, October 29, 1969); and Associated Press, "Satellite Fallout Said Not Harmful," *Japan Times*, April 12, 1970).

48. Steven Aftergood, "Nuclear Space Mishaps and Star Wars," *Bulletin of the Atomic Scientists*, October 1986, 40.

49. Kathleen Teltsch, "Canada to Continue Search for Fragments of Satellite that Fell," *New York Times*, July 6, 1978; and "Follow-Up on the News: 'Hot' Spy in the Cold," *New York Times*, April 26, 1981.

50. William J. Broad, "Despite Danger, Superpowers Favor Use of Nuclear Satellites," *New York Times*, January 25, 1983.

51. Tod S. Purdum, "Russian Mars Craft Falls Short and Crashes Back to Earth," *New York Times*, November 18, 1996; and Michael R. Gordon,

"Mystery of Russian Spacecraft: Where Did It Fall to Earth?, *New York Times*, November 19, 1996.

52. "Challenger Disaster," *Dallas Observer*, August 27, 1987; see also *The Nation*, February 22, 1986.

53. Ibid.

54. Major Mike Morgan, deputy director of the Space Control Center, U.S. Space Command, Cheyenne, Wyoming, in interview by Derek McGinty, National Public Radio, "All Things Considered," August 13, 1997.

55. William J. Broad, "New Plans on Reactors in Space Raise Fears of Debris," *New York Times*, October 18, 1988.

56. William J. Broad, "Debris Spews into Space After Satellites Collide," *New York Times*, February 12, 2009; Ron Cowen, "Two Satellites Collide in Earth Orbit," *Science News*, March 14, 2009; and BBC News, "Russian and U.S. Satellites Collide," February 12, 2009: http://news.bbc.co.uk/go/pr/fr/-/2/hi/sceince/nature/7885051.stm (accessed February 16, 2009).

57. Matthew L. Wald, "Nuclear Waste Tanks at Hanford Could Explode," *New York Times*, July 31, 1990.

58. Matthew L. Wald, "4 Decades of Bungling at Bomb Plant," *New York Times*, January 25, 1992.

59. Andrew Pollack, "After Accident, Japan Rethinks Its Nuclear Hopes," *New York Times*, March 25, 1997.

60. John H. Cushman, "Chemicals on Rails: A Growing Peril," *New York Times*, August 2, 1989.

61. Walt Bogdanich and Christopher Drew, "Deadly Leak Underscores Concerns About Rail Safety," *New York Times*, January 9, 2005; see also Ariel Hart, "8 Are Killed in Train Crash and Gas Leak," *New York Times*, January 7, 2005.

APPENDIX TABLES: MAJOR ACCIDENTS RELATED TO WEAPONS OF
MASS DESTRUCTION

Table 4.1 Major U.S. nuclear weapons-related accidents: A chronology of publicly reported events (1950–2009)

Date	Weapon System	Location	Description
*Feb. 13, 1950	B-36 Bomber	Pacific Ocean off Puget Sound, Washington	B-36 on simulated combat mission drops nuclear weapon from 8,000 feet then crashes. Chemical explosives detonate
*April 11, 1950	B-29 Bomber	Manzano Base, New Mexico	Minutes after takeoff from Kirtland AFB, plane crashes into mountain, killing crew. Nuclear warhead case destroyed, high explosive burned
*July 13, 1950	B-50 Bomber	Lebanon, Ohio	B-50 from Biggs AFB crashes, killing 16 crew. Chemical explosive of nuclear weapon detonates on impact
Fall 1950	B-50 Bomber	Quebec, Canada	B-50 bomber, returning from U.S. military exercise, accidentally drops nuclear warhead into the St. Lawrence. Conventional explosive detonates, dispersing HEU into river (*Christian Science Monitor Radio*, 2/18/97, KERA-FM Dallas)
*Aug. 5, 1950	B-29 Bomber	Fairfield-Suisun AFB California	B-29 with nuclear weapon crashes on takeoff, killing 19. Chemical explosive detonates
*March 10, 1956	B-47 Bomber	Mediterranean Sea	B-47 with "two capsules of nuclear weapon material" disappears

(*continued*)

Table 4.1 (Continued)

Date	Weapon System	Location	Description
*July 26, 1956	B-47 Bomber	U.S. Base at Lackenheath, United Kingdom	Unarmed B-47 crashes into "igloo" storing 3 nuclear bombs. USAF general said, had blazing jet fuel set off chemical explosive, "part of Eastern England [might] have become a desert."
*May 22, 1957	B-36 Bomber	New Mexico	Mark 17, 10-megaton H-bomb accidentally drops on desert near Kirtland AFB, Albuquerque, with conventional explosion
*July 28, 1957	C-124 Transport	Atlantic Ocean	C-124 from Dover AFB, loses power in two engines; jettisons two nuclear warheads over ocean. Weapons never located.
*Oct. 11, 1957	B-47 Bomber	Homestead AFB, FL	B-47—nuclear warhead and nuclear capsule aboard—crashes shortly after takeoff. Two low order detonations occurred
Dec. 12, 1957	B-52 Bomber	Fairchild AFB, Spokane, Washington	Crash on takeoff during "training mission"
*Jan. 31, 1958	B-47 Bomber	U.S. Strategic Air Command Base at Sidi Slimane, Morocco	B-47 crashes during takeoff and burns for 7 hours with nuclear warhead aboard "in strike configuration...." some contamination in area of crash

124

Table 4.1 (Continued)

Date	Weapon System	Location	Description
*Feb. 5, 1958	B-47 Bomber	Hunter AFB Savannah, Georgia	B-47 on simulated combat mission from Homestead AFB, Florida, collides midair with F-86. Nuclear weapon jettisoned after crash never found
Feb. 12, 1958	B-47 Bomber	Off coast near Savannah, Georgia	No details available
March 5, 1958	B-47 Bomber	Off coast of Georgia	Atomic weapon jettisoned after midair collision
*March 11, 1958	B-47 Bomber	Florence, SC	Bomb-lock system malfunction causes Hunter AFB bomber to accidentally jettison A-bomb. High explosive detonates
April 10, 1958	B-47 Bomber	12 miles south of Buffalo, New York	Plane explodes midair while approaching refueling tanker plane
*Nov. 4, 1958	B-47 Bomber	Dyess AFB, Texas	High explosive in nuclear warhead goes off when bomber crashes shortly after takeoff
*Nov. 26, 1958	B-47 Bomber	Chenault AFB, LA	Catches fire on flight line. Fire destroys one nuclear weapon aboard
*Jan. 18, 1959	F-100 Fighter	Pacific Base	F-100 loaded with one nuclear warhead catches fire on ground
*July 6, 1959	C-124 Transport	Barksdale AFB, Louisiana	Plane crashes and burns, destroying one nuclear weapon being transported

(continued)

Table 4.1 (Continued)

Date	Weapon System	Location	Description
*Sep. 25, 1959	Navy P-5M Aircraft	Whidbey Island, Washington	Antisubmarine aircraft ditches in Puget Sound, with nuclear antisubmarine weapon (*NY Times*, May 26, 1981)
*Oct. 15, 1959	B-52 Bomber	Near Glen Bean, Kentucky	B-52 and KC-135 tanker collide midair. Two nuclear weapons recovered undamaged
Before June 8, 1960	Nuclear Warhead	US AFB near Tripoli, Libya	No details (*NY Times* report, June 8, 1960)
Before June 8, 1960	Nuclear Warhead	England	No details (*NY Times* report, June 8, 1960)
*June 7, 1960	Bomarc Surface-to-Air Missile	McGuire AFB, New Jersey	Antiaircraft missile explodes and burns; nuclear warhead destroyed by fire
Before 1961	Corporal Missile	Tennessee River	Missile carrying nuclear warhead rolls off truck into Tennessee River
Jan. 19, 1961	B-52 Bomber	Monticello, Utah	B-52 explodes in flight
*Jan. 24, 1961	B-52 Bomber	Near Goldsboro, North Carolina	Bomber carrying two 24-megaton nuclear weapons crashes. Portion of one bomb never recovered; other fell into field without exploding, but 5 of 6 interlocking safety switches flipped by its fall
*March 14, 1961	B-52 Bomber	California	B-52 from Beale AFB carrying nuclear weapons crashes; training mission
June 4, 1962	Thor ICBM	Johnston Island, Pacific Ocean	A one-megaton nuclear warhead destroyed in flight when Thor explodes at U.S. Pacific missile test range

Table 4.1 (Continued)

Date	Weapon System	Location	Description
June 20, 1962	Thor ICBM	Johnston Island, Pacific Ocean	One-megaton warhead destroyed in attempt to launch high-altitude test shot
April 10, 1963	SSN *Thresher*	US Atlantic Coastline	Nuclear attack submarine sinks on maiden voyage, presumed carrying SUBROC nuclear-armed missile
*Nov. 13, 1963	Nuclear Weapons Storage Igloo	Medina Base, San Antonio, Texas	123,000 pounds of nuclear weapons conventional explosives detonates; "little contamination" from nuclear components stored elsewhere in same building
*Jan. 13, 1964	B-52 Bomber	Cumberland, Maryland	B-52 from Turner AFB, Georgia, crashes; two nuclear weapons on board
*Dec. 5, 1964	Minuteman I ICBM	Ellsworth AFB, South Dakota	LGM 30B missile on strategic alert seriously damaged when "retro-rocket" fires accidentally during repairs
*Dec. 8, 1964	B-58 Bomber	Bunker Hill AFB, Indiana	B-58 carrying at least one nuclear weapon catches fire on flight line; part of nuclear weapon burns
Aug. 9, 1965	Titan II ICBM	Little Rock AFB, Arkansas	Explosion and fire in missile silo
*Oct. 11, 1965	C-124 Transport	Wright-Patterson AFB, Ohio	Plane carrying nonexplosive components of nuclear weapons catches fire on flight line during refueling stop.

(*continued*)

Table 4.1 (Continued)

Date	Weapon System	Location	Description
*Dec. 5, 1965	A-4E Skywarrier Attack Jet	200 miles East of Okinawa	Jet with B43 nuclear warhead rolls off No. 2 elevator of aircraft carrier *Ticonderoga*; sinks in 16,000 feet of ocean
*Jan. 17, 1966	B-52 Bomber	Palomares, southern Spain	B-52 with four 20 to 25 megaton H-bombs collides with KC-135 tanker. One bomb lands undamaged; chemical explosives in two others detonate, scattering plutonium over fields; one recovered intact from Mediterranean after intensive 3-month underwater search.
*Jan. 21, 1968	B-52 Bomber	Thule Bay, Greenland	B-52 crashes, burns, melts through 7-foot-thick ice and sinks. Chemical explosives in all 4 H-bombs aboard go off, scattering plutonium. (S.Sagan, *Limits of Safety*, Princeton University Press, 1993, p. 180)
Feb. 12, 1968	B-52 Bomber	Near Toronto, Canada	Crash; nuclear weapons reportedly aboard; 20 miles north of Toronto
*May 27, 1968	SSN *Scorpion*	Mid-Atlantic Ocean	Mechanical problems; nuclear attack submarine sinks carrying two ASTOR nuclear-armed torpedoes
June 16-17, 1968	US Cruiser, *Boston*	Off Tonkin Gulf, Vietnam	*Boston*, carrying Terrier nuclear-armed antiaircraft missiles, accidentally attacked by U.S. fighter plane(s)

Table 4.1 (Continued)

Date	Weapon System	Location	Description
Jan. 14, 1969	Nuclear-powered aircraft carrier, *Enterprise*	75 miles south of Pearl Harbor, Hawaii	Series of conventional weapons explosions and fires aboard nuclear weapons-capable carrier; 25 dead, 85 injured (*NY Times*, Jan.30,1969) *Enterprise* armed with nuclear weapons.
April 3, 1970	B-52 Bomber	Ellsworth AFB, South Dakota	SAC bomber crashes on landing, coming to rest on fire atop 6 underground fuel storage tanks holding 25,000 gallons each (*Air Force Magazine*, Dec. 1970)
Nov. 29, 1970	Submarine Tender *Canopus*	Holy Loch, Scotland	*Canopus* burns; Polaris nuclear missiles aboard and two Polaris missile submarines moored alongside (*Daily Telegraph*, Nov. 30, 1970)
Nov. 1975	Guided Missile Cruiser, *Belknap*	70 miles east of Sicily	After crashing with aircraft carrier *J.F. Kennedy* during maneuvers in Mediterranean, *Belknap* suffers extensive fires and ordnance explosions. Both ships carrying nuclear weapons
Aug. 24, 1978	Titan II ICBM	Rock, Kansas	Propellant leak causes toxic cloud to spew from silo for 29 hours; two airmen die, several injured, hundreds evacuated (*Wichita Eagle-Beacon*, Aug. 26 & 28, 1978). See parallel Sep19, 1980 Titan II accident.

(continued)

Table 4.1 (Continued)

Date	Weapon System	Location	Description
1980	FB-111 Fighter Bomber	Off Coast of New England	Nuclear-armed fighter bomber crashes
Sep. 15, 1980	B-52 Bomber	Grand Forks AFB, North Dakota	Nuclear-armed B-52 burns on runway
*Sep. 19, 1980	Titan II ICBM	Damascus, Arkansas	Air Force repairman drops socket wrench; punctures Titan II's fuel tank, causing leak leading to explosion demolishing silo door; 9-megaton nuclear warhead hurled from silo later recovered intact
Jan. 11, 1985	Pershing 2 Ballistic Missile	Waldheide Army Base, near Heilbronn, Germany	First-stage engine of missile catches fire and burns during "routine training exercise;" 3 killed, 7 injured; truck and maintenance tent also ignited (*Associated Press* in *West Palm Beach Post*, Jan. 12, 1985)
*Sep. 28, 1987	B-1B Bomber	Colorado	Pelican strikes unreinforced part of right wing near engine intake, causing fire. Plane destroyed (*NY Times*, Jan. 21, 1988)
Nov. 8, 1988	B-1 Bomber	Dyess AFB Abilene, Texas	Two of four engines burn, crew ejects, plane crashes; explodes (*NY Times*, Nov. 10, 1988)
Nov. 17, 1988	B-1 Bomber	Ellsworth AFB, South Dakota	Bomber crashes landing in bad weather; explosion and fire. Crash 9 days after earlier crash causes grounding of all B-1s except those on alert and loaded with nuclear bombs (*NY Times*, Nov 10 & 19, 1988)

Table 4.1 (Continued)

Date	Weapon System	Location	Description
July 24, 1989	B-52 Bomber	Kelly AFB, San Antonio, Texas	Bomber on ground burns; explodes; 1 killed, 11 injured (*NY Times*, July 26, 1989)
June 24, 1994	B-52 Bomber	Fairchild AFB, Washington	Bomber crashes, killing all 4 crew; pilot avoided crashing into nuclear weapons storage area by pulling plane into fatal stalling turn (*NY Times*, Apr. 27, 1995)
Feb. 9, 2001	USS Greeneville	Off Honolulu, Hawaii	Nuclear attack submarine practicing emergency surfacing smashes and sinks Japanese fishing trawler Ehime Maru, 9 die; (*CNN.com/World*, Feb. 11, 2001)
Jan. 8, 2005	USS San Francisco	350 mile south of Guam	Nuclear attack submarine slams full-speed into underwater mountain, crushing ship's bow; 1 dead, 98 injured (*NY Times*, Jan. 15, 2005 & Mar. 13, 2005)
Jan. 9, 2007	USS Newport News	Strait of Hormuz, near Arabian Sea	Nuclear attack submarine hits Japanese tanker Mogami Gawa (*Bloomberg.com*, Jan. 9, 2007)
May 22, 2008	USS George Washington	Between Chile and San Diego, California	Fire aboard nuclear-powered aircraft carrier burns for 12 hrs; $70 million in damages (*Chicago Tribune*, July 31, 2008)

(continued)

Table 4.1 (Continued)

Date	Weapon System	Location	Description
May 23, 2008	Minuteman III missile complex	Warren AFB, Wyoming	Fire in wiring burns for 1 to 2 hrs; undiscovered for five days (*Global Security Newswire*, Oct 11, 2008)
March 20, 2009	USS Hartford	Strait of Hormuz, near Iran	Nuclear attack submarine collides with US Navy amphibious transport ship; damage to sub's tower (*BBC News*, March 20, 2009)

*Officially acknowledged by the Pentagon

Sources:

1. U.S. Department of Defense, "Nuclear Weapons Accidents, 1950-1980," as reprinted with comments by Center for Defense Information (Washington, DC), in *The Defense Monitor* (Vol. 9, No. 5, 1981).
2. "Accidents of Nuclear Weapons Systems" in *World Armaments and Disarmament Yearbook* (Stockholm International Peace Research Institute, 1977), pp. 65–71.
3. Arkin, William M. and Handler, Joshua M., "Nuclear Disasters at Sea, Then and Now," *Bulletin of the Atomic Scientists* (July/August 1989), pp. 20–24.
4. Department of Defense, *Summary of Accidents Involving Nuclear Weapons, 1950–1980 (Interim)*, as cited in Talbot, Stephen, "Nuclear Weapons Accidents: The H-Bombs Next Door," *The Nation* (February 7, 1981), pp. 145–147.
5. "A Deadly White Blur," *Science* (Jan. 29, 1988), p. 453.
6. Talbot, Stephen, "Broken Arrow: Can a Nuclear Weapons Accident Happen Here?," *KQED-TV* (San Francisco: Channel 9 Public Affairs, PBS, 1980).
7. As cited in table.

Table 4.2 Major nuclear weapons-related accidents, other than U.S. (publicly reported, 1957–2009)

Date	Weapon System	Location	Description
1957–1958	**Soviet** Nuclear Waste (from nuclear warhead production)	Kyshtym, Southern Ural Mountains, U.S.S.R.	Underground nuclear waste storage area explodes "like a volcano." 375 square miles contaminated: hundreds die, thousands affected.
Jan. 30, 1968	**British** Vulcan Bomber	Cottesmore, Rutland, United Kingdom	Strategic bomber crashes, burns; claim no nuclear warheads aboard, but RAF firemen sent equipped with radiation monitors (*Times*, Jan. 31, 1968)
Apr. 11, 1968	**Soviet** *Golf* Class Nuclear Missile Submarine	Pacific Ocean, 750 miles west of Oahu	Submarine, carrying three ballistic missiles and probably two nuclear torpedoes, sinks after explosions on board
Before 1970	**Soviet** Military Aircraft (unspecified)	Sea of Japan	American military recovers nuclear weapon from Soviet aircraft that crashed (*NBC Nightly News*, Mar. 19, 1975)
Early Feb. 1970	**Soviet** Nuclear Submarine Construction Facility	Gorki, U.S.S.R.	Large explosion rocks main Soviet nuclear submarine shipyard; several killed, Volga River radioactively contaminated (*Daily Telegraph*, Feb. 21, 1970, *Japan Times*, Feb. 22, 1970)

(continued)

133

Table 4.2 (Continued)

Date	Weapon System	Location	Description
Apr. 12, 1970	**Soviet** *November* Class Nuclear Attack Submarine	300 Nautical Miles North- west of Spain	Develops serious nuclear propulsion problem in heavy seas. Fails to rig towline to nearby merchant ship and ap- parently sinks. 2 nu- clear torpedoes likely aboard
Feb. 25, 1972	**Soviet** Nuclear Missile Submarine	North Atlantic Ocean, off Newfound- land	Submarine crippled by unknown causes, wal- lows in high seas (*Times*, March 1, 1972; *International Herald Tri- bune*, March 10, 1972)
Dec. 1972	**Soviet** Submarine	Off North America (East Coast)	Nuclear torpedo rup- tures, leaking radia- tion (*San Francisco Chronicle*, May 1, 1986)
Mar. 30, 1973	**French** Mirage IV Stategic Bomber	Atlantic Ocean, near France	Landing gear problem; told to ditch at sea; claim not nuclear- armed. Reluctance to try landing possibly because nuclear armed Mirage carries bombs under wing (*Le Monde*, April 1, 1973)
May 15, 1973	**French** Mirage IV Strategic Bomber	Luxueil, France	Strategic bomber crashes on takeoff (*Le Monde*, Sept. 28, 1973)
June 18, 1973	**French** Mirage IV Strategic Bomber	Near Bellegard, France	Bomber crashes; train- ing mission (*Times*, June 19, 1973)
Aug. 31, 1973	**Soviet** Nuclear Missile Submarine	Atlantic Ocean	Sub, carrying 12 nuclear-armed SLBMs, has missile tube accident
Sep. 1973	**Soviet** Nuclear Missile Submarine	Carribean Sea	U.S. aircraft spot sur- faced submarine with 8-foot gash in deck (*Daily Telegraph*, Sept. 6, 1973)

Table 4.2 (Continued)

Date	Weapon System	Location	Description
Sept. 27, 1973	**French** Mirage IV Strategic Bomber	Off Corsica, in Mediterranean Sea	Oil loss; engine seizure; bomber crashes (*Le Monde*, Sept. 28, 1973)
Sept. 1974	**Soviet** *Kashin* Class Guided Missile Destroyer	Black Sea	Ship explodes and sinks (*New York Times*, Sept. 27, 1974)
Sept. 8, 1977	**Soviet** *Delta* Class Nuclear Missile Submarine	Off Coast of U.S.S.R., near Kamchatka	Malfunction forces crew to open missile tube to correct dangerous pressure buildup and clear smoke filling missile compartment. 250-kiloton nuclear warhead thrown high into air; falls into sea (later recovered)
1979/1980	**Soviet** "Echo" Class Submarine	Pacific Ocean	Sub, armed with nuclear torpedoes badly damaged in collision with another sub (rumored Chinese and to have sunk).
Sept. 1981	**Soviet** Submarine	Baltic Sea	Powerful jolts rock sub; nuclear reactor rupture; radiation leaks; some crew contaminated (*San Francisco Chronicle*, May 1, 1986)
Oct. 27, 1981	**Soviet** Submarine	Near Karlskrona Naval Base, Sweden	Diesel-powered sub believed nuclear armed runs aground in Swedish waters, stuck for 10 days (*New York Times*, Nov. 6, 1987)
June 1983	**Soviet** Nuclear Powered Sub	North Pacific Ocean	Soviet submarine sinks, with about 90 people aboard (*New York Times*, Aug. 11, 1983)

(continued)

Table 4.2 (Continued)

Date	Weapon System	Location	Description
May 13, 1984	**Soviet** Northern Fleet Ammunition Depot	Severomorsk, Russia (on Barents Sea)	Huge explosion, then series of explosions and fires destroys major ammunition stocks, killing 200 to 300 people; over 100 nuclear-capable missiles destroyed; no evidence of nuclear explosion or radiation (*New York Times*, June 23, 26 & July 11, 1984)
Sept. 20, 1984	**Soviet** *Golf* 2-Class Nuclear Submarine	Sea of Japan	Missile fuel burns disabling nuclear-armed sub; goes adrift (*Dallas Times-Herald*, Sept. 21, 1984)
Aug. 10, 1985	**Soviet** *Echo*-Class Submarine	Chazma Bay (near Vladivostok)	One reactor explodes during refueling
Dec. 1985	**Soviet** *Charlie/ Victor*-Class Submarine	Pacific Ocean	Human error causes reactor meltdown
1986	**Soviet** *Echo*-Class Submarine	Cam Ranh Bay, Vietnam	On combat duty (with nuclear-armed torpedoes and cruise missiles) radiation detectors jump when crew adds wrong chemicals to reactor cooling system (possible meltdown)
Oct. 6, 1986	**Soviet** *Yankee*-Class Nuclear Missile Submarine	600 miles Northeast of Bermuda	Liquid missile fuel burns, causes explosion. Submarine sinks, carrying 16 ballistic missiles and two nuclear torpedoes; 34 nuclear warheads aboard (*New York Times*, Oct. 7, 1986)

Table 4.2 (Continued)

Date	Weapon System	Location	Description
June 26, 1987	**Soviet** *Echo* 2-Class Nuclear Submarine	Norwegian Sea	Pipes burst, crippling nuclear reactor; sub carries nuclear war-heads
Apr. 17, 1989	**Soviet** *Mike*-Class Nuclear Attack Submarine	Norwegian Sea	Short-circuit causes fire; submarine sinks; two nuclear torpedoes aboard
Dec. 5, 1989	**Soviet** *Delta IV*-Class Submarine	White Sea	Control of missile lost during test launch accident
Mar. 20, 1993	**Russian** Nuclear Missile Submarine	Barents Sea, Close to Kola Peninsula	Sub, carrying 16 nuclear-armed ballis-tic missiles, collides with U.S. nuclear-powered sub
Aug. 12, 2000	**Russian** Nuclear Attack Submar-ine "Kursk"	Barents Sea, Off Kola Peninsula	Torpedo explodes inside Kursk, sinking sub with 2 nuclear reactors and likely nuclear warheads aboard; 118 crew die (*BBC News Online*, Aug. 24, 2000)
Aug. 30, 2003	**Russian** Nuclear Submarine K-159	Barents Sea	Sub towed to scrap yard with nuclear reactor and 798 kg of nuclear fuel aboard spring leak and sinks, 9 die (*Time Europe*, Sept. 15, 2003)
Sept. 18, 2003	**Russian** TU-160 Long-Range Supersonic Bomber	Saratov Region of Russia, 450 miles southeast of Moscow	Onboard explosion and fire; plane crashes, killing all 4 crew on test mission; TU-160s have been backbone of Russian strategic nuclear bomber force since 1987 (*Moscow Times.com*, Sept. 19, 2003)

(continued)

Table 4.2 (Continued)

Date	Weapon System	Location	Description
November 2008	**Russian** Nuclear Submarine (Nerpa?)	Sea of Japan	20 killed, 21 more injured when "firefighting system went off unsanctioned; sub due to be leased to India (*AFP*, Nov. 8, 2008)
Feb. 3–4, 2009	**French** and **British** Nuclear Missile Submarines	"Deep below the surface of the Atlantic Ocean"	The French sub Le Triomphant and U.K. sub HMS Vanguard collide "at very low speed" while on routine patrol fully loaded with nuclear weapons; U.K. sub towed to port; French sub reaches port under escort; no injuries (Guardian.co.uk, Feb. 16, 2009)

Sources:

1. Medvedev, Zhores A., "Two Decades of Dissidence," *New Scientist* (November 4, 1976); see also Medvedev, Z. A., *Nuclear Disaster in the Urals* (New York: Norton, 1979).
2. Arkin, William M. and Handler, Joshua M., "Nuclear Disasters at Sea, Then and Now," *Bulletin of the Atomic Scientists* (July–August 1989), pp. 20–24.
3. "Accidents of Nuclear Weapons Systems" in *World Armament and Disarmament Yearbook* (Stockholm International Peace Research Institute, 1977), pp.74–78.
4. Handler, Joshua M., "Radioactive Waste Situation in the Russian Pacific Fleet, Nuclear Waste Disposal Problems, Submarine Decommissioning, Submarine Safety, and Security of Naval Fuel" (Washington, D.C.: Greenpeace, October 27, 1994; unpublished).
5. As cited in Table.

Table 4.3   Stray missiles and miscellaneous accidents related to actual or potential weapons of mass destruction (publicly reported)

Date	System	Location	Description
1945–1969	U.S. Biological Warfare Tests	United States	Army reports it carried out 239 secret "open air" tests of "simulated" germ warfare around the United States. Includes dumping bacteria into ocean near San Francisco (1950) resulting in 11 hospitalized for rare urinary infection; release of bacteria in NYC subways (1966) (*New York Times*, Dec. 23, 1976 and Mar. 9, 1977)
1951, 1958, 1964	U.S. Biological Warfare Laboratory	Fort Detrick, Maryland	Rare diseases being studied as bioweapons kill 3 employees: biologist and electrician (anthrax); caretaker (Bolivian hemorrhagic fever). Kept secret until 1975 (*New York Times*, Sep. 20, 1975)
Spring 1953	U.S. Nuclear Weapons Tests in Atmosphere	Lincoln County, Nevada a few miles to 120 miles east of Nevada Test Site	4,300 sheep grazing downwind of tests die after absorbing 1,000 times maximum allowable human dose of radioactive iodine (*New York Times*, Feb 15, 1979)
Dec. 5, 1956	U.S. Snark Guided Missile	Patrick AFB, Florida	Missile launched on "closed circuit" mission fails to turn, flies 3,000 miles, crashes into Brazilian jungle (*New York Times*, Dec. 8, 1956)

(continued)

Table 4.3 (Continued)

Date	System	Location	Description
1957	U.S. Nuclear Weapons Plant	Rocky Flats Plant, near Denver, Colorado	Fire causes release of plutonium and other radioactive material; plutonium contaminated smoke escapes plant for half a day. 30 to 44 pounds of plutonium may have burned up (*New York Times*, Sept. 9, 1981)
Early 1960	U.S. Matador Missile	Straits of Taiwan	Older version U.S. cruise missile launched from Taiwan accidentally turns and heads toward China; would not self-destruct (*NY World Telegram*, March 14, 1960)
1967	U.S. Mace Missile	Carribean Sea	Older version cruise missile accidentally overflies Cuba after Florida launch; would not self-destruct (*Miami Herald*, January 5, 1967)
March 1969	U.S. VX Nerve Gas	Skull and Rush Valleys Near Dugway Proving Grounds, Utah	6,000 sheep die when wind carries nerve gas sprayed from Air Force jet well beyond test site
Apr 30, 1970	French Masurca Missile	Le Lavandou, France	Accidental missile launch from French ship. Missile lands near Riviera resort beach with many bathers. Windows and doors blown out but claim cause was missile shock wave, not explosion (*Le Monde*, May 3, 1970)

Table 4.3 (Continued)

Date	System	Location	Description
July 11, 1970	**U.S.** Athena Missile (used to flight test warheads and other ICBM components)	White Sands, New Mexico	50-foot, 8-ton Athena flies into Mexico; crashes in remote area. Some release of radio-activity from small cobalt nose cone capsule. Third missile to crash into Mexico in 25 yrs. (*Japan Times* and UPI July 13, 1970; & *Daily Telegraph*, Aug. 5, 1970)
April 1979	**Soviet** Biological Warfare Facility	Military Compound 19, near Sverdlovsk, Ural Mountains., U.S.S.R.	200 to 1,000 near city die in anthrax epidemic resulting from explosion at secret biowarfare plant that released I-21 anthrax to air. (*New York Times*, Dec. 28, 1998; July 16, 1980); *Science News*, Aug. 2, 1980
Aug. 7, 1979	**U.S.** Factory Producing Nuclear Fuel for U.S. Navy Submarines	Tennessee	Highly enriched uranium dust (300 to 3,000 grams) accidentally released from plant's vent stack when pipe clogs; up to 1,000 people contaminated. Officials claim no health hazard. (*New York Times*, Oct. 31, 1980)
Oct. 18, 1982	**U.S.** GB Nerve Gas	Blue Grass Army Depot, Kentucky	Sensors register small nerve gas leak; 200 to 300 evacuated; army claims no danger; two dead cows found on post; army says cows hit by truck. (*Associated Press, West Palm Beach Post*, Oct. 20, 1982)

(*continued*)

Table 4.3 (Continued)

Date	System	Location	Description
Dec. 28, 1984	**Soviet** Cruise Missile	Barents Sea, off Northern Norway	Errant high-speed missile launched by Soviet Navy overflies Finland and Norway; Norway says it is submarine cruise missile; Denmark says it is 1954-type missile used as target in Navy exercises (*New York Times*, Jan. 4, 1985)
June 10, 1987	**US** NASA Rockets	Wallops Island, Virginia	Lightning causes accidental launch of three small NASA rockets; direction of flights not tracked because of unplanned firings (*Aviation Week & Space Technology*, June 15, 1987)
Sep. 12, 1990	**Soviet** Nuclear Fuel Plant	Ust-Kamenogorsk, Kazakhstan, U.S.S.R.	Explosion releases "poisonous cloud" of "extremely harmful" gas over much of city's 307,000 people; plant uses beryllium, which is toxic but not radioactive; Kazakh officials ask Moscow to declare city ecological disaster zone. (*New York Times*, Sept. 29, 1990)
April 2004	**Chinese** Biological Containment Lab	Beijing, China	Outbreak of severe acute respiratory syndrome (SARS) results from failing laboratory containment system; 1 dead, 9 infected, 600 quarantined

Table 4.3 (Continued)

Date	System	Location	Description
May 5, 2004	**Former Soviet** Biological Warfare Facility "Vector"	Siberia, Russia	Russian scientist infected with deadly Ebola virus after accidentally sticking herself; later dies (*New York Times*, May 25, 2004)
July 26, 2007	**Syrian** Al-Safir Chemical Weapons and Missile Facility	Aleppo, Syria	Explosion kills 15, injures 50; *Jane's Defense Weekly* claims VX and Sarin nerve gas and mustard blister agent dispersed outside/inside facility; chemical weapons involvement disputed (*Jane's Defense Weekly*, Sept. 26, 2007; and *WMD Insights*, Nov. 2007)

Sources:

1. Dugway nerve gas accident:
   a) Hearings on "Chemical and Biological Warfare: US Policies and International Effects," before Subcommittee on National Security Policy and Scientific Developments of the House Committee on Foreign Affairs, 91st Congress, 1st Session, December 2, 9, 18, and 19, 1969, p. 349.
   b) Hersh, Seymour, "Chemical and Biological Weapons: The Secret Arsenal," reprinted from *Chemical and Biological Warfare: America's Hidden Arsenal*, reprinted by Committee for a Sane Nuclear Policy, 1969.
2. As cited in table.

5

Accidental War with Weapons of Mass Destruction

In January 1987, the Indian army prepared for a major military exercise near the Pakistani border province Sind. Because the area was a stronghold of secessionist sentiment, Pakistan thought India might be getting ready to attack and moved its own forces to the Sind border. The two nations had already fought three wars with each other since 1947, and both were now nuclear-capable: India successfully tested a nuclear device in 1974; Pakistan was suspected of having clandestine nuclear weapons. The buildup of forces continued until nearly 1 million Indian and Pakistani troops tensely faced each other across the border. The threat of nuclear war hung in the air as they waited for the fighting to begin. Then, after intense diplomatic efforts, the confusion and miscommunication began to clear and the crisis was finally defused. India and Pakistan had almost blundered into a catastrophic war by accident.[1]

In May 1998, both India and Pakistan successfully carried out a series of nuclear weapons tests, removing any lingering doubt that they had joined the ranks of nuclear-armed nations. In May 1999, "over 1000 Pakistan-based militants and Pakistani regulars" launched a surprise attack across the so-called "Line of Control" in Kashmir, seizing Indian army outposts near the town of Kargil. The Indian Army counterattacked, driving back the Pakistani soldiers.[2] The conflict dragged on for two months, taking 1,300 to 1,750 lives, before outside diplomatic efforts finally persuaded Pakistan to withdraw. But it could easily have been much worse: During the 1999 Kargil war, Pakistan reportedly got its intermediate-range missiles ready for nuclear attack and "High level officials in both countries issued at least a dozen nuclear threats."[3]

In the midst of an active military conflict between two long-term rivals with both sides actively threatening each other with nuclear attack, it would not take all that much miscommunication, misinterpretation, systems failure, or simple human error for the conflict to escalate out of control and unintentionally precipitate an accidental nuclear war. According to an American intelligence assessment completed in May 2002—as tensions between India and Pakistan once again intensified—whatever its cause, "a full-scale nuclear exchange between the two rivals could kill up to 12 million people immediately and injure up to 7 million."[4] By 2009 both nations had made great progress in developing nuclear arsenals and little or no progress in resolving tensions that repeatedly brought them so close to accidental disaster.

More than a few times, nations have found themselves fighting wars that began or escalated by accident or inadvertence. World War I is a spectacular example. By 1914, two alliances of nations (the Triple Alliance and the Triple Entente), locked in an arms race, faced off against each other in Europe. Both sides were armed to the teeth and convinced that peace could be and would be maintained by the balance of power they had achieved: in the presence of such devastating military force, surely no one would be crazy enough to start a war. Then on June 28, 1914, a Serbian nationalist assassinated Archduke Francis Ferdinand of Austria-Hungary and his wife, an event that would have been of limited significance in normal times. In the presence of enormous, ready-to-go military forces, however, the assassination set in motion a chain of events that rapidly ran beyond the control of Europe's politicians and triggered a war that no one wanted.

Within two months, more than 1 million men were dead. By the time it was over, 9 million to 11 million people had lost their lives in a sickeningly pointless war, the worst the world had ever seen. The war itself had already taken a frightful toll. Now the terms of surrender added to the damage by demanding heavy reparations from the losers. This helped to wreck the German economy, sowing the seeds of the rise to power of one of the most brutal, aggressive, and genocidal regimes in all of human history. Within two decades, the Nazis and their allies once again plunged the world into war—a far more destructive war—that left some 40 million to 50 million people dead (half of them civilians) and gave birth to the atomic bomb. The American bomb project was born primarily out of fear that the Germans might develop such a weapon first.

There may be many links in the chain, but there *is* in fact a connected chain of events leading from the accidental war we call World War I to the ovens of Auschwitz, the devastation of World War II, and the prospect of nuclear holocaust that has cast a pall over us ever since. Yet the whole terrible chain of events might have been prevented but for a simple failure of communications. The Kaiser had sent the order that would have stopped the opening move of World War I (the German invasion of Luxembourg on August 3, 1914) before the attack was to begin. But the message arrived 30 minutes late. In a classic understatement, the messengers who finally delivered the belated order said, "a mistake has been made."[5]

NUCLEAR WAR STRATEGY AND ACCIDENTAL WAR

The American military's master plan for nuclear war, the Single Integrated Operating Plan (SIOP), is a complex and rigid set of integrated nuclear battle plans. SIOP bears a frightening resemblance to the mechanistic approach to war planning that led us to blunder into World War I. In 2004 Bruce Blair, a long-time analyst of (and one-time participant in) the U.S. nuclear response system, wrote of SIOP,

> It is so greased for the rapid release of U.S. missiles by the thousands upon the receipt of attack indications from early warning satellites and ground radar that the president's options are not all created equal. The bias in favor of launch on electronic warning is so powerful that it would take enormously more presidential will to withhold an attack than to authorize it.[6]

During the Cold War, the nuclear forces of the two superpowers were constantly watching each other, constantly ready to react. They were so "tightly coupled" that they were, in effect, one system. A serious glitch in either's forces could have led to a set of mutually reinforcing actions that escalated out of control and brought about a pointless war that neither wanted, just as it did in World War I. This time, the result would have been more than misery and carnage; it would have been mutual annihilation.

It is also widely assumed that with the end of the Cold War and the disappearance of the "Soviet threat," Russian and American missiles were taken off hair-trigger alert and no longer configured for launch-on-warning of attack. But that is simply not true.[7] Testifying before Congress on July 18, 2007—16 years after the demise of the Soviet Union—former Secretary of Defense William J. Perry said, "Both

American and Russian missiles remain in a launch-on-warning mode." Perry then added, "And the inherent danger of this status is aggravated by the fact that the Russian warning system has deteriorated since the ending of the Cold War."[8] In August 2007, "Russia declared . . . that it would begin regularly sending its strategic bombers within striking distance of the United States and allied nations for the first time since the end of the Cold War."[9] Since that time, Russian bombers have been intercepted by British and Norwegian fighter jets in NATO airspace, by Danish fighters close to Danish airspace, and repeatedly by U.S. and Canadian fighters as they approached North America.[10] All this was still going on when President Barak Obama took office in January 2009.

Five weeks after Obama's inauguration, the White House said that Obama intended "to make good on a campaign promise to 'work with Russia to take U.S. and Russian ballistic missiles off hair-trigger alert.'"[11] Launch-on-warning is an extremely dangerous policy. It is based on the tenuous assumption that through careful design and redundancy, human/technical warning systems can be created that are not subject to catastrophic failure. Yet it is painfully obvious that this belief is not true (see Chapters 6 to 10). Once a missile attack is launched based on a warning that turns out to be false, there is a real attack underway that cannot be recalled or destroyed en route. Even if the other side's warning systems did not detect the launch, they will soon know for certain they have been attacked, as the incoming warheads explode over their targets. Given the horrendous damage that will result, a contrite admission that "a mistake has been made" is unlikely to forestall retaliation. Some combination of human fallibility and technical failure will have brought about a disaster beyond history.

It is abundantly clear that the end of the Cold War did not put an end to the Cold War mentality that underlies the SIOP. The Cold War may be dead and gone, but the Cold Warriors—and the weapons and institutions of the Cold War—are still very much with us. As long as that is true, accidental nuclear war remains a real possibility.

GENERATING WAR BY MISTAKE: TRIGGERING EVENTS

Accidents[12]

During a time of confrontation and crisis, a weapons accident that resulted in a nuclear explosion on the territory of a nuclear-armed country or its allies could trigger an accidental nuclear war. It is even possible that an accident involving the weapons of friendly forces might be surrounded in fog long enough to be misread as an enemy attack. But a weapons accident does not have to involve a nuclear explosion to trigger nuclear war. There have been a number of publicly reported accidents in which the powerful conventional explosive in one or more nuclear weapons was detonated (see the Appendix to Chapter 4). Suppose one of these bombs had fallen onto an oil refinery, tank farm, or toxic chemical waste dump. Worse yet, if one fell into a nuclear waste storage area, the huge explosion and high levels of radioactivity that would result could easily be misinterpreted as an act of enemy sabotage or a deliberate attack—especially under the pressure and confusion of a crisis.

The crash of an aircraft or missile into a nuclear, chemical, or biological weapons storage area could also produce a provocative combination of explosion and contamination. Such incidents appear in the public record: in the summer of 1956, a B-47 bomber crashed into a storage igloo containing three nuclear weapons in England; on June 24, 1994, a B-52 bomber crashed as the pilot pulled the plane into a fatal stalling turn in a successful last-minute attempt to avoid crashing into a nuclear weapons storage area. For years, the flight path of civilian airliners in Denver took them close to or right over a nearby nerve gas storage area at the Rocky Mountain Arsenal. In the mid-1990s, commercial airliners flew directly over the Pantex nuclear weapons plant on their way to and from nearby Amarillo airport. More than 9,000 plutonium "pits" the size of bowling balls that had been removed from decommissioned nuclear weapons were being stored there at the time, in World War II vintage conventional weapons bunkers.[13]

There has been an elaborate set of controls in place intended to prevent accidental or unauthorized firing of American land-based ICBM's.[14] In addition to procedures that require a number of people to act together, since the 1960s the weapons have been fitted with special locks known as Permissive Action Links (PALs) that require higher command to send the proper enabling codes before on-site crews can physically launch the missiles. Of course, no system is ever failure proof or impossible to work around. Bruce Blair, who was a missile launch officer in the Strategic Air Command (SAC) in the early to mid-1970s, reports that SAC "quietly decided to set the 'locks' to all zeros in order to circumvent this safeguard.... And so the 'secret unlock code' during the height of the nuclear crises of the Cold War remained constant at 00000000." Blair reports that when he told former Secretary of Defense MacNamara (who ordered the PALs installed) about this in January 2004, MacNamara responded, "I am shocked, absolutely shocked and outraged. Who the hell authorized that?" The PALs were not "activated" until 1977.[15]

For decades a large part of the U.S. strategic arsenal has been carried on nuclear submarines, whose missiles are armed with thousands of city-destroying warheads. While submarine missile launch procedures require a number of people to act jointly, the U.S. Navy managed to stave off the installation of PALs until 1997.[16] For all those years, there was no external physical control against the launch of nuclear-armed missiles from American missile submarines. Furthermore, the BBC reported in 2007 that British strategic nuclear warheads, which are only carried on British missile submarines, were still not fitted with PALs.[17]

It is also important to note that the popular idea that only the president has the codes needed by launch crews to launch their nuclear weapons from land, sea, or air is simply incorrect. "Those authorizing or enabling codes . . . always remain in military custody at the national command posts in the Pentagon and alternate sites around the country. The president carries special codes to positively identify himself to the key nuclear commanders who might be ordered to launch a nuclear strike. But the military does not need these codes to launch the strike . . . "[18] Military commanders have had the authority to order the use of American nuclear weapons for decades. Presidents Eisenhower, Kennedy, and Johnson all pre-delegated the authority for emergency use of nuclear weapons to "six or seven three and four-star generals."[19] Presidents Carter and Reagan empowered senior military commanders "to authorize a SIOP retaliatory strike and to select the SIOP option to execute."[20] In short, there have been a lot more "fingers on the button" than most of us realize, making accidental nuclear war that much more likely.

False Warnings

The equipment and people of the nuclear attack warning systems are expected to detect and evaluate any real attack quickly and accurately, whenever it might come, while at the same time never failing to recognize that false warnings are false. The time available is a matter of minutes and the pressure is enormous.[21] Asking fallible human beings dependent on fallible technical systems to make decisions of such enormous importance in so little time is asking for trouble. It is virtually inevitable that system will fail catastrophically some day.

That day almost came on January 25, 1995, when Russian warning radars detected the launch of a rocket from the Norwegian Sea. About the size of a U.S. submarine-launched Trident missile, it seemed to be streaking toward Moscow, with projected impact in only about 15 minutes. The radar crew sent the warning of impending nuclear attack to a control center south of Moscow, where it was relayed up the chain of command. Soon the light flashed on the special "nuclear briefcase" assigned to Russian President Yeltsin. The briefcase was opened, revealing an electronic map showing the missile location and a set of options for nuclear strikes against targets in the United States. Alarms sounded on military bases all over Russia, alerting the nuclear forces to prepare to attack. Yeltsin consulted with his top advisors by telephone. Tensions rose as the stages of the rocket separated, making it look as though several missiles might be headed for the Russian capital. About eight minutes after launch—only a few minutes before the deadline for response—senior military officers determined that the missile was headed far out to sea and was not a threat to the Russian homeland. The crisis was over.[22]

Despite the end of the Cold War in 1991 and all of the dramatic changes that had occurred since that time, we were once again at the very brink of nuclear war with Russia. In fact, this was the first time that the famous "nuclear briefcase" had ever been activated in an emergency. The cause? The rocket the Russians detected, it turned out, was an American scientific probe designed to study the aurora borealis (northern lights). It was sent aloft from the offshore Norwegian Island of Andoya. Norway had notified the Russian embassy in Oslo in advance of the launch, but somehow "a mistake was made" and the message never reached the right people in the Russian military.[23]

False warnings of attack are probably the most likely trigger of accidental nuclear war. There have been many. Between 1977 and 1984 (the last year for which NORAD made these data publicly available), there were nearly 20,000 routine false alarms, an average of more than six a day.[24] Most were quickly recognized as false, but there have also been much more serious false warnings of attack (see Table 5.1).

From 1977 to 1984, *there were more than 1,150 serious false warnings of attack, an average of one every two and a half days for eight years.* These data were no longer made public after 1984, but we do have some credible information. We know, for example, that even as late as 1997, space junk that "might be construed as an ICBM warhead" continued to reenter the atmosphere at the rate of about one "significant object" a week, according to the deputy director of the U.S. Space

Table 5.1   Serious false warnings of attack

Year	Major Missile Display Conferences	Threat Assessment Conferences
1977	43	0
1978	70	2
1979	78	2
1980	149	2
1981	186	0
1982	218	0
1983	255	0
1984	153	0
1985	NORAD stopped releasing data on false warnings	
TOTAL	**1,152**	**6**

Source: Center for Defense Information, *Defense Monitor* (Vol. 15, No. 7, 1986), p. 6.

Control Center.[25] Writing about potential attack warnings in 2007, Bruce Blair affirmed, " . . . it happens practically on a daily basis, sometimes more than once per day, because there are many events involving apparent missile launches that require evaluation."[26] With that many potential attack warnings to evaluate, there are many opportunities to make mistakes.

Why do so many false warnings occur? All warning systems are subject to an inherent engineering tradeoff between sensitivity and false alarms. If a smoke alarm is designed to be so sensitive that it will quickly pick up the very first signs of a fire, it is bound to go off from time to time because someone is smoking a cigarette or a little smoke from a backyard barbecue blows in the window. If the smoke alarm system is made less sensitive, it will not go off as often when it should not. However, it also will not go off as quickly when there is a real fire. Modern nuclear weapons delivery systems are so quick and accurate that a military not prepared to ride out an attack requires very sensitive warning systems to get maximum warning time. Such systems are virtually certain to generate many false alerts.

A few examples of the kinds of false attack warnings that have been publicly reported follow:

- During the 1962 Cuban Missile Crisis, U.S. forces were on high alert. Near midnight, a guard at an air force facility in Duluth saw someone climbing the base security fence. Fearing sabotage, he fired at the intruder and set off a sabotage alarm that was linked to alarm systems at other bases nearby. A flaw in the alarm system at one of those bases (Volk Field in Wisconsin) triggered the wrong warning. Instead of sounding the sabotage alarm, the klaxon that signaled the beginning of nuclear war went off. Fighter pilots rushed to their nuclear-armed aircraft, started their engines, and began to roll down the runway, believing a nuclear attack was underway. The base commander, having determined that it was a false alarm, sent a car racing onto the tarmac, lights flashing, signaling the aircraft to stop. It turned out that the suspected saboteur shot climbing the fence was a bear.[27]

- At 10:50 AM on November 9, 1979, monitors at NORAD in Colorado, the National Command Center at the Pentagon, Pacific headquarters in Honolulu, and elsewhere simultaneously lit up. A nuclear missile attack was underway. The profile of the attack was realistic, a salvo of submarine-based missiles against just the kind of targets that the American military might expect the Soviets to strike. Ten jet interceptors were ordered into the air from bases in the United States and Canada. Six minutes later, it was determined that this was a false alarm. The attack profile seemed realistic to the U.S. military because they created it. An operator had fed a test tape for a simulated attack into the computer as part of an exercise. According to Scott Sagan, the " . . . software test information . . . was *inexplicably* transferred onto the regular warning display at Cheyenne Mountain and simultaneously sent to SAC, the Pentagon and Fort Richie."[28] Through some unknown combination of human and technical error, the realistic test attack data had gone out through the wrong channels and showed up as a "live" attack warning.[29]

- In the early morning hours of June 3, 1980, the displays at the Strategic Air Command suddenly showed a warning from NORAD that two submarine-launched ballistic missiles (SLBMs) had been fired at the United States. Eighteen seconds later, the warning indicated more SLBMs were on the way. SAC sent its bomber crews racing to their B-52s. The engines were fired up and 100 B-52s, loaded with more than enough nuclear warheads to devastate any nation on earth, were prepared to take-off.[30] Launch crews in land-based missile silos across the United States were told to get ready for launch orders. Ballistic missile submarines at sea were alerted. Special battle control aircraft, equipped to take over if normal command centers failed, were prepared for takeoff. One took off from Hawaii.[31] The American military got ready for Armageddon.

Just then, the warning disappeared. SAC contacted NORAD and was told none of the satellites or radars had detected any missiles at all. Bomber crews were told to shut down their engines, but stay in their planes. Then, suddenly the displays at SAC again showed an attack warning from NORAD, this time indicating a barrage of Soviet land-based missiles launched against the United States. At the same time, the National Military Command Center at the Pentagon also received a warning from

NORAD—but it was a different warning. Their displays showed Soviet submarine-launched missiles, not land-based ICBMs speeding toward America.[32] Something was clearly wrong. What turned out to be wrong was that a faulty computer chip (costing 46 cents) was randomly generating 2's instead of 0's in the transmissions from NORAD to each of the different command centers. Rather than 000 sea or land-based missiles heading toward the United States, there would be 020 or 002 or 200. The faulty chip was the immediate problem, but the underlying cause was a basic design error. The NORAD computer was not programmed to check the warning messages it was sending against the data it was receiving from the attack warning sensors to make sure they matched—an elementary mistake. This kind of error checking capability was routine in the commercial computer systems of the day. But the software that controlled a key part of the world's most dangerous technological system did not have this basic safety precaution.[33]

- On the night of September 26, 1983, Lt. Colonel Stanislav Petrov was operations duty officer at the Soviet Union's Serpukhov-15 Ballistic Missile Early Warning System (BMEWS). It was a time of high tension between the United States and the U.S.S.R. President Reagan had famously called the Soviet Union the "Evil Empire," and the Russian military had shot down a South Korean passenger jet only three weeks earlier, an act vociferously condemned by the United States and its allies. Suddenly, an alarm went off at Petrov's command and control post. A huge screen displaying the United States showed the missile launch that a Soviet satellite had detected—one missile, then a second, third, and fourth—from the same American base headed toward Russian territory. In Petrov's words, "We did what we had to do. We checked the operation of all systems—on 30 levels, one after another. Reports kept coming in: All is correct; the probability factor [of an real missile attack] is two. . . . The highest."[34]

  Without any indications to the contrary and in the face of computer confirmations that this was a real attack, and despite all the tensions and hostile rhetoric of the day, Petrov told his superiors in the Kremlin that it was a false alarm, and the Russian military stood down its missiles. In a 2004 interview with the *Moscow News*, he said, "You can't possibly analyze things properly within a couple of minutes. All you can rely on is your intuition. . . . The computer is by definition brainless. There are lots of things it can mistake for a missile launch."[35]

There have also been incidents in which a number of false warnings occurred at the same time, reinforcing each other and making it easier to believe that a real attack was under way. Warning systems can and do fail in many different ways. Sensors sometimes accurately detect real events, only to have computers or people misinterpret them.

Sometimes computers or people generate false warnings, without the sensors having detected anything. Sometimes the alarms themselves go off without sensors, computers, or people generating any indication of a problem. And sometimes warning systems fail because they miss events they are supposed to detect.

Failures of Detection

Knowing that alarm systems have failed to detect what they are supposed to detect can lead to a state of hypervigilance in critical situations. Hypervigilance makes it easy for people to see danger where there is none. They are so determined not to miss the critical event that they are primed to overreact. That overreaction can lead to tragedy, even to the tragedy of accidental war. Striking examples of warning system detection-related failures include:

- The frigate Stark was part of the U.S. Navy fleet patrolling the Persian Gulf during the Iran-Iraq War. On May 17, 1987, an Iraqi air force pilot flying a French Mirage fighter bomber began to approach the Stark. Suddenly, the plane fired one Exocet missile, then another at the ship. The attack came as a total surprise. According to the Navy, the ship's electronic warning system detected that the Mirage had switched on a special radar used to pinpoint targets a minute before the first missile struck the ship—but no one told the captain. The Stark's state-of-the-art warning equipment did not pick up the missile firings. The only warning came from the oldest, lowest technology warning system on board, a lookout watching from the deck. He spotted the 500-mile-per-hour missiles seconds before they slammed into the ship, setting it ablaze and killing 37 men. Iraq later said the attack was a mistake, and apologized.[36]
- On May 28, 1987 a West German teenager named Matthias Rust filed a flight plan for Stockholm and took off from Helsinki, Finland flying a single-engine Cessna 172. He had removed some seats and fitted extra fuel tanks to the plane to increase its range, but it was otherwise a perfectly ordinary small civilian aircraft. Rust diverted from his flight plan, turned south, and flew into Soviet airspace over the coast of Estonia. About 7:00 PM, he reached Moscow, circled low over Red Square, and landed near the Kremlin Wall. He had flown through sophisticated radar systems, across more than 400 miles of heavily defended territory to the heart of the Soviet Union, without triggering a single alarm. Rust climbed out of the plane, and was immediately surrounded—not by police or the military, but by a crowd of curious Russians and foreign tourists. Children climbed on the tail of the plane. He chatted with the crowd until the authorities arrived. Rust was

arrested, imprisoned, then freed and expelled from the country a year later. The United States had spent many billions of dollars developing "stealth" fighters and bombers to sneak past Soviet radars and dense defenses such as those surrounding Moscow, but the whole system had been penetrated with ease by a boy flying an inexpensive, run-of-the-mill airplane he had rented from a flying club in his hometown. To add to the irony, he did this as the Soviets were celebrating Border Guards Day.[37]

- On October 17, 2000 the USS Kitty Hawk, a nuclear-powered aircraft carrier, was overflown by Russian SU-24 and SU-27 fighter planes, "which were not detected until they were virtually on top of the carrier. The Russian aircraft . . . caught the ship completely unprepared." The fighters flew within several hundred feet of the carrier. The Russians not only took very detailed photos of the carrier's flight deck, they emailed them directly to the Kitty Hawk's commander. Such an event should not have been all that big a surprise since Russian aircraft had flown over the carrier less than a week before this incident. And they did it again a few weeks later.[38]

- On May 10, 2001, a major fire broke out at Serpukhov-15, an important ground relay station in the Russian satellite control system. The fire soon engulfed the complex. Four Russian early-warning military satellites controlled by the Space Forces command center were designed to immediately detect any ICBM launches from the territory of the United States and flash a warning to that relay station, which would send it on to higher command. With the relay station out of action, the satellites, still in orbit, were out of control. According to a Russian news report, the loss of contact left the satellites in " 'sun-orientation mode' entirely on their own. Essentially, they are blind. We have a week to ten days to make contact again, or the satellites may enter the upper atmosphere. That would be the end of them."[39] Geoffrey Forden of MIT expressed concern that Russia's deteriorating space-based early warning system created a risky situation: "Russia no longer has the working fleet of early warning satellites that reassured its leaders that they were not under attack during the [1995] false alert."[40]

By the end of 2001, only four communications satellites of what had been a Soviet-era fleet of more than 100 early-warning, communications, and intelligence satellites were still operational.[41] According to a Russian news report in 2003, "The Russian Defense Ministry has been left without any eyes or ears in space" when the last orbiting Russian spy satellite "simply fell apart." The so-called Don reconnaissance satellite had been launched from the Baikonur spaceport in August 2003 to compensate for the failure of two other early warning satellites earlier that year. But it too had failed prematurely.[42]

In light of all this, it is interesting that by 2006 the Pentagon had come up with a scheme to develop a conventional version of the Trident missile warhead and began to deploy "24 conventionally armed

Trident missiles aboard submarines that also carry a look-alike nuclear version of the D-5 weapon." Members of Congress had steadfastly refused to fund the program because they were quite rightly concerned "that Russian early-warning systems—and perhaps future Chinese technology—would be unable to distinguish which kind of warhead the missile was carrying during its 12 to 24 minute flight time." That would be quite a task for a fully functional warning system, let alone one that had fallen into an advanced state of disrepair. Yet in 2008, the vice-chairman of the Joint Chiefs of Staff "branded as 'facetious' congressional concerns that the launch of a proposed long-range conventional missile might be mistaken for a nuclear salvo."[43]

FAILURES OF VERIFICATION AND COMMUNICATION

No sane national leader would launch a retaliatory nuclear strike until an attempt has been made to verify that the triggering event really did mark the beginning of nuclear war. The most dangerous failure of such an attempt at verification is a false positive (i.e., the system confirms that the triggering event is real when it is not). This would sharply increase the likelihood that a nuclear strike would be launched in retaliation for an enemy attack that never actually happened. Fortunately, the kind of compound error that would be necessary to get false positive confirmation, though not impossible, is relatively unlikely. It is much more likely that verification will fail because there will not be enough time to find out if the triggering event is real. Those in authority will then have to decide whether or not to retaliate without being sure that the attack is actually under way. In such a situation, people tend to believe what they have been predisposed to believe.

Only a year after the Stark's failure to react cost the lives of three dozen men, the U.S. Navy cruiser Vincennes found itself in what appeared to be a similar situation. The Vincennes was engaged in a firefight with several Iranian gunboats in the Persian Gulf on the morning of July 3, 1988. Suddenly, the cruiser's combat information center detected an aircraft rising over Iran, headed toward the Vincennes. The combat center concluded that the oncoming plane was one of the 80 F-14 fighters that the United States had earlier sold Iran. The ship warned the plane to change course, but it kept coming and did not respond. The ship's crew saw the aircraft descending toward the Vincennes from an altitude of 7,800 feet at a speed of 445 knots, as if to attack. Seven minutes after the plane was first detected, the cruiser fired two surface-to-air missiles and blew it out of the sky.[44]

But the plane was not an F-14, and it was not attacking. The Vincennes had destroyed a European-built Airbus A-300 civilian airliner with nearly 300 people on board. Iran Air Flight 655 was a regularly scheduled flight on a well-known commercial route. Its transponder was automatically broadcasting a signal that identified it as a civilian aircraft. The A-300 has a very different shape than an F-14. It is nearly three times as long and has more than twice the wingspan. And it was climbing, not descending.[45] How could such a thing happen?

The Vincennes was in combat in a war zone at the time the plane was detected. American commanders in the Gulf had been warned that Iran might try to attack U.S. forces around the Fourth of July holiday. There were only seven to eight minutes after detection to react, only three minutes after the plane was declared hostile. There was also the specter of the tragic loss of life on the Stark, whose crew failed to fire at an attacking aircraft they could not believe was attacking. Under so much stress, with so little time, and a predisposition to believe they were being attacked, the crew was primed for error. In the process of verifying the threat posed by the blip on the radar screens, they misinterpreted the electronic identification signal coming from the plane, misread the data which correctly showed that the approaching plane was gaining altitude (not descending to attack), and failed to find Flight 655, clearly listed in the civilian airline schedule they hastily consulted.

Five years before the destruction of Iran Air 655, at a time when the Cold War was still alive and well, the Soviets had made a similar mistake with similarly tragic results. On the night of August 31/September 1, after a long flight across the Pacific, Korean Airlines (KAL) Flight 007 drifted far off course and flew directly over Kamchatka Peninsula and Sakhalin Island, the site of a secret Soviet military complex.[46] Soviet radars detected the intruder and scrambled fighter planes to identify and intercept it. The radars tracked the plane for more than two hours, while the interceptors struggled to find it. They mistook the Boeing 747 civilian airliner with 269 people aboard for a U.S. Air Force RC-135 reconnaissance plane. A Soviet SU-15 fighter finally caught up with the plane, following it through the night sky. When the airliner did not respond to a warning burst of canon fire, the fighter pilot fired an air-to-air missile that blew the plane apart.

The Reagan Administration's immediate reaction was to condemn the U.S.S.R. as a brutal nation that had knowingly shot down a civilian airliner out of a combination of barbarism and blood lust. Within five weeks, however, it was reported that "United States intelligence experts say they have reviewed all available evidence and found no indication that Soviet air defense personnel knew it was a commercial plane before the attack."[47] There had been many previous instances of American spy planes intruding on Soviet airspace since 1950. In fact, there was an American RC-135 spy plane in the air at the time of this incident that had crossed the flight path of KAL 007. The Soviet military was preparing for a missile test in the area, which they knew the United States would want to monitor. As in the case of the Vincennes, all of this predisposed the Soviet armed forces to see an enemy military plane when they were actually looking at an ordinary civilian airliner.[48]

It is anything but reassuring that—with interceptor planes in the air and radars tracking KAL 007 for some two and a half hours—the Soviets still could not determine the nature of the intruder they had detected.[49] If they could not tell the difference between a harmless jumbo jet airliner and a much smaller military aircraft in two and a half hours, what reason is there to believe that their deteriorating military warning system today can verify that a missile attack warning is false in less than 30 minutes?

Despite modern equipment and well-trained military forces, it is amazing how easy it is to mistake completely harmless intruders for hostile forces when you are looking for and expecting to find an enemy.

Problems with Command and Control

It is never easy to assure that military commanders, let alone political decisionmakers, have a clear and accurate picture of what is happening in critical situations. In the words of one senior military commander:

> Even in an ordinary war, you learn to discount the first report, only give partial credence to the second report, and only believe it when you hear it the third time.... The chances for misappreciation of information, for communications foul-ups, for human errors are enormous. Even in tactical exercises—with people not under emotional stress.... I've never seen a commander who was not surprised by what had actually happened versus what he thought had happened.[50]

If that is true in conventional war and war games, it is that much more likely to be true when decision time is as limited and stakes are as high as when nuclear war hangs in the balance.

On the morning of September 11, 2001, the Pentagon's domestic air defense system faced a challenge for which, as it turned out, it was disastrously unprepared—a major terrorist airborne strike on American soil. According to the independent commission set up to investigate the events of that day, its response and that of the Federal Aviation Administration were confused and poorly coordinated. With the world's most powerful and best funded military at its command and close to an hour to respond after the first plane hit the World Trade Center, the military command could not get fighter pilots in the air fast enough to shoot down the airliner that ultimately crashed into the Pentagon building itself. For some reason, the emergency order authorizing the shooting down of the commandeered jetliners ultimately came from Vice President Cheney, not President Bush. With all that time available (as compared to the 15 to 30 minutes to respond to a warning of nuclear attack), it did not reach the military pilots until after the fourth hijacked plane had crashed into a field in Pennsylvania.[51]

High-level military communications do not always work smoothly even in normal times. When writer Daniel Ford was being taken around NORAD's command center by Paul Wagoner, the general in charge of NORAD's combat operations, he was shown five telephones next to the commander's battle station desk. The NORAD fact sheet Ford had been given said of their communications system, "When we pick up a telephone, we expect to talk to someone at the other end—right now." So he asked General Wagoner for a demonstration:

> starting with the special black phone that would be used to talk to the President. He picked up the phone briskly and punched the button next to it, which lit up right away ... About twenty seconds elapsed and nothing happened. The General ... hung up, mumbled something about perhaps needing 'a special operator,' and started to talk about other features of the Command Post. I interrupted and pointed to another of the five telephones ... a direct line to the Joint Chiefs of Staff. He picked it up ... Once again, nothing happened. ... I looked at the other three phones, but General Wagoner decided to terminate his testing of the communications system.[52]

Virtually every part of the military communications system has failed at one time or another. For example:

- The famous "hot line"—the emergency communications link between Moscow and Washington set up to prevent accidental war—failed six times during the 1960s.[53]

- NORAD's backup National Attack Warning System headquarters near Washington failed during a routine test in February 1984, and it took over half an hour for officers to even notify the main Colorado headquarters that the system was down.[54]
- In April 1985, the Pentagon flew a group of reporters to Honduras to test its system for news coverage of emergency military operations. Their reports were delayed 21 hours because technicians could not make Navy telephones work.[55]
- The major earthquake that rocked San Francisco in October 1989 and disrupted the World Series also knocked out the U.S. Air Force center that manages America's early warning and spy satellites for 12 hours.[56]

Military communications have often been plagued with co-ordination and compatibility problems.[57] It was not until the American assault on Grenada in 1983 that the Army discovered its invasion forces were sometimes unable to communicate with the Navy's because their radios operated on different frequencies. Yet when the Army later bought new communications systems specifically so that its battlefield commanders could coordinate battle plans with other units, it bought devices that were not compatible with those used by the Air Force. An additional $30 million had to be spent for adapters. What makes this story even more peculiar is that the Army had acted as purchasing agency for the Air Force equipment! An army spokesman said that he could not explain why the two services wound up with incompatible systems.[58]

In 1971, the House Armed Services Committee found that the global U.S. defense communication system took an average of 69 minutes to transmit and receive a "flash" (i.e., top-priority) message. In a real nuclear attack, the missiles would have arrived well before the message did. While there have been revolutionary improvements in the technology of communications since then, it is important to note that only about five of the 69 minutes was actual transmission time. The remainder was due to a variety of largely organizational and individual human foul-ups.[59]

On May 12, 1975, the Cambodian Navy seized a U.S. merchant ship, the Mayaguez, in the Gulf of Thailand. The subsequent recovery of the ship and its crew by the American military has often been cited as an example of the effectiveness of improvements in communications and response made as a result of earlier incidents.[60] Actually, it is quite the opposite. While U.S. military command could speak directly to the captains of Navy ships in the Gulf, U.S. forces in the area could not talk to each other. The Air Force had to fly in compatible equipment so that everybody in the area could communicate. The communications link with the Pentagon was not exactly trouble free either. A key military satellite failed to function properly, requiring some messages to be routed through the Philippines where, for some unknown reason, they were sent out to both the Air Force and the Navy uncoded over open channels. Because of this, the Cambodians knew the size of the U.S. Marine assault force coming to rescue the crew and where they were going. When the Marines landed by helicopter, they took nearly 100 casualties, 41 of them fatal, boarding a ship that had already been deserted.

Not only was that assault costly, it was also unnecessary. On May 14, *before* the Marines attacked, Cambodian radio announced that the crew and the ship were being released. As dozens of Marines were paying the ultimate price attempting to rescue the crew of the Mayaguez, the Cambodians were taking that same crew to the U.S. Navy destroyer Wilson for release. Either military command in the area had not been monitoring Cambodia broadcasts—an inexcusable and sophomoric error—or it had been unable to communicate with the Marines quickly enough to cancel the assault. Upon release of the crew, the president ordered an immediate halt to the bombing of the Cambodian mainland. Despite a direct presidential order, however, U.S. forces continued to bomb Cambodia for at least another half hour.[61] If nuclear war had been hanging in the balance, this slow a response would have been terminal.

It is important to emphasize that although there have been truly revolutionary improvements in both computer systems and communication equipment since these events transpired, much of what went wrong here was at least partly—if not primarily—the result of human and organizational problems, not purely technical failures. Furthermore, huge advances in computer and communications technology between these events and the events of September 11, 2001 did not prevent the American military response on 9/11 itself from being sluggish to the point of impotency in the face of attacks on the American homeland that took thousands of lives.

THE CRISIS CONTEXT

Crises increase the chances of accidental war. During a crisis, forces armed with weapons of mass destruction are kept at a high level of readiness. If a nation under attack wants to react before the other side's warheads reach their targets and explode, they must be able to move quickly (within 5 to 30 minutes). Since normal peacetime controls will get in the way, security is tightened and peacetime controls are removed in layers, as forces are brought to progressively higher levels of readiness.

In addition to the heightened danger inherent in removing peacetime barriers to unauthorized use, crises increase the chance of accidental war because false warnings are easier to believe when real attack is more likely—especially when verification fails. Each side will be looking for any sign that the other is preparing to attack. They will be predisposed to see what they expect to see, and—as in the case of the Vincennes—that can lead to tragedy, this time on a catastrophic scale.

Blair uses a simple Bayesian statistical model to illustrate the effect of initial expectations of attack on the credibility of the attack warning system. He assumes a warning system that has only a 5 percent chance of generating a false alarm, and only a 5 percent chance of missing a real attack. If the pre-warning estimate of the likelihood of a real attack was, say, 1 in 1,000—as it might be in normal times— even two corroborating warnings of attack would result in an estimated probability of only 27 percent that the attack was real; after a third positive warning, the probability would still be under 90 percent. But if a crisis raised the pre-warning expectation of a real attack to 1 in 10 (still low), two warnings would result in near certainty (98%) that a real attack was underway. As the crisis deepened and the pre-warning estimate of the probability of a real attack rose to 30 percent, even one warning would result in near 90 percent belief that the attack was real. If the chance of war were judged to be 50–50, a *single* warning would yield a 95 percent estimate that a real attack was underway; two warnings would make it seem virtually certain (99.7%).[62]

During the Cuban Missile Crisis, President Kennedy reportedly believed that the chances of war between the United States and U.S.S.R. were between 30 percent and 50 percent.[63] If the American attack warning system were as effective as Blair assumes, a single false warning would have led decisionmakers having Kennedy's assessment of pre-attack warning probabilities to believe that there was a 90 percent to 95 percent chance that a real attack was underway. Add to this the fact that four of the six Soviet nuclear-tipped intermediate range missile sites in Cuba had become operational in the middle of the crisis[64] and the fact that President Kennedy had publicly threatened to launch a "full retaliatory" nuclear strike against the U.S.S.R. if even one nuclear missile was launched from Cuba.[65] What you have is a recipe for accidental nuclear war on a global scale.

To make matters worse, after a high-flying U2 spy plane was shot down over Cuba in the midst of the crisis, Robert Kennedy told the Soviet ambassador that if even one more American plane was shot down, the United States would immediately attack all anti-aircraft missile sites in Cuba and probably follow that with an invasion.[66] At the time, Cuban anti-aircraft guns, not under Soviet control, were shooting at low-flying American spy planes entering Cuban airspace.[67] Without realizing it, the United States had given Fidel Castro, or possibly even an unreliable Cuban soldier operating an anti-aircraft gun, the power to trigger an all-out nuclear war.

Crises also make accidental war more likely because they put a great deal of stress on everyone involved, increasing the chances of mistakes, poor judgment, and other forms of human error. High stress can cause serious deterioration in performance. Sleep deprivation, disruption of circadian rhythms (the "biological clock"), fatigue, and even physical illness (common in national leaders) can magnify that deterioration (see Chapter 7). Crises also stimulate group pathology that can lead to very poor decisionmaking (see Chapter 8). All of this gets worse as the crisis drags on.

The stress of a crisis can also multiply the dramatic and sometimes dangerous side effects of prescribed and over-the-counter medication and increase the tendency to abuse drugs and alcohol (see Chapter 6). In the short term, chemical stimulants such as amphetamines may actually help decisionmakers to overcome stress and fatigue-induced impairment of performance. However, there is no way of knowing early on how long the crisis will last, and prolonged use of stimulants can lead to impulsiveness, highly distorted judgment, and even severe paranoia—a deadly combination in the midst of a confrontation between nuclear-armed adversaries.[68]

The Cuban Missile Crisis was not the only Cold War crisis that could have served as a backdrop for accidental nuclear war. In 1954 to 1955, fighting between China and Taiwan over the islands of Quemoy and Matsu threatened to draw the United States into war with China; in 1969, a confrontation erupted over the border between the Soviet Union and China during which these two nuclear-armed adversaries were shooting at each other; and by at least one account, in Israel's darkest days during the "Yom Kippur War" in October 1973, the Israeli military had put its nuclear-armed Jericho missiles on high alert.[69] That same 1973 Middle East war also came close to bringing the superpowers into direct conflict.

Despite the end of the Cold War, international crises continue to be fertile ground for accidental war with weapons of mass destruction. During the Persian Gulf War in 1991, Iraq, which had already used chemical weapons in war (against Iran and the Iraqi Kurds), threatened to launch a nerve gas missile attack against Israeli cities. Israel, which had already attacked an Iraqi nuclear reactor in the past, could certainly have struck back. Never before had Israel sat out an Arab attack on its territory, military or terrorist, without retaliating. This time they made a commitment to show uncharacteristic restraint, and held back a counterattack even as Iraqi Scud missiles rained down on Israel. Had they instead adopted a launch-on-warning strategy like that of the United States, they might have launched a nuclear counterattack before the Scuds even reached Israel and exploded—before they knew that the missiles were only carrying conventional warheads.

In 1991, President Bush had sent a message to Saddam Hussein implying that the United States would retaliate with nuclear weapons if Iraq used chemical or biological weapons in the war. This is often credited with having prevented Iraq from using these weapons. The same letter reportedly attached the same veiled threat against burning the oil fields in Kuwait, however, and Iraq did that anyway.[70]

There is no doubt that the dramatic lessening of tensions that followed the end of the Cold War reduced the chances of full-scale accidental war between nuclear-armed superpowers. Yet none of the critical elements that can combine to generate accidental nuclear war have disappeared: vast arsenals of weapons of mass destruction still stand at the ready; serious international crises and small-scale wars that threaten to escalate continue to occur; and human fallibility and technical failure still conspire to produce triggering events and failures of communication that might lead to nuclear war in the midst of conventional war or crisis.

Beyond this, both the United States and Russia continue provocative Cold War practices in the post-Cold War era. Thousands of strategic warheads are still on launch-on-warning alert. During the 1990s, American strategic nuclear bombers still flew up to the North Pole on simulated nuclear attack missions against Russia. In 2007, Russia began actively flying its nuclear bombers toward the territory of the U.S. and its allies, provoking repeated interceptions, and in 2009 the Russian Navy seems to have resumed continuous nuclear-armed submarine patrols for the first time in a decade.[71] The U.S. Navy continues to operate its nuclear missile submarines at Cold War rates, with about eight subs on patrol at any given time.[72]

Even more frightening, Russia may still be operating a "doomsday" system it had allegedly put in place some 25 years ago to insure a devastating retaliation if America attacked first. Switched on by the Russian general staff during a crisis, the almost completely automated system was designed to fire communications rockets if nearby nuclear explosions were detected and communications with military command were lost. These rockets would overfly Russian land-based missiles and transmit orders causing them to fire, without any action by local missile crews.[73] Given all the false warnings and other human and technical failures that have plagued the nuclear military, simply having such a system greatly increases the probability of accidental nuclear war. What justification can there possibly be for continuing these extraordinarily dangerous practices today?

In 2003, the RAND Corporation issued a report analyzing the risks inherent in the nuclear state-of-affairs between the United States and Russia: "[D]espite the steps taken by both countries to put Cold War hostilities behind them, an important nuclear risk remains—specifically that of accidental and unauthorized use of nuclear weapons. . . . [A]lthough both countries have significantly reduced their nuclear forces, they still retain nuclear postures and deterrence doctrines formulated when tension between them was much higher than it is today."[74] In addition to launch-on-warning, RAND pointed to a number of other reasons for the heightened risk of unintentional war, including human reliability problems aggravated by the problematic social and economic state of Russia and the deterioration of the Russian early warning system. In releasing the report, former Senator Sam Nunn "called for urgent action to address the increased danger of an accidental or unauthorized launch of a nuclear weapon between the United States and Russia."[75] Four years later, in July 2007, former Secretary of Defense William Perry reportedly stated that he believed the risk of an accidental nuclear war had increased since the end of the Cold War.[76]

After every major accident, false warning, or communications disaster, we are always assured that changes have been made that reduce or eliminate the chances of it ever happening again. Yet, perhaps in a different form, it *always* seems to happen again.

We have often failed to learn from even our most serious mistakes. The corrective measures that were taken were too frequently too little and too late. We certainly can do better. But the problem goes much deeper. The fallibility inherent in human beings, their organizations, and every system they design and build makes it impossible to do well enough.

As long as any nation continues to have arsenals of weapons of mass destruction, we must do everything possible to reduce the threat of blundering into cataclysmic war. Yet the reality is, we will never remove that threat unless and until we find a way to get rid of those arsenals. In human society, failure is an unavoidable part of life, and yet we still continue to bet our future on the proposition that the ultimate failure will never occur.

NOTES

1. Harry Anderson, Sudip Mazmazumdar, and Donna Foote, "Rajiv Gandhi Stumbles: Has India's Young Leader Become 'Directionless?' " *Newsweek*, February 23, 1987, 38.

2. Benjamin Friedman, "India and Pakistan: War in the Nuclear Shadow," *Nuclear Issues*, Center for Defense Information, June 18, 2002: http://www.cdi.org/nuclear/nuclearshadow-pr.cfm: (accessed July 1, 2009).

3. M. V. Ramana and A. H. Nayyar, "India, Pakistan and the Bomb", *Scientific American*, December 2001, 74.

4. Thom Shanker, "12 Million Could Die at Once in an India-Pakistan Nuclear War", *New York Times*, May 27, 2002.

5. Daniel Ford, *The Button: America's Nuclear Warning System - Does It Work?* (New York: Simon and Shuster, 1985), 52–53.

6. Bruce G. Blair, "Keeping Presidents in the Nuclear Dark: The SIOP Option that Wasn't," *The Defense Monitor*, Center for Defense Information, March/April 2004, 2.

7. Frank Von Hippel, "De-Alerting," *Bulletin of the Atomic Scientists*, May/June 1997, 35.

8. Russian Information Agency Staff Writers, "Current Nuclear Threat Worse Than During Cold War," *RIA Novosti*, Washington: July 20, 2007.

9. "Russia Resumes Regular Strategic Bomber Flights," *Global Security Newswire*, August 20, 2007, http://www.nti.org/d_newswire/ (accessed August 23, 2007).

10. "Russian Bombers Enter NATO Airspace," September 14, 2007; "Danish Jets Intercept Russian Bomber," October 31, 2007; "Russia Boosts Bomber Drills Near North America," October 2, 2007; "Russian Bombers Irk Canadian Leaders," March 2, 2009, *Global Security Newswire*, http://www.nti.org/d_newswire/ (accessed various dates).

11. Elaine M. Grossman, "Top U.S. General Spurns Obama Pledge to Reduce Nuclear Alert Posture," *Global Security Newswire*, February 27, 2009:

http://gsn.nti.org/siteservices/print_friendly.php?ID=nw_20090227_8682 (accessed July 1, 2009).

12. Unless specifically noted, the incidents discussed in this section are discussed in Chapter 4 or its Appendix.

13. Don Wall, "The Secret Arsenal," investigative report on the WFAA-TV *Evening News*, Channel 8: Dallas, Texas, November 2, 1995.

14. For a more detailed description, see op. cit. Daniel Ford, 118–119.

15. Bruce Blair, "Another Episode: The Case of the Missing 'Permissive Action Links,' " *The Defense Monitor*, Center of Defense Information: March/ April 2004, 2.

16. Stephen I. Schwartz, ed., *Atomic Audit: The Costs and Consequences of U.S. Nuclear Weapons Since 1940* (Washington, DC: Brookings Institution Press, 1998), 220.

17. U.K. Defends Nuclear Security Following BBC Report," *Global Security Newswire*, November 16, 2007; http://www.nti.org/d_newswire/ (accessed November 19, 2007).

18. Op. cit. Stephen Schwartz, 222.

19. The source of this information is Daniel Ellsberg, who was closely involved with American nuclear war planning at the Pentagon in the 1960s, as noted in op. cit. Paul Bracken, 198–199.

20. Bruce Blair, *The Logic of Accidental Nuclear War* (Washington, DC: The Brookings Institution, 1993), 50.

21. The following brief discussion is drawn from Bruce Blair, *The Logic of Accidental Nuclear War* (Washington, DC: The Brookings Institution, 1993), 188–191. For a detailed description of the warning systems, see also op. cit. Daniel Ford; Paul Bracken, *The Command and Control of Nuclear Forces* (New Haven, CT: Yale University Press, 1983); Bruce Blair, *Strategic Command and Control: Redefining the Nuclear Threat* (Washington, DC: Brookings Institution, 1985).

22. Bruce W. Nelan, "Nuclear Disarray," *Time* (May 19, 1997), 46–47; and Bruce G. Blair, Harold A. Feiveson, and Frank N. von Hippel, "Taking Nuclear Weapons Off Hair Trigger Alert," *Scientific American* (November 1997), 74–76.

23. Ibid.

24. Center for Defense Information, *Defense Monitor* (Vol. xv, No.7, 1986), 6. NORAD officials have said that the figures for the years prior to 1977 are "unavailable." These data were classified beginning in 1985 by the Reagan Administration, though it is hard to imagine any legitimate security reason why these data could not be made public. As far as I have been able to determine, they are still classified.

25. Major Mike Morgan, deputy director of the Space Control Center of the U.S. Space Command, in interview by Derek McGinty, National Public Radio, "All Things Considered," August 13, 1997.

26. Bruce Blair, "Primed and Ready," *Bulletin of the Atomic Scientists* (January/February 2007), 35.

27. Op. cit. Scott Sagan, 99–100.

28. Scott Sagan, *The Limits of Safety: Organizations, Accidents and Nuclear Weapons* (Princeton, NJ: Princeton University Press, 1993), 238.

29. A. O. Sulzberger, "Error Alerts U.S. Forces to a False Missile Attack," *New York Times*, November 11, 1979; Richard Halloran, "U.S. Aides Recount Moments of False Missile Alert," *New York Times*, December 16, 1979.

30. Senator Gary Hart and Senator Barry Goldwater, "Recent False Alerts from the Nation's Missile Attack Warning System," *Report to the Committee on Armed Services of the United States Senate: October 9, 1980* (Washington, DC: U.S. Government Printing Office, 1980), 5–6.

31. Richard Halloran, "Computer Error Falsely Indicates A Soviet Attack," *New York Times*, June 6, 1980; Richard Burt, "False Nuclear Alarms Spur Urgent Effort to Find Flaws," *New York Times*, June 13, 1980. See also op. cit. Scott Sagan, 231.

32. Op. cit. Daniel Ford, 78. See also footnotes 23 and 24.

33. Op. cit. Daniel Ford, 79; op.cit., Scott Sagan, 232.

34. Yuri Vasilyev, "On the Brink: More than 20 Years Ago, Stanislav Petrov Saved the World From a Thermonuclear War," *Moscow News* (May 12–18, 2004). See also Mark McDonald, "U.S. and Russian Nuclear Missiles Are Still on Hair Trigger Alert," *Knight Ridder Newspapers*, December 16, 2004.

35. Op. cit. Yuri Vasilyev.

36. John Kifner, "U.S. Officer Says Frigate Defenses Were Turned Off," *New York Times*, May 20, 1987; John Kifner, "Captain of Stark Says Ship Failed to Detect Missiles," *New York Times* (May 21, 1987); John Cushman, "Blind Spot Left Stark Vulnerable, U.S. Officials Say," *New York Times*, June 1, 1987.

37. Felicity Barringer, "Lone West German Flies Unhindered to the Kremlin," and Reuters News Service, "Pilot Said to Have Passion for Flying" *New York Times*, May 30, 1987; Philip Taubman, "Moscow Frees Young German Pilot," *New York Times*, August 4, 1988.

38. Roger Thompson, "Is the U.S. Navy Overrated?, *A Knightsbridge Working Paper* (Knightsbridge University: May 2004). See also "Pentagon Admits Russians Buzzed Kitty Hawk, Then E-mailed Pictures," *Agence France-Presse*, December 7, 2000.

39. Viktor Baranets, "Americans Can Now Make the First Nuclear Strike," *Komsomolskaya Pravda*, May 15, 2001, as reprinted in *Johnson's Russia List*, May 15, 2001, #14.

40. Jon Boyle, "Russia Loses Control of Army Satellites After Fire," *Reuters*, May 10, 2001, as reprinted in *Johnson's Russia List*, May 11, 2001, #1.

41. "Military Radar 'Blind' over 2/3 of Russia—General," *Reuters*, December 10, 2001, as reprinted in *Johnson's Russia List*, December 11, 2001, #2.

42. Konstantin Osipov, "Deaf, Dumb, and Blind? The Defense Ministry Lost Its Last Satellite: It Simply Fell Apart," *Tribuna*, November 26, 2003, as reprinted in *CDI Russia Weekly*, November 26, 2003, #10.

43. Elaine Grossman, "General Plays Down Concerns Over Ambiguous Missile Launches," *Global Security Newswire*, April 22, 2008; http://www.nti.org/d_newswire/ (accessed April 22, 2008).

44. Richard Halloran, "U.S. Downs Iran Airliner Mistaken for F-14; 290 Reported Dead; A Tragedy, Reagan Says," *New York Times*, July 4, 1988.

45. Stephen Engleberg, "Failures Seen in Safeguards on Erroneous Attacks," *New York Times*, July 4, 1988; Michael Gordon, "Questions Persist on Airbus Disaster," *New York Times*, July 5, 1988; "U.S. Pushes Inquiry on Downing of Jet; Questions Mount," Richard Halloran, "Team Goes to Gulf," and Julie Johnson, "No Shift in Policy," *New York Times*, July 5, 1988; John Hess, "Iranian Airliner," *New York Observer*, July 18, 1988; Les Aspin, "Witness to Iran Flight 655," *New York Times*, November 18, 1988.

46. David Shribman, "Korean Jetliner: What Is Known and What Isn't," *New York Times*, September 8, 1983.

47. David Shribman, "U.S. Experts Say Soviet Didn't See Jet Was Civilian," *New York Times*, October 7, 1983.

48. David Shribman, "U.S. Experts Say Soviet Didn't See Jet Was Civilian," *New York Times*, October 7, 1983, and "Korean Jetliner: What Is Known and What Isn't," *New York Times*, September 8, 1983; David Pearson, "KAL 007: What the U.S. Knew and When We Knew It," *The Nation*, April 18–25, 1984, 105–124; Steven R. Weisman, "U.S. Says Spy Plane Was in the Area of Korean Airliner," *New York Times*, September 5, 1983; Dick Polman, "Challenging the U.S. on Flight 007," *Philadelphia Inquirer*, October 18, 1984; Bernard Gwertzman, "A New U.S. Transcript Indicates Soviet Pilot Fired 'Canon Bursts,' " *New York Times*, September 12, 1983. See also, Seymour M. Hersh, *The Target Is Destroyed: What Really Happened to Flight 007 & What America Knew About It* (New York: Vintage, 1987).

49. Steven Weisman, "U.S. Says Spy Plane Was in the Area of Korean Airliner," *New York Times*, September 5, 1983; Steven Erlanger, "Similarities with KAL Flight Are Rejected by U.S. Admiral," *New York Times*, July 4, 1988.

50. Op. cit. Daniel Ford, 87.

51. Philip Shenon, "Panel Investigating 9/11 Attacks Cites Confusion in Air Defense," *New York Times*, June 16, 2004.

52. Op. cit. Daniel Ford, 20–21.

53. Richard Hudson, "Molink Is Always Ready," *New York Times Magazine*, August 26, 1973.

54. "National Warning Plan Fails," *Sacramento Bee*, February 18, 1984.

55. Bill Keller, "Pentagon Test on News Coverage Hurt by Communications Lapses," *New York Times*, April 27, 1985.

56. *Sacramento Bee*, November 19, 1989.

57. "Coordination Plans of Military Services Attacked," *New York Times*, March 22, 1984.

58. "Army Phones Need $30 Million Repair," *New York Times*, March 2, 1985.

59. U.S. Congress, Armed Services Investigating Subcommittee, Committee on Armed Services, House of Representatives, *Review of Department of Defense World-Wide Communications* (Washington, DC: U.S. Government Printing Office, 1971).

60. Such as the 1967 Israeli attack on the U.S. ship Liberty and the 1968 North Korean attack on the U.S. ship Pueblo.

61. Op. cit. Daniel Ford, 173–174, and Desmond Ball, "Can Nuclear War Be Controlled?," *Adelphi Papers* (London: International Institute for Strategic Studies, No.169), 12–13; see also "Resistance Met: All 39 From Vessel Safe— U.S. Jets Hit Cambodian Base," *New York Times*, May 15, 1975.

62. Bruce Blair, *The Logic of Accidental Nuclear War* (Washington, DC: The Brookings Institution, 1993), 224–228. Bayesian analysis is a probabilistic technique for calculating the extent to which additional information causes one to revise earlier estimates of the likelihood of an event.

63. Lynn Rusten and Paul C. Stern, *Crisis Management in the Nuclear Age*, Committee on International Security and Arms Control of the National Academy of Sciences, and Committee on Contributions of Behavioral and Social Science to the Prevention of Nuclear War of the National Research Council (Washington, DC: National Academy Press, 1987), 26.

64. B. J. Bernstein, "The Week We Almost Went to War," *Bulletin of the Atomic Scientists*, February 1976, 17.

65. John F. Kennedy, Presidential Address to the Nation (October 22, 1962), reprinted in Robert F. Kennedy, *Thirteen Days: A Memoir of the Cuban Missile Crisis* (New York: W. W. Norton, 1969), 168.

66. Daniel Ellsberg, "The Day Castro Almost Started World War III," *New York Times*, October 31, 1987.

67. Ibid.

68. Derek Paul, Michael Intrilligator, and Paul Smoker, ed., *Accidental Nuclear War: Proceedings of the 18th Pugwash Workshop on Nuclear Forces* (Toronto: Samueal Stevens and Company and University of Toronto Press, 1990), 43–51.

69. Avner Cohen, "The Last Nuclear Moment," Op-Ed, *New York Times*, October 6, 2003.

70. Brian Hall, "Overkill Is Not Dead," *New York Times Magazine*, March 15, 1998, 84.

71. "Russia Restores Nuclear-Armed Submarine Patrols," *Global Security Newswire*, February 18, 2009; http://gsn.nti.org/siteservices/print _friendly.php?ID=nw_20090218_4795 (accessed July 7, 2009).

72. "U.S. Missile Submarines Sustain Active Schedule," *Global Security Newswire*, March 16, 2009; http://gsn.nti.org/siteservices/print_friendly.php ?ID=nw_20090316_6606 (accessed July 7, 2009).

73. Bruce Blair, "Hair-Trigger Missiles Risk Catastrophic Terrorism," Center for Defense Information, as published in *Johnson's Russia List*, April 29, 2003, #1. See also Ron Rosenbaum, "The Return of the Doomsday Machine?" *Slate.com*,

August 31, 2007; http://www.slate.com/toolbar.aspx?action=print&id =2173108 (accessed July 7, 2009); and "Russia's Doomsday Machine," and William J. Broad, "Russia Has a Nuclear 'Doomsday' Machine," *New York Times*, October 8, 1993.

74. David E. Mosher, Lowell H. Scwartz, David R. Howell, and Lynn E. Davis, "Beyond the Nuclear Shadow: A Phased Approach for Improving Nuclear Safety and U.S. Russian Relations," RAND Corporation, released May 21, 2003, xi.

75. "Nunn Urges Presidents Bush and Putin to Address Nuclear Dangers," *Nuclear Threat Initiative Press Release*, May 21, 2003, 1. See also op. cit. David Mosher et al, xii.

76. Russian Information Agency Staff Writers, "Current Nuclear Threat Worse Than During Cold War," *RIA Novosti*, Washington: July 20, 2007.

Part III

Why It Can Happen Here: Understanding the Human Factor

6

Substance Abuse
and Mental Illness

We all make mistakes every day, most of them so trivial and fleeting that we pay little attention to them. We misdial telephone numbers, misplace notes, and misspell words. Our brains do not always stay focused tightly on the issue at hand. Attention drifts, and something that should be noticed goes unnoticed. Just as the complex technical systems we create and operate are subject to breakdown and less-than-perfect performance from time to time, so is the complex biological system that creates and operates them: the human mind and body. We are very powerful and very capable, and yet we are prone to error, subject to malevolence, and all the other limitations of being human. We are fallible in our actions and in our essence.

Our failures do not have to be continuous in order to be dangerous. Failures that are of little consequence most of the time can be catastrophic if they occur in critical situations at critical moments. The pilot of a transatlantic airliner could be asleep for most of the flight without causing any real problem. But if he/she is not fully reliable at the moment a problem requiring quick corrective action develops, hundreds of lives could be lost. Similarly, most times it is not a problem if a nuclear weapons guard is drug or alcohol impaired because most times nothing happens. But if that guard is not alert and ready to act the moment that terrorist commandos try to break into the compound, a disaster of major proportions could result. The problem is that it is impossible to predict when those critical moments will occur. So when dangerous technologies are involved, there is little choice but to take even fleeting failures of reliability seriously.

Human error is a frequent contributor to commonplace technological mishaps. Some 29 percent of 1,300 fatal commercial aircraft accidents worldwide over the more than half century from 1950 to 2008 were caused by pilot error unrelated to weather or mechanical problems.[1] Federal Aviation Administration (FAA) data show that 2,396 people died between 1987 and 1996 in one of the strangest categories of commercial aircraft accident, "controlled flight into terrain." These crashes, in which a malfunctioning crew flies a properly operating aircraft into the ground, accounted for 25 percent of world commercial aircraft accidents during that period.[2] In a 1998 study of 10 nuclear power plants, representing a cross section of the U.S. civilian nuclear industry, the Union of Concerned Scientists concluded that nearly 80 percent of reported problems resulted from flawed human activity (35% from "workers' mistakes" and 44% from poor procedures).[3] In 2000, the Institute of Medicine of the U.S. National Academy of Sciences published a report finding that medical errors cause more deaths each year in the United States than breast cancer or AIDS.[4]

The use and abuse of chemical substances is one of the most common ways that individuals make themselves unreliable. Even in moderation, alcohol and drugs can at least temporarily interfere with full alertness, quickness of reaction, clarity of the senses, and sound judgment. When used to excess or at inappropriate times, they seriously threaten reliability. Mental illness and emotional trauma also impair performance. These problems are, unfortunately, all too common in the human population. Since every team designing dangerous technologies and every workforce operating and maintaining them is recruited from that population, the problems of human fallibility can never be avoided entirely.

ALCOHOL

Alcohol is the most widely used drug. In 2006, the *average* U.S. resident drank 1.6 gallons of pure alcohol annually (roughly one shot of hard liquor per day).[5] Since an estimated one-third of the U.S. population does not drink at all, the average *drinker* drank 2.4 gallons of pure alcohol per year, or roughly one and a half shots of hard liquor per day. Even those in critical positions are sometimes less careful than they should be. The FAA and all commercial airlines have strict rules concerning alcohol use by flight crews. That did not stop all three members of the cockpit crew of a March 1990 Northwest Airlines flight from Fargo, North Dakota to Minneapolis-St. Paul from flying the plane under the influence of alcohol. The pilot reportedly "had so much alcohol in his blood ... that he would have been too drunk to legally drive an automobile in most states."[6] The crewmembers were arrested in Minnesota, and their licenses were later revoked.[7]

About 1,200 airline pilots were treated for alcoholism and returned to flying from the mid-1970s to 1990, under a government-run rehabilitation program. According to the psychiatrist who helped begin that program, "Every pilot I treat, it's clear ... should have been treated 10 or 15 years ago ... [and] there are five more who just sneak into early retirement."[8] A 2008 FAA civil aviation study analyzed 2,391 pilots involved in fatal accidents from 2000 to 2007, finding that 215 (9%) "had a documented alcohol or drug related offense in their past."[9]

Alcohol in the American Military

The armed forces are intimately involved with a wide variety of dangerous technologies, from weapons of mass destruction to nuclear power reactors to highly toxic chemicals. Focusing on the military should help avoid exaggerating the problem, since the military has more power to regulate the behavior of the people who live and work within its confines than do most organizations that use dangerous technologies. That should make it easier to minimize these human reliability problems and take quick corrective action when they do occur. Furthermore, the American military is a public institution, prominent in a society democratic enough to make more data available.

In 1980, the Pentagon began a series of worldwide surveys in an effort to collect data on substance abuse and related problems within the armed forces. There has been a survey about every three years since 1980, with 2005 the latest publicly available at this writing. Interestingly enough, after the 1988 survey was released, it was discovered that human fallibility had struck: a programming error had caused substantial understatement of the "heavy drinkers" category for each military branch individually and all branches combined in the 1982, 1985, and 1988 surveys.[10] Data presented below are corrected for this error.[11]

The surveys defined heavy drinking as "consumption of five or more drinks per typical drinking occasion at least once a week." In 1980, nearly 21 percent of all active duty military personnel were classified as heavy drinkers during the 30 days previous to the survey. By 1982, that number was 24 percent. In 1998, it hit a low of 15.4 percent, but by 2005 it has climbed to 19 percent—more than a third of the way back to the 1982 peak. Keeping in mind that these numbers are limited to recent heavy drinking, it is clear that active-duty American military personnel continue to consume substantial quantities of alcohol.[12] The surveys indicate a substantial decline in the percentages of personnel reporting negative effects of alcohol use since 1980. Still, in 2005 serious negative consequences were reported by more than 8 percent of military personnel. Among heavy drinkers, it was three times as high; even for "infrequent/light" drinkers, it was still 5 percent. That year the figures for "alcohol-related productivity loss" were higher: 13 percent for all military personnel, 6 percent for "infrequent/light" drinkers, and nearly 36 percent for heavy drinkers.[13]

These data may be understated. It is difficult enough to collect accurate information on sensitive issues from the general public. People who live and work in an authoritarian "total institution" like the military are less likely to respond accurately.[14] While the data-collection teams "assured the respondents of anonymity, and informed participants of the voluntary nature of participation,"[15] what incentive would anyone have to accurately answer questions that could potentially cause them a major problem, even ruin their careers, if anonymity broke down? What would serious abusers lose by understatement that compares to what they could lose if they answered honestly and their answers were traced back to them somehow? Verbal assurances of survey takers are likely not strong enough to overcome the powerful incentive to give answers that are more acceptable than accurate.[16]

There is no way of directly assessing the degree of bias in the data. Since there is certainly no reason that respondents would *overstate* their substance abuse or its consequences, however, these data can be considered conservative estimates. Even so, it is clear that alcohol use is a widespread and continuing problem in the U.S. military.

Alcohol in the Soviet/Russian Military

Russia has the only other military that has arsenals of weapons of mass destruction comparable to those of the United States and comparable involvement with other dangerous technologies. There are few systematic data available on Soviet/Russian military alcohol use, but the extent of alcohol abuse among the general population makes clear that it is a problem.

Vladimir Treml of Duke University tried to unravel the confusion and distortions of official data on alcoholism among the population of the former USSR and present day Russia. He estimated that the average person (15 years of age or older) consumed about 3.85 gallons of pure alcohol during 1984—45 percent more than average American consumption that year. By 1989, average Soviet alcohol consumption had fallen to about the U.S. per capita level.[17] By 1993, per capita alcohol consumption in Russia had climbed to roughly the same peak it had reached in the Soviet Union (3.81 gallons).[18] Treml's figures include state-produced alcohol and "samogon," a popular homemade alcohol, but not homemade wine or beer or "stolen industrial alcohol and alcohol surrogates."[19] Since these forms of alcohol were widespread, these estimates should be regarded as conservative. Treml concludes, "The Soviet Union . . . [was] a country of heavy drinkers. . . ."[20]

Combining analysis of Soviet military journals with a survey of several hundred former Russian soldiers, Richard Gabriel concluded in 1980:

> Russian soldiers at all ranks have at least the same high rates of alcoholism and alcohol-related pathologies as the society at large . . . probably considerably higher. . . . The impression one gathers . . . is that heavy drinking and chronic alcohol abuse are common characteristics of Soviet military life.[21]

Furthermore:

> drunkenness seems to cut across all rank levels in the army. . . . Reports of alcoholism are just as common in the elite technical units, such as strategic rocket forces [*the mainstay of the Soviet nuclear military*] and the air defense corps, as they are in infantry and tank units.[22]

For most, economic life in Russia (and some other Soviet successor states) today is not much better than before. Russian military life deteriorated even more dramatically. Military industry workers and soldiers lost much of the status they once enjoyed. Many have fallen farther than the average citizen, and that can be a source of great stress.

DRUG USE

In the mid-1990s, the Substance Abuse and Mental Health Services Administration took over a national survey series on drug use begun in 1971. The population targeted all U.S. civilian residents, including civilians living on military installations. Active duty military personnel, persons living in institutions such as hospitals and jails, and children under 12 were excluded. The data were collected through personal visits to each residence. An attempt was made to assure anonymity. Table 6.1 presents summary data from the 2006 and 2007 survey.[23]

The 2007 survey estimated that more than 45 percent of all Americans have used some illicit drug some time in their lives. More than 1 in 7 admitted to using an illicit drug within the preceding year; about 1 in 12 admitted to doing so within the past month. The most commonly used drugs were marijuana and hashish, with about 6 percent reporting recent use and more than 10 percent reporting use within the past year.

From the human reliability point of view, cocaine and hallucinogens are among the most troubling drugs. For each, about 14 percent reported using the drug at some point, and some 1.5 to 2.5 percent reported use within 12 months; 0.4 percent reporting using hallucinogens within the past 30 days and at least twice that percentage using cocaine. Considering the total 2007 U.S. population (excluding children) of 245 million,[24] these numbers mean that more than 35 million Americans have used cocaine and close to 34 million have used hallucinogens as of 2007: almost 6 million used cocaine and nearly 4 million used hallucinogens within the last year, and close to 2 million used cocaine and 1 million used hallucinogens within the last month. There is no category of drug in the table used by less than 0.1 percent over the previous 30 days, and even 0.1 percent is almost 250,000 people.

The use of illicit drugs by males of prime military age (age 18 to 25) is far more common, as shown in Table 6.2. In 2007, the rate of recent drug use (within 12 months) for young males was more than one and a half times that for the population; for current drug use (within 30 days), it was more than three times as high.

During the 1980s, cocaine was very much at the forefront of public attention in the United States. Cocaine was so pervasive it even managed to penetrate two of America's most prominent institutions, baseball and high finance. By 1985, scores of major league baseball players had been implicated as users and sometimes as sellers in criminal investigations. According to the *New York Times*, "One owner . . . said an agent had told him . . . that he could field an all-star team in each league with players who were using cocaine."[25] Today we seem to worry more about steroids and other performance-enhancing drugs in sports. As to high finance, on April 16, 1987, "Sixteen brokers and a senior partner of Wall Street firms were arrested by federal agents . . . and charged with selling cocaine and trading the drug for stocks, information and lists of preferred customers." The special agent in charge said of the raid, "We don't believe this case is an aberration."[26]

Table 6.1 Types of illicit drug use in lifetime, past year, and past month among persons aged 12 or older: Percentages, 2006 and 2007

	Time Period					
	Lifetime		Past Year		Past Month	
Drug	2006	2007	2006	2007	2006	2007
Illicit Drugs[1]	45.4	46.1	14.5	14.4	8.3	8.0
Marijuana and Hashish	39.8	40.6	10.3	10.1	6.0	5.8
Cocaine	14.3	14.5	2.5	2.3	1.0	0.8
Crack	3.5	3.5	0.6	0.6	0.3	0.2
Heroin	1.5	1.5	0.2[a]	0.1	0.1[a]	0.1
Hallucinogens	14.3	13.8	1.6	1.5	0.4	0.4
LSD	9.5	9.1	0.3	0.3	0.1	0.1
PCP	2.7	2.5	0.1	0.1	0.0	0.0
Ecstasy	5.0	5.0	0.9	0.9	0.2	0.2
Inhalants	9.3	9.1	0.9	0.8	0.3	0.2
Nonmedical Use of Psychotherapeutics[2,3]	20.7	20.3	6.7	6.6	2.9	2.8
Pain Relievers	13.6	13.3	5.1	5.0	2.1	2.1
OxyContin®	1.7	1.8	0.5	0.6	0.1	0.1
Tranquilizers	8.7	8.2	2.1	2.1	0.7	0.7
Stimulants[3]	9.1	8.7	1.5[b]	1.2	0.6[a]	0.4
Methamphetamine[3]	5.8[a]	5.3	0.8[b]	0.5	0.3	0.2
Sedatives	3.6	3.4	0.4	0.3	0.2	0.1
Illicit Drugs Other than Marijuana[1]	29.6	29.7	8.6	8.5	3.9	3.7

[a]Difference between 2006 estimate and 2007 estimate is statistically significant at the 0.05 level.

[b]Difference between 2006 estimate and 2007 estimate is statistically significant at the 0.01 level.

[1]Illicit Drugs include marijuana/hashish, cocaine (including crack), heroin, hallucinogens, inhalants, or prescription-type psychotherapeutics used nonmedically. Illicit Drugs Other Than Marijuana include cocaine (including crack), heroin, hallucinogens, inhalants, or prescription-type psychotherapeutics used nonmedically. The estimates for Nonmedical Use of Psychotherapeutics, Stimulants, and Methamphetamine incorporated in these summary estimates do not include data from the methamphetamine items added in 2005 and 2006. See Section B.4.6 in Appendix B of the *Results from the 2007 National Survey on Drug Use and Health: National Findings.*

[2]Nonmedical Use of Prescription-Type Psychotherapeutics includes the nonmedical use of pain relievers, tranquilizers, stimulants, or sedatives and does not include over-the-counter drugs.

[3]Estimates of Nonmedical Use of Psychotherapeutics, Stimulants, and Methamphetamine in the designated rows include data from methamphetamine items added in 2005 and 2006 and are not comparable with estimates presented in prior NSDUH reports. See Section B.4.6 in Appendix B of the *Results from the 2007 National Survey on Drug Use and Health: National Findings.*

Source: SAMHSA, Office of Applied Studies, National Survey on Drug Use and Health, 2006 and 2007.

Table 6.2 Illicit drug use in lifetime, past year, and past month among persons aged 18 to 25 (percentages, 2006 and 2007)

Demographic Characteristic	Time Period					
	Lifetime		Past Year		Past Month	
	2006	2007	2006	2007	2006	2007
TOTAL	59.0	57.4	34.4	33.2	19.8	19.7
GENDER						
Male	64.4	60.5	38.4	37.2	23.7	24.1
Female	56.5	54.4	30.3	29.1	15.8	15.3

Source: SAMHSA, Office of Applied Studies, National Survey on Drug Use and Health, 2006–2007. (http://www.oas.samhsa.gov/NSDUH/2k7NSDUH/tabs/Sect1peTabs1to46.htm# Tab1.21B; extracted from Table 1.21B; accessed September 4, 2009).

Bad behavior among the captains of high finance has certainly not gone away. But today—with good reason—we have become more concerned about their creativity in finding ways to misinform, mislead, and sometimes out-and-out defraud investors who are their actual or potential clients than we are about the extent to which they might be using drugs (although there could be a connection between the two).[27]

Psychedelics

We have highlighted psychedelics (also called hallucinogens) because they are particularly relevant to reliability. The premier psychedelic, d-lysergic acid diethylamide (LSD), was first synthesized in 1938,[28] but it was not brought to widespread public attention in America until the 1960s. LSD is one of the most powerful psychoactive drugs known. Whereas the effective dose of most such drugs is measured in the tens or hundreds of milligrams, as little as 10 micrograms (0.01 milligram) of LSD produces some euphoria, and 50 to 100 micrograms (0.05 to 0.1 milligram) can give rise to the full range of its psychedelic effects.[29]

LSD magnifies and intensifies sensory perception, " ... ordinary objects are seen ... with a sense of fascination or entrancement, as though they had unimagined depths of significance ... colors seem more intense ... music more emotionally profound .... "[30] It can alter the drug takers perceptions of their own body parts, cause the field of vision to quiver and swell, and distort perspective. It can also produce "synesthesia," a kind of crossing of the senses in which colors are heard, sounds are seen, and so forth. Time may seem to slow or even stop completely. LSD has even more profound emotional effects," . . . feelings become magnified to a degree of intensity and purity almost never experienced in daily life; love, gratitude, joy . . . anger, pain, terror, despair or loneliness may become overwhelming." These powerful emotions can produce intense fear of losing control, even to the point of paranoia and panic. The drug taker may lose sight of where his/her body ends and the environment begins. The worst experience, the "bad trip," can be truly terrible" . . . a fixed intense emotion or distorted thought that can seem like an eternity of hell . . . [with] remorse, suspicion, delusions of persecution or of being irreversibly insane."[31]

Which of these effects occur and at what level of intensity is extremely difficult to predict. "One person may feel only nervousness and vague physical discomfort from a dose that plunges another into paranoid delusions and a third into ecstasy . . . [T]he drug . . . is so unpredictable that even the best environment and the highest conscious expectations are no guarantee against a painful experience."[32] Furthermore, the same individual may experience radically different effects from the same dose taken at different times. The nature and unpredictability of the drug's effects render the user completely unreliable while he/she is "under the influence."

LSD normally takes effect 45 to 60 minutes after the drug is taken orally. Its effects peak in two or three hours and can last as long as half a day. Users are also subject to spontaneous recurrences, called "flashbacks." Without having actually taken the drug again, the user may suddenly be on a complete or partial "LSD trip" at some unpredictable future time. A flashback" . . . can last seconds or hours . . . it can be blissful, interesting, annoying, or frightening. Most . . . are episodes of visual distortion, time distortion, physical symptoms, loss of ego boundaries, or relived intense emotion. . . . [I]n a small minority of cases, they turn into repeated frightening images or thoughts."[33]

About a quarter of psychedelic drug users experience flashbacks. They can occur up to several months after the last dose, but rarely longer than that. Their timing is difficult if not impossible to predict, but flashbacks are known to be most likely at a time of high emotional stress, fatigue, or drunkenness. Marijuana can trigger LSD flashbacks.[34] This illustrates the importance of considering drug interactions as well as the user's physical and emotional state, in evaluating how substance abuse affects reliability.

Because LSD is so powerful, it is normally taken in extremely small doses. What little is taken is almost completely metabolized in the body—very little is excreted unchanged. Urine (and blood) tests *can* detect LSD use, but the tests required are more sophisticated, time consuming, and expensive than the tests for detecting other major illicit drugs.[35] This and the fact that LSD does not linger in the body for days makes it easier for a determined drug user to avoid detection in an environment where random testing is common.

LSD is not the only psychedelic. Psilocyn, Psilocybin, DMT, MDA, MMDA, mescaline (peyote), DOM, ketamine, a class of drugs called harmala alkaloids, DOET, and THC (the active ingredient in marijuana and hashish) can all have effects like those of LSD.[36]

Legal Drugs

Most people understand that using illicit drugs can affect mood, clarity of thought, judgment, and reaction time in ways that compromise reliability. But it is not widely understood that commonly prescribed legal drugs—even over-the-counter medications—have interaction and side effects that can also seriously degrade performance and reliability. These range from mild drowsiness that often accompanies the use of antihistamines to such dramatic effects as depression, hallucinations, convulsions, and rage.[37]

Among the 50 drugs most frequently prescribed in the United States in 2005 were Tramadol, Lyrica, Toprol, Xanax, and Ativan. Tramadol, a pain reliever, can cause dizziness, weakness, anxiety, agitation, and emotional instability. Lyrica, for nerve pain, has common side effects that include tremors, memory loss, and accidental injury, and less commonly can cause nerve pain (paradoxically), difficulty swallowing, hostility, hallucinations, and heart failure. Toprol, prescribed for high blood pressure and abnormal heart rhythms, can produce breathing difficulties, depression, confusion, disorientation, memory loss, and emotional instability. Xanax, used for anxiety, can cause depression, lethargy, disorientation, dizziness, blurred or double vision, and stupor. Ativan, prescribed for anxiety, tension, fatigue, and agitation can cause depression, lethargy, nervousness, hysteria, and even psychosis.[38]

Because over-the-counter medications are presumed safe enough to be taken without medical oversight and are relatively cheap, they are used more casually. Yet they too have reported serious side effects. Among over-the-counter medications are such common household names as Contac, Alka Seltzer Plus, Drixoral, Sudafed, and Advil.

The advertising that made these brand names household words also conveyed the impression that they are completely benign. Yet Contac, used to temporarily relieve sinus and nasal congestion, can cause dizziness, trembling, and insomnia. It can interact with prescription antidepressants known as monoamine oxidase inhibitors (MAOI) to produce dangerous increases in blood pressure. Alka Seltzer Plus has side effects including nervousness, drowsiness, dizziness, and insomnia; and more rarely fatigue, changes in vision, agitation, and nightmares. It can interact with alcohol, antidepressants, and antihistamines to cause excessive sedation.[39]

Drixoral reduces the symptoms of allergies. Its common side effects include insomnia and agitation; its rarer side effects include nightmares, seizures, and hallucinations. Drixoral has interaction effects similar to those of Alka Seltzer Plus. Sudafed, a decongestant, can cause agitation and insomnia, and on rare occasions hallucinations and seizures. Finally Advil, a pain reliever, has been known to produce drowsiness and depression, and in rare cases, blurred vision, confusion, and convulsions.[40]

Many of the more striking and dangerous prescription and nonprescription drug side effects are temporary and occur as the result of interaction with alcohol or other drugs. They also tend to be rare. Nevertheless, rare events do occur.[41] In anyone working with dangerous technologies in a job subject to sudden criticality, temporary failure of reliability resulting from a rare side effect or interaction can have permanent, catastrophic consequences.

Drug Use in the American Military

For each major category of illicit drug, Table 6.3 presents data on self-reported abuse among American military personnel. The data are from worldwide surveys conducted by Defense Department contractors. These are essentially the same surveys that measured alcohol abuse, and hence they have the same potential biases discussed earlier.

In 1995, almost 6.5 percent of U.S. military personnel worldwide reported using an illicit drug within 12 months and 3 percent reported doing so within 30 days. The most widely used drug was marijuana: 4.6 percent reported its use within the year; 1.7 percent within the month. Almost 4 percent used some other drug in the past year; 2.0 percent in the past month.

Table 6.3 Nonmedical drug use in the U.S. military (by estimated percentage of combined military population)

	Used in Past Year	Used in Past Month
Any Illicit Drug[*]		
1995	6.5	3.0
2005	10.9	5.0
Any Drug Except Marijuana		
1995	3.9	2.0
2005	9.5	4.4
Marijuana		
1995	4.6	1.7
2005	4.2	1.3
Cocaine		
1995	0.9	0.3
2005	1.9	0.6
Hallucinogens (LSD, PCP)		
1995	1.7	0.7
2005	1.9	0.8
Amphetamines/Stimulants		
1995	0.9	0.5
2005	1.4	0.6
Barbiturates/Sedatives		
1995	0.3	0.1
2005	2.0	1.0
Tranquilizers/Depressants		
1995	0.6	0.3
2005	2.0	0.9
Analgesics/Other Narcotics		
1995	1.0	0.6
2005	7.3	3.3
Inhalants		
1995	0.7	0.4
2005	2.1	0.9
Heroin/Other Opiates		
1995	0.2	0.1
2005	0.9	0.5

*Nonmedical use of any drug or class of drugs listed in the table. (Plus, 1995 data include combinations of individual legal drugs created specifically for their psychoactive effects, known as "designer drugs"; 2005 data do not include "designer drugs" as such, but do include steroids and sexual enhancers).

Sources: Bray, R. M. et al., Worldwide Survey of Substance Abuse and Health Behaviors Among Military Personnel, 1995 edition (Research Triangle Institute), Table 5.1, p. 85 and Table 5.3, p. 90 within the past 30 days; and Bray, R. M., et al., 2005 Department of Defense Survey of Health Related Behaviors Among Active Duty Military Personnel (RTI International [formerly, Research Triangle Institute], December 2006), Table 5.3, p. 100 (30 days), and Table 5.4, p. 101 (12 months).

By 2005, a substantial increase in reported drug use had been recorded. Use of any illicit drug during the past year increased to almost 11 percent, and use within the month increased to 5 percent. Marijuana use decreased slightly, but every other illicit drug listed increased, including a quadrupling of heroin use, a doubling of cocaine use, and a 10 percent increase in hallucinogens.

These percentages seem relatively low and may be underestimates, but to be conservative, suppose we use them. Since there were 1,143,000 active duty military personnel in the U.S. Armed Forces in 2005,[42] applying the estimated percentages to this overall force implies that more than 124,500 U.S. military personnel used illicit drugs within 12 months, and more than 57,000 used them within the preceding month. Excluding marijuana still leaves more than 108,500 active duty military who used illegal drugs within the year and over 50,000 who used them within the past 30 days. Nearly 22,000 active duty military had been tripping on LSD or another hallucinogen in the year; more than 9,100 within the past 30 days. Thus, thousands to many tens of thousands of American military personnel abuse all sorts of illicit drugs, despite the military's ongoing anti-drug efforts.

In 1988, over 40 percent reported they did not seek help for drug abuse because "seeking help for a drug problem will damage military career"; nearly 50 percent believed they "can't get help for drug problem without commander finding out"; and more than 60 percent were kept from seeking help because they believed "disciplinary action will be taken against a person (with a drug problem)."[43] But distrust of the system is not the only problem. For many, especially in the nuclear military, the conditions of life and work are a breeding ground for substance abuse (see Chapter 7). There are also technical and human problems in implementing anti-drug efforts themselves.

There is a frightening possibility that at least some illegal drug users may shift from more easily detected, relatively benign drugs like marijuana to less easily detected, more dangerous reliability-reducing drugs like LSD. Table 6.3 shows falling marijuana use and increasing LSD use between 1995 and 2005. Beyond this, the potency of LSD means much smaller amounts are needed. It would be relatively easy to smuggle enough on board nuclear missile submarines, for example, by soaking part of a T-shirt in LSD, then cutting it into little patches that could be swallowed whole.

There are no publicly available systematic data on drug and alcohol abuse in military industry or other weapons-related facilities. Yet periodic reports make it clear that these workplaces are not immune.

For example, in 1988 a congressional oversight committee reported that an undercover agent investigating Lawrence Livermore nuclear weapons labs two years earlier collected information suggesting drugs were being bought, sold, or used on the job by more than 100 Livermore employees. These included scientists and staff with high security clearances.[44]

Drug Use in the Military of the Former Soviet Union/Russia

There is little reliable information available concerning use and abuse of drugs in the former Soviet Union. For a long time, the problem was completely denied. In the mid-1980s, Soviets claimed that " 'serious drug addiction does not exist' ... and 'not a single case' of addiction to amphetamines, cocaine, heroin, or LSD has been recorded."[45] While they no longer deny the problem outright, sources in Russia and other former Soviet states concede that they still do not have an accurate picture of the extent of the drug problem.[46]

Two surveys in Georgia, in the mid-1970s and mid-1980s, confirm high drug use. "Over 80 percent of the individuals in each sample consumed drugs at least once a day and a majority of them ... at least twice a day."[47] There were vast areas in the USSR where opium poppies and hemp (used for hashish) were cultivated or grew wild. Many households, perhaps hundreds of thousands, grew them illegally, for their own use or to earn extra income. According to a deputy minister of agriculture in the former USSR, more than half the legally grown hemp and poppy crop on collective farms was left in the fields after harvest and became "a fertile source of drugs for local users and illegal drug producers."[48] Alternative sources of supply included theft from medical stores and smuggling from neighboring countries (such as Afghanistan). Available data suggest that many individuals spent more on illicit drugs per month than the average monthly wage, creating the conditions for drug-related crime.[49] When the rigid controls of their police states were relaxed, the Russian Federation and other former Soviet states experienced rising crime rates amidst economic and political chaos. Drugs played a part in this as well.

There is less information available concerning drug abuse in the Soviet/Russian military than for alcoholism. It seems to have been relatively minor until the Soviets entered Afghanistan in the late 1970s. Like America's Vietnam War, the Soviet War in Afghanistan was unpopular, ill defined, and fought against guerrilla forces in an area that has long been a center of drug production. Though the evidence is anecdotal and very spotty, the Afghan War apparently did increase drug abuse among the Soviet armed forces. (It is similarly possible that U.S. military involvement in Afghanistan and Iraq after 2003 may be part of the reason for rising U.S. military drug abuse between 1995 and 2005). There is little information available about the drug problem today in the militaries of the Russian Federation and other republics of the former Soviet Union. But the armed forces are certainly not drug free.

Mental Illness

Mental illness is disturbingly common. In 2008, the National Institute of Mental Health (NIMH) estimated that more than one-quarter (26.2%) of the U.S. population age 18 or older suffered from a diagnosable mental disorder.[50] Applied to the U.S. 2007 population age 18 or older (227,719,000),[51] that means that almost 60 million American adults had some form of mental illness. Fortunately, only about 6 percent suffered from a *serious* mental disorder. Still, that amounted to an estimated 14 million people. According to NIMH, "mental disorders are the leading cause of disability in the U.S. and Canada for ages 15–44. Nearly half (45%) of those with any mental disorder meet criteria for 2 or more disorders, with severity strongly related to comorbidity."[52]

Even though the disorders in these statistics are not restricted to the most extreme behaviors, they are still relevant. Many subtler forms of mental illness can pose a serious threat to reliability—without being as easily noticed. No human life is without trauma. From time to time, life events produce emotional shock waves in even the strongest and most solid among us, which sometimes give rise to prolonged anguish or trigger other mental or physical disorders. Sometimes it passes quickly and has little or no long-term effect on emotional stability. But even when trauma leaves no deep or permanent scars, it can still render us less reliable when that wave of troubling emotion is washing over us.

One common mental aberration is the dissociative experience. Dissociation occurs when any group of mental processes are split off from the rest of the mind's functions and operate almost independently of the rest of the psyche. A frequent and mild form occurs when the mind "wanders"—for example, when a person is listening to someone else talk and suddenly realizes that part of what was just said did not register. At critical moments even a wandering mind can cause major problems. But dissociation can also take the form of amnesia, depersonalization (a feeling of loss of identity, of unreality, frequently accompanied by bewilderment, apathy, or emotional emptiness), auditory hallucinations (e.g., "hearing voices inside your head"), or even multiple personality.[53]

Ross, Joshi, and Currie found that some form of dissociative experience was very common among the general population. They administered the Dissociative Experiences Scale to more than 1,000 adults in Winnipeg, Canada. The subjects were specifically instructed not to report any experiences that occurred when they were under the influence of drugs or alcohol. Nearly 13 percent reported a substantial number of dissociative experiences. Some 26 percent of those tested sometimes heard "voices inside their head which tell them to do things or comment on things they are doing," and over 7 percent heard such voices frequently. The results of that study, combined with previous clinical studies, led the researchers to tentatively predict that significant dissociative problems—including multiple personality disorder—may affect as much as 5 to 10 percent of the general population.[54]

Post-traumatic stress disorder (PTSD) has received considerable attention since the Vietnam War. In 2006, it was estimated that nearly 20 percent of Vietnam veterans had experienced PTSD at some point since the war ended.[55] A 2004 report in the *New England Journal of Medicine* estimated that almost 17 percent of soldiers returning from the Iraq war were suffering from PTSD or other emotional problems.[56] (As of 2008, there were at least 121 cases of troubled Iraq/Afghanistan veterans charged with committing homicide after returning to the United States.)[57] Though war is a fertile breeding ground for PTSD, it can also be triggered by violent personal assaults such as rape, mugging, or severe domestic violence, as well as serious accidents, acts of terrorism, and natural or human-caused disasters. PTSD can also develop at any age, including childhood, although research indicates the median age of onset is 23 years.[58] In 2005, it was estimated that about 3.5 percent of the U.S. population 18 years old or older have PTSD in any given year. That amounts to 7.7 million American adults.[59]

In its most extreme forms, PTSD can be debilitating.[60] Symptoms of PTSD range from emotional detachment and extreme suspicion to nightmares and intrusive thoughts about past traumatic events that can verge on flashbacks. A national survey conducted at the University of Michigan in the mid-1990s estimated PTSD has struck nearly one adult in 12 in the United States at sometime in his/her life.

In more than one-third of the cases, symptoms have persisted for at least a decade.[61]

It is therefore not surprising that mental disorders and emotional disturbances are a significant problem for the American military. In a sample of 11,000 naval enlistees, roughly 1 in 12 (8.7%) was discharged during first enlistment because of psychological problems.[62] During the 1980s, nearly 55,000 in the Army were diagnosed as having psychiatric disorders not specifically involving drugs or alcohol. Of these, more than 6,300 were schizophrenic, and nearly 6,400 exhibited other psychotic disorders.[63] Over the same period, 129,000 navy personnel were diagnosed as having psychiatric problems.[64] According to the 2005 worldwide survey, close to one-fifth of all active duty personnel reported experiencing considerable family stress and almost one-third felt highly stressed at work;[65] 4.3 percent of those in the armed forces (more than 49,000 people) reported considering hurting or killing themselves in reaction to life stress.[66]

ALCOHOL, DRUGS, AND MENTAL ILLNESS IN AMERICA'S NUCLEAR MILITARY

Across all branches of the military, it is the nuclear forces that consistently deal with the most dangerous technologies. The Pentagon has a special Personnel Reliability Program (PRP) that covers anyone in the military who has direct access to nuclear weapons or components, control over the access of others to such weapons or components, or both. In practice, the PRP covers everyone who works with nuclear weapons, guards them, has access to the authenticator codes needed to fire them, or is part of the chain of command that releases them. That is, anyone but the President, a rather important exception.

There is an initial PRP screening intended to prevent anyone who is physically or mentally unreliable from ever being assigned to nuclear duty. It includes a security investigation, a medical evaluation, a review of personnel records, and a personal interview. Once an individual is certified by the PRP and assigned to nuclear duty, he/she is subject to temporary or permanent de-certification and removal from nuclear duty at any time if his/her subsequent condition or behavior fails to meet PRP standards. This part of the program is the most effective, but those who are decertified were *on* nuclear duty for some time *before* being removed. It is impossible to know how much of that time they were unreliable.

Although the PRP continues to be key to maintaining reliability among the nuclear forces today, after 1990 the Defense Department stopped making data on the number of permanent PRP de-certifications publicly available. The data that we do have shows that every year from 1975 until 1990, permanent removals ranged from 2.6 to 5.0 percent of the personnel subject to the program, averaging nearly 4 percent (see Table 6.4). That means for a decade and a half, at least 1,900 people each year—and in some years as many as 5,800—were judged unreliable enough to be *permanently* removed from nuclear weapons duties they were already performing.

Reasons for removal range from poor attitude to substance abuse and mental breakdown. Drug abuse was by far the most common reason from 1975 through 1985; from 1986 through 1989, it was "poor attitude or lack of motivation." In 1990, the PRP category that includes mental disturbances and aberrant behavior became the largest, accounting for 27 percent of that year's removals. Alcohol abuse grew in importance as a cause for removal, rising from 3 percent in 1975 to nearly 18 percent by 1990.

Since the military has no reason to exaggerate human reliability failures, assuming the numbers in Table 6.4 are accurate will give a conservative estimate of this problem from 1975 to 1990. According to these data, *more than 66,000 people were permanently removed from nuclear duties* they had already been performing because they were found unreliable—an average of more than 4,100 removals every year for a decade and a half!

The 1975 to 1990 record of permanent PRP de-certifications reveals a general pattern of reliability problems that is clearly not only historically important but also still relevant today. After all, if the problem had largely evaporated after 1990, it is extremely unlikely that the data would have been withdrawn from public view. Furthermore, we have concrete evidence (though not directly on the PRP program) from the highest levels of the Defense Department that general personnel reliability issues continue to be a significant problem within the nuclear military. In 2008, the Pentagon released the *Report of the Secretary of Defense Task Force on DoD Nuclear Weapons Management, Phases I and II*, motivated by two serious failures in handling nuclear weapons and related equipment. The first occurred in two parts, in October and again in November 2006, when "four forward-section assemblies used on the Minutemen III intercontinental ballistic missile," a mainstay of the land-based nuclear missile forces of the United States, were inadvertently shipped to Taiwan, despite the fact that

Table 6.4   Permanent removals from military nuclear duty under the personnel reliability program (PRP), by reason (1975–1990)

Reason	1975	1976	1977	1978	1979	1980	1981	1982	1983	1984	1985	1986	1987	1988	1989	1990
Alcohol abuse	169	184	256	378	459	600	662	645	621	545	500	395	415	388	365	337
Drug abuse	1,970	1,474	1,365	1,972	2,043	1,728	1,702	1,846	2,029	1,007	924	555	477	257	363	151
Negligence or delinquency	703	737	828	501	234	236	236	252	220	160	365	170	158	103	113	130
Court conviction or contemptuous behavior toward the law	1,067	1,333	1,235	757	747	694	560	605	604	580	327	447	473	486	437	340
Mental or physical condition, character trait, or aberrant behavior, prejudicial to reliable performance	1,219	1,238	1,289	1,367	1,233	941	1,022	882	704	646	550	408	437	486	481	510
Poor attitude or lack of	NA	NA	NA	822	996	1,128	1,053	980	904	828	627	556	564	609	633	432

(continued)

Table 6.4 (Continued)

Reason	1975	1976	1977	1978	1979	1980	1981	1982	1983	1984	1985	1986	1987	1988	1989	1990
motivation																
TOTAL REMOVALS																
Number	5,128	4,966	4,973	5,797	5,712	5,327	5,235	5,210	5,085	3,766	3,293	2,530	2,524	2,294	2,392	1,900
As percentage of people in PRP	4.3%	4.3%	4.2%	5.0%	4.8%	4.7%	4.8%	4.9%	4.9%	3.6%	3.2%	2.6%	2.7%	2.8%	3.1%	2.9%
Number of people subject to PRP	119,624	115,855	118,988	116,253	119,198	114,028	109,025	105,288	104,772	103,832	101,598	97,693	94,321	82,736	76,588	66,510

Sources: Department of Defense, Office of the Secretary of Defense, *Nuclear Weapon Personnel Reliability Program, Annual Disqualification Report,* RCS: DDCOMP(A) 1403, Calendar Years 1975–1977 (Washington, D.C.); and *Nuclear Weapon Personnel Reliability Program, Annual Status Report,* RCS: DDPOL(A)1403, Calendar Years 1978–1990 (Washington, D.C.).

they are considered "sensitive missile components" . . . [T]he components were not properly recovered until March 2008."[67] The second incident involved the unintentional transport of six nuclear-armed cruise missiles across the United States in a B-52 bomber:[68] "the unauthorized [nuclear] weapons transfer from Minot Air Force Base (AFB) in North Dakota to Barksdale AFB in Louisiana in August 2007 . . . was due to a breakdown in procedures in the accounting, issuing, loading, and verification processes."[69]

According to the report, "The . . . investigations revealed a serious erosion of focus, expertise, mission, readiness, resources, and discipline in the nuclear weapons enterprise within the Air Force."[70] Furthermore, "the Task Force found that the lack of interest in and attention to the nuclear mission . . . go well beyond the Air Force. This lack of interest and attention has been widespread throughout the DoD . . ."[71]

It is quite clear that serious problems of human reliability still trouble the American nuclear military today, and will continue. Furthermore, there is no evidence to indicate, and no reason to believe, that human reliability is any less a problem in the militaries of any other nuclear weapons country.

Some Incidents

Statistics and sanitized reports can take us only so far in understanding this kind of a problem. A few short illustrative stories should help.

- In August 1969, an Air Force major was suspended after allowing three men, with "dangerous psychiatric problems," to continue guarding nuclear weapons at a base near San Francisco. One of the guards was accused of going berserk with a loaded carbine at the base. The major testified that he had received unfavorable psychiatric reports on the three guards, but he had not removed them because he was short of staff and without them the "hippies" in San Francisco would try to steal the weapons.[72]
- In March 1971, three airmen with top security clearance working with U.S. nuclear war plans were arrested for possession of marijuana and LSD. The men worked at the computer section that maintains the war plans for all U.S. nuclear armed forces, at the top secret underground SAC base near Omaha, Nebraska.[73]
- On May 26, 1981, a Marine EZ-6B Prowler jet crashed and burned on the flight deck of the nuclear-powered aircraft carrier Nimitz. Fourteen people died, 44 were injured, and 20 other airplanes were damaged. The crew of

the Nimitz and the Prowler's pilot were certified reliable by the PRP. Yet autopsies revealed six of the deckhands who died had used marijuana, at least three of them either heavily or shortly before the crash. The pilot had six to 11 times the recommended level of antihistamine (brompheniramine) in his system, enough to cause sedation, dizziness, double vision, and tremors.[74]

- On November 7, 1986, petty officer John Walker, Jr. was convicted in one of the U.S. Navy's most serious spy cases. Walker, who had been a communications specialist on a Polaris nuclear missile submarine, was convicted of selling classified documents to the USSR from 1968 on. He had high-level security clearance and was PRP-certified reliable for nuclear weapons duty.[75]

  The Bangor Submarine Base in Washington State was home to some 1,700 nuclear weapons (about 1,500 of them aboard Trident nuclear missile submarines) at the time of the following incidents. There were additional nuclear weapons stored at Bangor's Strategic Weapons Facility. All of these events took place in a single year:[76]

- On January 14, 1989, an 18-year-old Marine, Lance Corporal Patrick Jelly, was on duty in the guard tower at the Bangor Strategic Weapons Facility. At 9:30 PM, he shot himself in the head with his M-16 rifle. For weeks prior to committing suicide, "He had talked about killing himself, punctured his arms with a needle and thread and claimed to be the reincarnation of a soldier killed in Vietnam." Yet despite such obviously aberrant behavior, Jelly remained PRP certified until the night he died.

- Tommy Harold Metcalf had direct responsibilities in maintaining, targeting, and firing the 24 ballistic missiles on the Trident submarine Alaska, each carrying several city-destroying nuclear warheads. On July 1, 1989, Metcalf went to the home of an elderly couple, bound them and taped plastic bags over their heads, and murdered them by suffocation. The keys to their motor home were found in his pocket when he was arrested. Metcalf was PRP certified as reliable at the time.[77]

- In early August 1989, Commander William Pawlyk was arrested after stabbing a man and a woman to death. Pawlyk had been commander of Submarine Group 9 at Bangor and served aboard the nuclear submarine James K. Polk for five years. He headed a reserve unit in Portland, Oregon at the time of the murders.[78]

- Shyam Drizpaul, like Tommy Metcalf, was a PRP-certified fire control technician, serving on the nuclear submarine Michigan. On January 15, 1990, he shot and killed one crew member in the lounge at his living quarters, then another in bed. Later, while attempting to buy a 9mm pistol, he grabbed the gun from the clerk, shot her to death, and critically wounded her brother. Fleeing the scene, Drizpaul checked into a motel near Vancouver and used the same weapon to kill himself. A subsequent Navy investigation discovered that Drizpaul drank excessively "and claimed to have been a trained assassin."

All of these people were PRP certified as reliable, and relatively recent reviews had uncovered no behavioral or attitudinal problems that might have caused them to be removed from nuclear duty.

The only reasonable conclusion to draw from all this is that there are serious limits to the effectiveness of the PRP. It cannot guarantee either the stability or reliability of those in America's nuclear forces.

Reliability in the Russian Military

It is clear that the Russian nuclear military has parallel problems. In October 1998, the Chief Military Prosecutor's Office indicated that 20 soldiers in the strategic rocket forces were removed from duty during 1997 and 1998 because of serious psychiatric problems. Some had been nuclear weapons guards. In a crime-ridden country, the strategic rocket forces had also achieved the distinction of having the highest increase in crime of any branch of the Russian armed forces between 1996 and 1997.[79]

The following illustrative incidents all reportedly occurred within a three-week period:[80]

- Novaya Zemlya is the only nuclear weapons test facility in Russia. On September 5, 1998, five soldiers there murdered one guard and took another hostage. They then tried to hijack an airplane, taking more hostages before they were finally captured and disarmed.
- On September 11, 1998, a young sailor aboard a nuclear attack submarine in Murmansk went berserk, killing seven people on the sub. He then barricaded himself in the torpedo bay, threatening "to blow up the submarine with its nuclear reactor." Sometime during the 20 hours he was in the torpedo bay, the distraught teenager reportedly killed himself.
- More than 30 tons of weapons-grade plutonium separated from civilian reactors is stored at Mayak. On September 20, 1998, a sergeant from the Ministry of Internal Affairs stationed there "shot two of his comrades and wounded another." He then managed to get away.

The alcohol and drug abuse and the mental and emotional disturbances that seriously compromise reliability are often born in the wide variety of traumas and frailties to which human beings are heir. Some lie deep within the past of the people involved, but the environment in which people live and work can also cause or bring to the surface these and other forms of unreliable behavior. That is certainly true of life in the military in general and the nuclear forces in particular. It is also true of some of those who work with other dangerous technologies. It is to these matters that we now turn.

NOTES

1. PlaneCrashInfo.com database, "Causes of Fatal Accidents by Decade," http://planecrashinfo.com/cause.htm (accessed August 27, 2009).

2. Matthew L. Wald, "Crew of Airliner Received Warning Just Before Guam Crash," *New York Times*, March 24, 1998.

3. David Lochbaum, *The Good, The Bad and the Ugly: A Report on Safety in America's Nuclear Power Industry*, Union of Concerned Scientists (June 1998), v.

4. L. T. Kohn, J. M. Corrigan, and M. S. Donaldson, ed. *To Err Is Human: Building a Safer Health System* (Washington, DC: National Academy Press, 2000).

5. Bureau of the Census, U.S. Department of Commerce, *Statistical Abstract of the United States, 2009*, Table 207, http://www.census.gov/prod/2008pubs/09statab/health.pdf (accessed August 27, 2009). The per capita data are based on all U.S. residents. These data were converted assuming alcohol content of 3.5 percent for beer, 12 percent for wine, and 40 percent for distilled spirits.

6. J. H. Cushman, "Three Pilots Dismissed in Alcohol Abuse," *New York Times*, March 17, 1990.

7. Nearly 20 years later, one of the crew members of that infamous 1990 flight released a "tell-all" book called *Flying Drunk*, chronicling the events leading up to that flight. Jon Hilkevitch, "One of the 'Drunken Pilots' Now Writes of Redemption, "*Chicago Tribune*, August 8, 2009.

8. E. Weiner, "Drunken Flying Persists Despite Treatment Effort," *New York Times*, July 14, 1990.

9. When specimens from those 215 pilots were analyzed in a toxicology lab, it was found that slightly more than 10 percent of them (23) had consumed alcohol shortly before the fatal flight, and 70 percent of those (16) had blood alcohol levels above the legal limit. See Sabra R. Botch and Robert D. Johnson, "Alcohol-Related Aviation Accidents Involving Pilots with Previous Alcohol Offenses," Federal Aviation Administration, Office of Aerospace Medicine, Washington, DC: October 2008.

10. D. Newhall, Acting Assistant Secretary of Defense (Health Affairs), Memorandum to the Assistant Secretaries of the Army, Navy and Air Force (M&RA), Subject: "Error in 'Heavy Drinkers' Category in the 1982, 1985, and 1988 Worldwide Survey of Substance Abuse and Health Behaviors Among Military Personnel," August 15, 1989.

11. Op. cit., R. M. Bray. Note that because the corrections were distributed some time after the 1988 report was issued, the figures cited here may differ

from those given in the text of the original reports of the 1982, 1985, and 1988 surveys.

12. R. M. Bray, et al., *2005 Department of Defense Survey of Health Related Behaviors Among Active Duty Military Personnel* (RTI International (formerly, Research Triangle Institute), December 2006), Table 4.2: "Trends in Heavy Alcohol Use, Past 30 Days, Unadjusted and Adjusted for Socio-Demographic Differences, 1980–2005," 70. The numbers discussed in the text are unadjusted.

13. Ibid., Table 4.9: "Negative Effects of Alcohol Use, Past 12 Months, by Drinking Level," 83.

14. Sociologists use the term "total institution" to refer to an organizational setting in which the institution encompasses all aspects of the "inmate's" life. Examples include prisons, mental institutions, and the military.

15. R. M. Bray, et al., *1988 Worldwide Survey of Substance Abuse and Health Behaviors Among Military Personnel* (Research Triangle Institute: December 1988), 17.

16. To their credit, the surveyors do, in fact, recognize that "self-reports may sometimes underestimate the extent of substance abuse," Ibid., 21–22.

17. V. G. Treml, "Drinking and Alcohol Abuse in the USSR in the 1980's," in *Soviet Social Problems*, ed. A. Jones, W. D. Connor, and D. E. Powell (Boulder, CO: Westview Press, 1991), 121.

18. Vladimir G. Treml, "Soviet and Russian Statistics on Alcohol Consumption and Abuse," in *Premature Death in the New Independent States*, ed. J. L. Bobadilla, C. A. Costello, and F. Mitchell (Washington, DC: National Academy Press, 1997), 222–224.

19. V. G. Treml, "Drinking and Alcohol Abuse in the USSR in the 1980s," in *Soviet Social Problems*, ed. A. Jones, W. D. Connor, and D. E. Powell, (Boulder, Colorado: Westview Press, 1991), 121.

20. This pattern is all the more striking considering the cost of alcohol in the USSR at the time. Treml estimates the average Soviet industrial worker had to work 19 hours to earn enough money to buy a liter of vodka. Ibid., 125 and 131.

21. R. A. Gabriel, *The New Red Legions: An Attitudinal Portrait of the Soviet Soldier* (Westport, CT: Greenwood Press, 1980), 153.

22. Ibid., 154.

23. Substance Abuse and Mental Health Services Administration (SAMHSA), Office of Applied Studies, *2007 National Survey on Drug Use and Health: Detailed Tables*, http://www.oas.samhsa.gov/NSDUH/2k7NSDUH/tabs/Sect1peTabs1to46.htm#Tab1.1B (accessed August 29, 2009), Table 1.1B "Types of Illicit Drug Use in Lifetime, Past Year, and Past Month among Persons Aged 12 or Older: Percentages, 2006 and 2007."

24. This value for the resident U.S. population (244,926,000) is probably conservative, since it uses the available data for population over 14 (rather than 12) years of age in 2007. Economics and Statistics Administration,

U.S. Census Bureau, U.S. Department of Commerce, *Statistical Abstract of the United States: 2009* (Washington, DC: U.S. Government Printing Office, 2008), Table 7, "Resident Population by Age and Sex: 1980–2007," http://www.census.gov/compendia/statab/tables/09s0007.pdf (accessed September 4, 2009).

25. M. Goodwin and M. Chass, "Baseball and Cocaine: A Deepening Problem," *New York Times*, August 19, 1985.

26. P. Kerr, "17 Employees of Wall Street Firms Are Arrested on Cocaine Charges," *New York Times*, April 17, 1987.

27. The late twentieth/early twenty-first century antics of top managers at companies such as Enron, WorldCom, HealthSouth, and CUC International are good examples. Their "creative accounting" and financial gamesmanship turned the financial statements on which investors and employees depended into flights of fantasy, covering a crumbling edifice of poor decisions and asset stripping. Then there were those like financier Bernard Madoff, who committed out and out fraud, leaving his clients tens of billions of dollars poorer.

28. LSD (d-lysergic acid diethylamide) was first synthesized by the Sandoz drug company in Basel, Switzerland.

29. Even so, it is difficult if not impossible to become physically addicted to LSD. The body builds up resistance to its effects in two to three days. Resistance fades just as quickly. See L. Grinspoon and J. B. Bakalar, *Psychedelic Drugs Reconsidered* (New York: Basic Books, 1979), 11.

30. Ibid., 12–13 and 158.

31. Ibid.

32. Ibid., 13–14.

33. Ibid., 159.

34. Ibid., 160.

35. See, for example, Sarah Kerrigan and Donald Brooks, "Immunochemical Extraction and Detection of LSD in Whole Blood," *Journal of Immunological Methods*, Volume 224, Issues 1–2 (April 22, 1999): 11–18.

36. Ibid., 14–35.

37. Of course, illegal drugs also have a variety of interaction and side effects that can magnify their already troubling effects on reliability and performance. For example, the side effects of methamphetamine include paranoia and extremely violent behavior. See Dirk Johnson "Good People Go Bad in Iowa," *New York Times*, February 22, 1996.

38. Harold Silverman, ed., *The Pill Book* (New York: Bantam Books, May 2006), 55–57, 344–346, 699–700, 873–875, and 1,096–1,097.

39. H. W. Griffith, *Complete Guide to Prescription and Non-Prescription Drugs* (The Body Press, 1990), 116, 236, 336, and 790.

40. Ibid., 156, 514, and 856.

41. Strange as it seems, in Chapter 11 we will see extremely rare events occur more often than one might think.

42. Bureau of the Census, U.S. Department of Commerce, *Statistical Abstract of the United States, 2009*, Table 489 "Military and Civilian Personnel and Expenditures: 1990–2006," http://www.census.gov/compendia/statab/tables/09s0489.pdf (accessed September 5, 2009).

43. Op. cit., R. M. Bray, Table 10.4, 209.

44. M. Barinaga, "Drug Use a Problem at U.S. Weapons Laboratory?," *Nature* (June 23, 1988): 696.

45. J. M. Kramer, "Drug Abuse in the USSR," in *Soviet Social Problems*, ed. A. Jones, W. D. Connor, and D. E. Powell (Boulder, CO: Westview Press, 1991), 94.

46. Ibid., 99.

47. Ibid., 100–101.

48. Ibid., 103.

49. Ibid.,104.

50. National Institute of Mental Health, National Institutes of Health, "Numbers Count: Mental Disorders in America," Washington, DC: 2008; http://www.nimh.nih.gov/health/publications/the-numbers-count-mental -disorders-in-america/index.shtml#Intro, introductory paragraph (accessed September 5, 2009).

51. Economics and Statistics Administration, U.S. Census Bureau, U.S. Department of Commerce, *Statistical Abstract of the United States: 2009* (Washington, DC: U.S. Government Printing Office, 2008), Table 7, "Resident Population by Age and Sex: 1980–2007," http://www.census.gov/compendia/statab/tables/09s0007.pdf (accessed September 5, 2009).

52. National Institute of Mental Health, National Institutes of Health, "Numbers Count: Mental Disorders in America," Washington, DC: 2008; http://www.nimh.nih.gov/health/publications/the-numbers-count-mental -disorders-in-america/index.shtml#Intro, introductory paragraph (accessed September 5, 2009),

53. L. E. Hinsie and R. J Campbell, *Psychiatric Dictionary* (New York: Oxford University Press, 1970), 200 and 221.

54. C. A. Ross, S. Joshi, and R. Currie, "Dissociative Experiences in the General Population," *American Journal of Psychiatry* (November 11, 1990): 1550 and 1552.

55. B. Dohrenwend, J. B. Turner, N. A. Turse, B. G. Adams, K. C. Koen, and R. Marshall, "The Psychological Risk of Vietnam for U.S. Veterans: A Revisit with New Data and Methods," *Science* (2006: 313[5789]): 979–982.

56. Anahad O'Connor, "1 in 6 Iraq Veterans Is Found to Suffer Stress-Related Disorder," *New York Times*, July 1, 2004.

57. In January 2008, the *New York Times* published a series of articles concerning violence committed by American soldiers after they returned from the wars in Iraq or Afghanistan. As described in Lizette Alvarez and Dan Frosch, "A Focus on Violence by Returning G.I.'s," *New York Times*, January 2, 2009.

58. R. C.Kessler, A. Berglund, O. Demler, R. Jin, and E. E. Walters, "Lifetime Prevalence and Age-of-Onset Distributions of DSM-IV Disorders in the National Comorbidity Survey Replication" (NCS-R), Archives of General Psychiatry (June 62(6)), 2005): 593–602.

59. R. C. Kessler, W. T. Chiu, O. Demler, and E. E Walters, "Prevalence, Severity, and Comorbidity of Twelve-Month DSM-IV Disorders in the National Comorbidity Survey Replication (NCS-R). Archives of General Psychiatry (2005: June 62(6)), 617–627.

60. Daniel Goleman, "Severe Trauma May Damage the Brain as Well as the Psyche," *New York Times*, August 1, 1995.

61. B. Bower, "Trauma Disorder High, New Survey Finds," *Science News*, December 23 and 30, 1995.

62. J. A. Plag, R. J. Arthur, and J. M.Goffman, "Dimensions of Psychiatric Illness Among First-Term Enlistees in the United States Navy," *Military Medicine* (Vol. 135, 1970), 665–673.

63. Department of the Army, "Disposition and Incidence Rates, Active Duty Army Personnel, Psychiatric Cases, Worldwide, CY1980–1984 and CY1985–1989, "U.S. Army Patient Administration Systems and Biostatistics Activity" (Washington, DC: 1985, 1990).

64. Department of the Navy, "Distribution of Psychiatric Diagnoses in the U.S. Navy (1980–1989), Naval Medical Data Services Center (Bethesda, MD: 1990).

65. R. M. Bray, et al., *2005 Department of Defense Survey of Health Related Behaviors Among Active Duty Military Personnel* (RTI International [formerly, Research Triangle Institute], December 2006), Table 9.1, "Levels of Perceived Stress at Work and in Family Life, Past 12 Months, by Service," 196.

66. Ibid., Table 9.4, "Behaviors for Coping with Stress, By Service," 201.

67. "Section 1. Background," *Report of the Secretary of Defense Task Force on DoD Nuclear Weapons Management, Phase I: The Air Force's Nuclear Mission* (September 2008): 13.

68. Jon Fox, "Inattention Sent Nukes Airborne, Air Force Says," *New York Times*, October 22, 2007.

69. "Section 1. Background," *Report of the Secretary of Defense Task Force on DoD Nuclear Weapons Management, Phase I: The Air Force's Nuclear Mission* (September 2008): 13.

70. Ibid., "Executive Summary," 1.

71. "Executive Summary," *Report of the Secretary of Defense Task Force on DoD Nuclear Weapons Management, Phase II: Review of the DoD Nuclear Mission*, December 2008, iii.

72. "3 Atom Guards Called Unstable: Major Suspended," *New York Times*, August 18, 1969.

73. "Three at Key SAC Post Are Arrested on Drug Charges," *International Herald Tribune*, March 29, 1971.

74. H. L. Abrams, "Human Instability and Nuclear Weapons," *Bulletin of the Atomic Scientists* (January/February 1987): 34.

75. P. Earley, *Family of Spies: Inside the John Walker Spy Ring* (New York: Bantam, 1988), 358. See also H. L. Abrams, "Sources of Human Instability in the Handling of Nuclear Weapons," in *The Medical Implications of Nuclear War*, Institute of Medicine, National Academy of Sciences (Washington, DC: National Academy Press, 1986), 512.

76. Herbert L. Abrams, "Human Reliability and Safety in the Handling of Nuclear Weapons," *Science and Global Security* (Vol.2, 1991), 325–327.

77. Ibid. See also Associated Press, "Man Pleads Guilty to Two Murders," *Seattle Post-Intelligence*, June 20, 1990.

78. Op. cit., Herbert L. Abrams. See also "Courts," *Seattle Post-Intelligencer*, October 5, 1993.

79. Kenneth L. Luongo, and Matthew Bunn, "Some Horror Stories Since July," *Boston Globe*, December 29, 1998.

80. Ibid.

7

The Character of Work;
the Conditions of Life

Error is pervasive in human activity, even when lives and treasure hang in the balance. In the airline industry, in the space program, in the world of high finance, in the field of medicine, in the criminal justice system—nowhere are we completely insulated from its consequences. Witness:

- In May 1995, the chief of neurosurgery at the world-renowned Memorial Sloan-Kettering cancer center in New York operated on the wrong side of a patient's brain. Six months later, the New York State Department of Health reported that "systemic deficiencies" at the hospital, such as failure to always follow medical practices as basic as reviewing diagnostic reports and medical records prior to surgery, had played a role in the incident.[1]
- On a clear night in December 1995, the pilots of American Airlines Flight 965—both with thousands of hours of flying time and spotless records— made a series of "fatally careless mistakes" as they approached Cali, Colombia. After flying past the locational beacon 40 miles north of the airport, they programmed their navigational computers to fly toward it and steered their plane into the side of a mountain. Flight recorders showed no evidence that they had discussed approach procedures, as is mandatory before every landing. It was the worst accident involving an American air carrier in seven years.[2]
- A Boeing 757 crashed off the coast of Peru in early October 1996, killing all 70 people on board. According to the National Transportation Safety Board, the plane went down "because maintenance workers forgot to remove tape and paper covers they had put over sensors while polishing the plane."[3]
- In April 1997, the Inspector General of the Justice Department found that flawed scientific practices and sloppy performance were common at the FBI's world famous crime laboratory. These "extremely serious and

significant problems" jeopardized dozens of criminal cases, including the bombing of the Murrah Federal Building in Oklahoma. In that case, lab workers accused superiors of engaging in sloppy, improper, or unscientific practices that so compromised bomb debris evidence that none of it could be tested.[4]

- On October 2, 2002, during the last 20 minutes of trading a clerical error at Bear Stearns resulted in an order to sell 1,000 times the value of stock the correct order had actually specified ($4 billion instead of $4 million worth). Although Bear Stearns said that it was able to cancel all but $622 million of the erroneous sell order, the company still sold 155 times as much stock as they should have, helping to accelerate the decline of a market that was already tumbling.[5]

- On December 23, 2003, workers drilling into the Chuandongbei natural gas field near the major city of Chongqing, China accidentally punctured a store of natural gas and hydrogen sulfide that was being kept at high pressure. The potentially lethal mixture of gas and toxic chemicals exploded, spewing 100 feet into the air. As the toxic plume spread, more than 190 people were killed, 10,000 sought medical help, and in excess of 40,000 people had to be evacuated from the area of 10 square miles that the state media referred to as a "death zone," strewn with the bodies of dead people and dead animals.[6]

- On April 25, 2005, a young train driver, trying to make up for being 90 seconds late, approached a curve in Amagasaki, Japan at 62 miles per hour on a section of track with a 44-miles-per-hour speed limit. The train jumped the tracks and slammed into a nine-story apartment building only about 12 feet away. More than 90 people died in the deadliest accident in Japan in 40 years.[7]

- On September 12, 2005, large areas of Los Angeles lost electricity when utility workers accidentally cut a power line. Although the outage lasted less than an hour in most places, it sent the Police Department into high alert and caused an unusual amount of tension across Los Angeles since the outage came just one day after September 11 and amidst reports that Al Qaeda had made threats against the city. The situation was not helped by TV station broadcasts of images of what appeared to be some area refineries on fire. The towering flames at those refineries that had looked so frightening turned out to be the intentional flaring of gas as a routine precautionary measure after those refineries had lost power.[8]

- On July 20, 2006, the National Institute of Medicine released a report indicating that 1.5 million people are hurt and several thousand are killed every year in the United States as a result of errors in medication. The financial cost to the nation was estimated at no less than $3.5 billion. According to the *New York Times*, "Drug errors are so widespread that hospital patients should expect to suffer one every day they remain hospitalized."[9]

It was not drug addiction, alcoholism, or mental illness that caused these problems. People are prone to make mistakes even under the best of circumstances. However, the physical, psychological, and sociological circumstances in which we live and work are often not the best. They have powerful effects on our state of mind and thus our behavior on (and off) the job. Even people who are emotionally, physically, and mentally healthy and are not abusing drugs or alcohol have limits when subjected to the stress, boredom, and isolation that are so often a part of working with dangerous technologies. Spending endless hours interacting with electronic consoles, repeating essentially the same lengthy and detailed routine over and over, watching lighted panels and screens, flipping switches, checking and double checking, sitting hour after hour in the control room of a nuclear power plant or missile silo, sailing for months in a submerged submarine isolated from most of humanity yet poised to destroy it—these working conditions are bound to aggravate the already strong human tendency to make mistakes.

The nuclear military has among the most difficult of dangerous technology work environments. It is isolating because it is enveloped in secrecy. No one on nuclear duty is permitted to talk about the details of their work with anyone lacking the proper security clearance. They cannot share what they do with friends and family. For much of the nuclear military, the work is also isolating because it requires long periods away from friends and loved ones.

Because of the constant repetition of routines and because it is so isolating, life in the nuclear forces is boring too. It is stressful as well for at least five reasons:

1. Boredom itself is stressful.
2. Isolation also creates stress (which is why solitary confinement is such a severe punishment for prisoners).
3. For safety and security reasons, people on nuclear duty are always "on-call" even when they are "off-duty."
4. Being highly trained to perform a task, but never being able to take it to completion, is frustrating and stressful.
5. Many on nuclear duty are aware that if nuclear war comes, they will be part of the largest-scale mass murder in human history.

The nuclear military may be the maximum case of a stressful, boring, and isolating dangerous technology work environment, but it is far from the only case. One or more of these reliability-reducing characteristics are found in many other dangerous technology workplaces.

Nuclear power plant operators do work that is boring most of the time and stressful some of the time, though it is not all that isolating. Working around highly toxic chemical or dangerous biological agents has significant levels of background stress, punctuated by periods of intense stress when something goes really wrong. Because it combines most of the reliability-reducing elements relevant to dangerous technology workplaces in general, the military is worth special attention.

BOREDOM AND ROUTINE

In 1957, Woodburn Heron described laboratory research funded by the Canadian Defense Research Board in which subjects lived in an exceedingly boring environment. There were sounds, but they were constant droning sounds; there was light, but it was constant, diffuse light. There was no change in the pattern of sensory stimuli. The subjects became so eager for some kind of sensory stimulation that they would whistle, sing, or talk to themselves. Some had a great deal of difficulty concentrating. Many lost perspective, shifting suddenly and unpredictably from one emotion to another. Many also began to hallucinate, seeing or hearing things that were not there.

People vary widely in their sensitivity to boredom. It is difficult to predict the threshold of monotony that will trigger these reactions in "real-world" situations, yet there is ample evidence that grinding boredom and dulling routine can produce such problems. For example, in 1987 it was reported that "Congressional committees, watchdog groups and the [Nuclear Regulatory] commission have repeatedly found operators of nuclear plants asleep or impaired by alcohol and drugs." Attempting to explain such behavior, a representative of the Atomic Industrial Forum (an industry group) said, "The problem is that it's an extremely boring job. It takes a great deal of training. Then you sit there for hours and hours and take an occasional meter reading."[10]

Boredom can be so painful that people feel compelled to escape, sometimes taking refuge in drugs and alcohol. In interviews of Vietnam veterans conducted by the Psychiatry Department of the Walter Reed Army Institute of Research during the Vietnam War (1971), soldiers often cited boredom as the main reason they used drugs. "Descriptions of work activities invariably included statements like, 'There was nothing to do, so we smoked dope. . . . You had to smoke dope, or drink, or go crazy doing nothing. . . . It was boring until I started smoking skag [heroin]; then I just couldn't believe just how fast the time went.'"[11]

A sailor who served as helmsman on the nuclear aircraft carrier USS Independence claimed he regularly used LSD *on duty* during the late 1970s and early 1980s. It was the only way, he explained, to get through eight hours of extremely boring work. He said that there was almost never a day in his whole tour of duty that he was not on LSD or marijuana, most often LSD.[12]

The dulling effects of routine can create great danger in systems subject to sudden criticality. According to Marrianne Frankenhaeuser of the Karolinska Institutet Department of Psychiatry and Psychology in Stockholm:

> An early sign of under stimulation is difficulty in concentrating ... accompanied by feelings of boredom, distress and loss of initiative. One becomes passive and apathetic ... [then] when a monotonous situation all of a sudden becomes critical ... the person on duty must switch instantaneously from passive, routine monitoring to active problem solving. ... The sudden switch ... combined with the emotional pressure, may cause a temporary mental paralysis. ... The consequences of such mental paralysis—however brief—may be disastrous. ... [13]

Eastern Airlines Flight 401 was on a routine final approach to Miami International Airport on a dark night in December 1972. The crew included three able-bodied and experienced pilots. Weather conditions were fine. As they prepared to land, the crew noticed that the nose landing gear light was not working, so they could not tell whether the gear was extended and locked. An emergency landing with a possible nose gear problem is not all that risky and is not particularly rare, but because of the otherwise routine conditions, the crew was so removed from the primary job of flying the plane that they became fixated on the light bulb. They did not notice that the autopilot had disengaged and the plane was slowly descending. They ignored the altimeter and did not even react when the altitude alert sounded. By the time they realized what was happening, they were "mentally paralyzed," unable to react fast enough to prevent disaster. The jumbo jet crashed into the Florida Everglades, killing 99 people.[14]

In 1904, Sigmund Freud published a very popular book, *The Psychopathology of Everyday Life*. He had collected hundreds of examples of inconsequential everyday errors, such as misreadings, misquotes, and slips of the tongue. Freud interpreted these not as meaningless accidents but as unintended revelations of the unconscious mind, the famous "Freudian slip." Having analyzed many slips reported by 100 subjects over a number of years, psychologist James Reason of

Manchester University reported that nearly half the absent-minded errors involved deeply ingrained habits: "The erroneous actions took the form of coherent, complete sequences of behavior that would have been perfectly appropriate in another context. In each case the inappropriate activity, more familiar to the subject than the appropriate one, had been carried out recently and frequently, and almost invariably its locations, movements, and objects were similar to those of the appropriate action."[15]

Under normal circumstances these errors are easily corrected and of little consequence. Under abnormal circumstances, following familiar routines can lead to disaster. It is important to understand that *the difference between a trivial and a catastrophic error is situational, not psychological.* What creates the disaster is the context within which the error occurs, not the mental process that caused it. In 1977, the experienced Dutch pilot of a Boeing 747 jumbo jet departing from Tenerife in the Canary Islands failed to wait for takeoff clearance, roared off, and crashed into another 747 that was still taxiing on the runway. How could a well-trained, experienced pilot who was head of KLM Airlines' flight training department for years make such an elementary error? He had spent some 1,500 hours in flight simulators over the preceding six years and had not flown a real aircraft for three months. To save money, pilots in flight simulators were not required to hold position while waiting for takeoff clearance. Apparently, the pilot simply reverted to the routine of the simulator with which he was so familiar. What in a different context might have been a trivial error instead cost 577 lives.[16]

STRESS

A little stress "gets the juices flowing," increasing alertness, effectiveness, and reliability. As the pressure continues to mount, however, performance tends to level off then decline, sometimes very sharply. Excessive stress can create all sorts of physical, mental, and emotional problems that affect reliability, ranging from irritability to high blood pressure to complete mental breakdown. On the physical side, there is evidence that high levels of mental stress can strain the heart, damage brain cells, and increase vulnerability to aging, depression, rheumatoid arthritis, diabetes, and other illnesses.[17]

In late 1991, a team of cardiologists showed that stress can cause abnormal constriction of blood vessels in patients whose coronary arteries are already clogged with atherosclerotic plaque. The narrowing of the arteries further impedes blood flow to the heart, raising the chances of heart attack.[18]

Psychiatrists widely accept the notion that stress can play a significant role in triggering episodes of severe depression. In the late 1980s, National Institute of Mental Health psychiatrist Philip Gold suggested how this linkage might work. When confronted by a threat, we naturally experience a "fight or flight" response—a complex biochemical and behavioral mobilization of the mind and body that includes increased respiration rate, a general sense of alertness, and a feeling of released energy. The "fight or flight" response may work well as a reaction to an acute, short-lived physical threat, the kind of threat for which it evolved. However, the emotional stress so common to modern life tends to be ongoing, longer term, and cumulative. Extended periods of high stress (particularly emotional stress) can overload the system, leading to maladaptive reactions like severe melancholic depression.[19]

The effects of chronic stress may be temporary, subsiding when the sources of stress are finally removed, or they may have a very long reach, possibly even affecting the physical structure of the brain.[20] Chronically high levels of the stress hormone cortisol can shrink the hippocampus, the part of the brain apparently responsible for turning off the body's stress response when the threat has subsided.[21] Acute stress from emotional traumas such as job loss, divorce, and the death of a loved one can have both powerful short-term effects and long-term impacts that can last from a few years to a lifetime. But the most extreme, longest-term reactions to stress seem to occur where the level of stress is both high and prolonged.

Stress also increases the likelihood of so-called "ironic errors." Writing in *Science* magazine in 2009, Harvard psychologist Daniel Wegner defines an ironic error as "when we manage to do the worst possible thing, the blunder so outrageous that we think about it in advance and resolve not to let that happen." He goes on to explain, "[T]he person with the sore toe manages to stub it, sometimes twice. . . . [R]esearch traces the tendency to do precisely the wrong thing to . . . mental . . . monitoring processes [that] keep us watchful for errors . . . and enable us to avoid the worst thing in most situations, but . . . increase the likelihood of such errors when we attempt to exert control under mental load," such as when we are under severe stress.[22]

In recent years, post-traumatic stress disorder (PTSD) has been one of the more celebrated stress effects. An immediate or delayed aftermath of trauma, the disorder involves recurring dreams, memories, and even flashbacks of the traumatic events sometimes triggered by a sight, sound, smell, or situation with some relationship to the original crisis. PTSD typically involves emotional detachment from loved ones, extreme suspicion of others, and difficulty concentrating.

At least 500,000 of the 3.5 million American soldiers who served in Vietnam were diagnosed with PTSD.[23] An estimated 30 percent suffer from such a severe version of the disorder that they will never lead a normal life without medication and/or therapy.[24] One-quarter of those who saw heavy combat were involved in criminal offenses after returning to the United States, and only 4 percent of those had had prior psychological problems.[25] A 1997 study of Vietnam veterans with PTSD found that they were also more likely to be suffering from serious physical ailments, such as heart disease, infections, and digestive and respiratory disorders.[26]

The diagnostic manual of the American Psychiatric Association indicates that PTSD is induced by events "outside the range of usual human experience." However, there is evidence that PTSD may be triggered by experiences that are not nearly as unusual as the manual implies. A study of young adults in metropolitan Detroit revealed that 40 percent of the more than 1,000 randomly selected subjects had experienced one or more traumatic events defined as "PTSD stressors." These included sudden serious injury or accident, physical assault or rape, seeing someone seriously hurt or killed, and receiving news of the unexpected death of a friend or close relative. About one-quarter of those who reported these experiences developed PTSD.[27] If these data are generalizable, PTSD would rank fourth among the most common psychiatric disorders troubling young urban adults.[28]

As of 2008, there were at least 121 cases of troubled Iraq/Afghanistan veterans charged with committing homicide after they returned to the United States.[29] In August 2009, the *New York Times* reported "the number of suicides reported by the Army has risen to the highest level since record-keeping began three decades ago."[30]

More than half of the Korean War prisoners of war (POWs) who developed PTSD also suffered from other forms of anxiety disorder, such as panic attacks, and about one-third experienced severe depression. POWs seem far more likely to develop PTSD than combat veterans.[31] The traumas that they experienced were both prolonged and extreme, involving random killings, forced marches, months of solitary confinement, torture, and the like. Yet traumas need not be this extreme to have very long-term effects on emotional health and mental stability.

From a human reliability perspective, two characteristics of trauma-induced disorders are particularly relevant. First, since the onset of the problems that result from stress disorders may be delayed days, months, or even years, someone who appears to be completely recovered from trauma and untouched by such disorders may still harbor them. Psychiatrist Andrew Slaby used the general term "aftershock" to describe "any significant delayed response to a crisis, whether this reaction is anxiety, depression, substance abuse or PTSD."[32] According to Slaby, " . . . everyone, even the calmest, most levelheaded person, has a breaking point that a trauma, or a series of traumas, can set off and bring on aftershock."[33] Second, while someone who has been severely traumatized in the past does not necessarily appear to be dysfunctional or unreliable, his or her ability to cope with stress can be severely compromised. Either chronic stress or the acute stress of a future crisis might overwhelm him or her well before it became severe enough to render a person without such problems unreliable. Because these disorders are common in the human population and can be difficult or impossible to detect, it is impossible to completely avoid them when recruiting a large dangerous technology workforce.

CIRCADIAN RHYTHMS: DISRUPTING THE BIOLOGICAL CLOCK

The behavior and metabolism of most biological organisms seems to be partly regulated by an internal biological clock. For centuries, it was believed that this rhythmic time pattern was simply the result of plants or animals responding passively to natural cycles in their environment (such as the day/night cycle). But even when an organism is deprived of all environmental time cues (keeping light, sound, temperature, and food availability constant), the majority of its time patterns continue—with a period close to but not exactly 24 hours.[34] Such patterns are called "circadian" rhythms. Apparently, most organisms on this planet have internalized the 24-hour period of an earth day.

The part of the human circadian pattern most important to reliability is the sleep/wake cycle, or more generally the variation in alertness and psychomotor coordination over the course of a day.[35] When time patterns are abruptly shifted, the internal biological clock is thrown out of phase with the external time of day. For example, flying at jet speed across a number of time zones causes "jet lag" that disrupts sleep, dulls awareness, reduces attention span, and produces a general feeling of disorientation and malaise. It even takes several days for most people to completely adjust to the simple twice a year, one-hour time shift between standard time and daylight saving time.[36]

The fundamental problem with being out of phase with external time is that the world may be demanding highest alertness and capability just when the internal cycle is at its lowest ebb. A business-person flying from New York to London crosses five time zones. When it is 9:00 AM in London, he or she will want to be bright and alert to deal with typically high demands at the beginning of a new business day. However, his or her internal clock will be set at 4:00 AM, a time when the level of alertness and psychomotor performance tends to be at or near its daily minimum.[37]

Many dangerous technology workers must staff all critical duty stations throughout the 24-hour day, every day. That kind of round-the-clock shift work inevitably plays havoc with the biological clock. There appears to be an underlying circadian rhythm that reaches its lowest levels at night, regardless of sleep/wake schedules. Swedish studies showed that the normal performance of night shift workers was similar to that of day shift workers who had lost an entire night's sleep.[38] Rotating the work schedules of shift workers aggravates the problem and spreads it to the day shift. Yet a survey by the National Center for Health statistics in the United States showed that by the late 1970s, more than 27 percent of male workers and 16 percent of female workers rotated between day and night shifts. Over 80 percent of these shift workers suffered from insomnia at home and/or sleepiness at work, and there is evidence that their risk of cardiovascular problems and gastrointestinal disorders also increased.[39]

Keeping the same workers on the night shift permanently would not solve the problem. Night shift workers usually try to function on something approaching a day schedule on their days off, so they will not be completely out-of-step with the world around them. Consequently, their circadian rhythms are in a continual state of disruption. If they do not try to follow a more normal schedule on their days off, they will be much more socially isolated, and that will subject them to increased stress that will also degrade their performance.

In the early 1980s, a group of applied psychologists in England studied sleep and performance under "real-life" conditions.[40] They monitored and recorded the sleep of a dozen male factory shift workers in their own homes, then measured their performance at the factory where they all worked. The men were followed over one complete three-week cycle during which they rotated between morning, after-noon, and night shifts. As compared to night sleep, sleep during the day was lighter and more "fragile." The normal pattern of sleep stages was disrupted. Not only were reaction times slower on night shift work, but performance tended to get worse as the week progressed. Performance of morning and afternoon shift workers remained nearly stable. Even taking more or longer naps, night shift workers could not compensate for the lower quantity and poorer quality of day sleep.[41] "The night shift worker must sleep and work at times when his or her body is least able to perform either activity efficiently. The body is programmed to be awake and active by day and asleep and inactive by night, and it is extremely difficult to adjust this program in order to accommodate artificial phase shifts in the sleep-wake cycle. . . . "[42]

Circadian rhythms also play a role in disease. Studies of ongoing disease processes in animals show real circadian variation in the average dose of toxins and the severity of injuries that prove fatal. Thus, shift workers may also be more vulnerable to health-related reliability problems. Further, the effectiveness or toxicity of a variety of drugs also follows a circadian rhythm, apparently because of circadian rhythms in drug absorption, metabolism, and excretion.[43] Drug and alcohol abuse might therefore reduce the reliability of dangerous technology shift workers more than it reduces the reliability of workers on a stable day schedule.

The timing of the accident at the Three Mile Island nuclear power plant in 1979 strongly suggests that circadian factors played a significant role. The operators at Three Mile Island had just passed the middle of the night shift when the accident occurred at 4:00 AM, and they had been on a six-week slow shift rotation cycle.[44] In general, the incidence of work errors seems to be much higher in the early morning hours for rotating shift workers whose circadian rhythms have not been fully synchronized to their work schedules. According to a study by the National Transportation Safety Board, truck drivers falling asleep at the wheel are a factor in 750 to 1,500 road deaths each year. Fatigue is a bigger safety problem for truckers than drugs or alcohol.[45] According to another study, they are three times as likely to have a single-vehicle accident at 5:00 AM than during usual daytime hours.[46]

Poor circadian adjustment is also an important problem in aviation. Late one night, a Boeing 707 jetliner whose crew had filed a flight plan to land at Los Angeles International Airport passed over the airport at 32,000 feet headed out over the Pacific. The aircraft was on automatic pilot and the *whole crew* had fallen asleep! When local air traffic controllers could not get a response from the aircraft, they managed to trigger a series of cockpit alarms. One of the crew woke up. The plane, which had flown 100 miles over the Pacific, still had enough fuel to turn around and land safely in Los Angeles. Not all such incidents have a happy ending. Pilot error was cited as the main cause of the 1974 crash of a Boeing 707 in Bali that killed 107 people. The crew had flown five legs on this flight since it began in San Francisco, combining night and day flying across 12 time zones. Disrupted circadian rhythms were likely a major contributor to "pilot error," officially listed as the cause of this accident.[47]

Forcing workers to follow an average day length radically different from 24 hours can also interfere with the normal functioning of the biological clock. As of the late 1990s, American nuclear submarine crews normally operated on an 18-hour day. Each sailor was at work for six hours, off-duty for twelve hours, then back at work for another six-hour shift. Short-term studies have shown that the 18-hour day can cause insomnia, impaired coordination, and emotional disturbance. This schedule is probably one reason for the extremely high turnover rate of American submarine crews. After each voyage, as much as 30 percent to 50 percent of the enlisted crew failed to sign up for another tour. Only a small number of sailors undertake more than two or three of the 90-day submarine missions.[48] The high turnover rate means that a large fraction of the crew on any given voyage are not fully experienced in the operation and maintenance of the ship on which they are sailing, and not used to being confined to a tube sailing under water for three months. This too is a potential source of unreliability.

Finally, disturbances of circadian rhythms are associated with certain forms of mental illness.[49] Waking up early in the morning, unable to fall back asleep, for example, is one of the classic symptoms of depression.[50] There is evidence that sleep-wake disorders in bipolar individuals may result from misfiring of their circadian pacemakers. It is even possible that those malfunctions may help cause manic-depressive syndrome. In general, studies suggest that disturbances of the biological clock caused by abnormalities in circadian pacemakers may contribute to psychiatric illness. If so, the circadian disruptions so common among dangerous technology workers may create reliability problems through this route as well.[51]

LEADERS AND ADVISORS

There is nothing in the exalted positions of political leaders (or the advisors on whose counsel they depend) that makes them immune to any of the reliability problems we have already discussed. There is also nothing in the process of achieving those exalted positions that insures that failures of reliability will not occur. Quite the opposite is more likely, as most political leaders in nearly all modern governments follow a long and arduous path to the pinnacles of power. By the time they get there, they have been exposed to a great deal of physical and mental stress and have often reached an advanced age. By the time they relinquish power, they are older still and have been subjected to even greater stress. On the positive side, those who reach high political positions have lived enough years to accumulate valuable experience and learned to cope with stress. On the negative side, stress and advanced age take their toll.

Some people show significant physical or psychological effects associated with aging before they leave their fifties, while others suffer little or no deterioration in intellectual and creative ability well into their seventies or eighties. Bernard Baruch authored the post-WWII American plan for controlling atomic energy at age 76; Goethe completed *Faust* at age 80; Michelangelo's Pieta was finished at age 84.[52] With the right lifestyle, nutrition, and exercise, it is possible for individuals to maintain their critical physical capacities much longer as well. Still, the mental and physical ability to cope successfully with the pressures of an acute crisis tend to diminish with age.[53]

A variety of physical and psychological problems relevant to reliability are more common among older people; when these problems occur, aging tends to aggravate them. Psychiatrist Jerrold Post listed psychological difficulties that tend to grow worse with age "once the march of symptomatic cerebral arteriosclerosis or other pre-senile cerebral degeneration has begun":

1. Thinking tends to become more rigid and inflexible, with things seen more in terms of black and white, right or wrong.
2. Concentration and judgment are impaired, and behavior becomes more aggressive and less tolerant of provocations.
3. There is less control of emotions, with anger, tears, and euphoria more easily triggered and a greater tendency to depressive reactions.
4. Rather than mellowing, earlier personality traits can become exaggerated (e.g., someone who has been distrustful can become paranoid).
5. The ability to perform mental tasks is degraded, but wide day-to-day fluctuations in mental function can lead others to underestimate the deterioration that has taken place.
6. There is a marked tendency to deny the seriousness and extent of disabilities—a failing leader may therefore "grasp the reins of power more tightly at the very time when he should be relinquishing them."[54]

Political leaders who have manifested these difficulties are as diverse as Joseph Stalin and Woodrow Wilson. Stalin was never a trusting soul, but according to Post, "Joseph Stalin in his last years was almost surely in a clinically paranoid state."[55] Woodrow Wilson became quite ill while he was president, yet refused to acknowledge the extent of his illness. He " . . . suffered a major cerebrovascular accident [a stroke] in September 1919 which left him paralyzed on the left side of his body and was manifested by severe behavioral changes. The manner in which Wilson stubbornly persisted in fruitless political causes was in part to sustain his denial of disability. . . . "[56] The European political leaders of the 1930s are often condemned for failing to stop the rise to power of aggressive dictators in Germany and Italy short of World War II. Yet "All the evidence suggests that . . . [the leaders of Europe] were sick men rather than sinners."[57]

During the final days of his presidency, Richard Nixon was under enormous pressure as a result of the Watergate scandal. Investigative journalists Bob Woodward and Carl Bernstein reported that General Alexander Haig, White House Chief of Staff, said Nixon was a battered man, strained to his limit, and he was afraid the president might try to kill himself. Woodward and Bernstein described a meeting between Nixon and Secretary of State Kissinger on August 7, 1974—two days before Nixon resigned in disgrace: "The President was drinking. He said he was resigning . . . broke down and sobbed. . . . And then, still sobbing, Nixon leaned over, striking his fist on the carpet, crying, 'What have we done? What has happened?'. . . . [T]he man . . . was curled on the carpet like a child. The President of the United States."[58] Later that night, Nixon telephoned Kissinger. "The President . . . was drunk. He was out of control. . . . He was almost incoherent."[59]

It is possible to empathize with the specter of a man whose life was coming apart at the seams. Nevertheless, that same man was at this very time the only person authorized to order the use of American nuclear weapons on his own judgment. He was at the head of the nuclear chain-of-command as his life slowly descended into chaos. As long as there are nuclear weapons, we cannot insure that no leader of a nuclear-armed nation will ever get into an emotionally tortured state like this again.

Fourteen of the eighteen U.S. presidents that held office during the twentieth century had significant illnesses during their terms. Four had strokes; five suffered from various kinds of chronic respiratory illness; six underwent major surgery at least once; seven suffered from serious gastrointestinal disorders; and nine had heart disease. Wilson,

Franklin Roosevelt, Eisenhower, Johnson, and Reagan were medically incapacitated while president; Harding, FDR, McKinley, and Kennedy died in office (the latter two by assassination). There were some reports that by 2004, the first president of the twenty-first century, George W. Bush, was displaying "increasingly erratic behavior and wide mood swings. . . . In meetings with top aides, the President goes from quoting the Bible in one breath to obscene tantrums against the media, Democrats and others that he classifies as 'enemies of the state.' "[60] In an off-the-record interview, one White House aide was quoted as saying, "We seem to spend more time trying to destroy John Kerry [the Democratic candidate for president in 2004] than al Qaeda and our enemies list just keeps growing and growing."[61]

Four of the seven leaders of the U.S.S.R. suffered from serious heart conditions and five died while in office.[62] After the collapse of the Soviet Union, Boris Yeltsin became the first President of Russia. He was first hospitalized for heart trouble in the late 1980s. In July 1995, he was rushed to the hospital with "acute heart problems." It appears he had yet another heart attack in the spring of 1996, and underwent major coronary bypass surgery. During the first six months after his re-election, he was only able to work in his office for two weeks.[63] By 1999, it was clear that Yeltsin remained a very sick man, still holding the reigns of power in an economically and politically deteriorating nation, armed with the world's second largest nuclear arsenal.

The kinds of illness and trauma that have so frequently plagued political leaders impair both physical and psychological function. Heart attacks, for example, are often followed by anxiety, depression, and difficulties in sleeping and concentrating. In more than half the patients, these psychological disturbances persist for months after the attack. According to one study, more than 30 percent suffer from irritability, fatigue, impaired memory, inability to concentrate, and emotional instability for six months to two years after their heart attack.[64]

Strokes, another common problem in aging leaders, cause many patients to suffer depression, anxiety, and emotional volatility. Forty to sixty percent are cognitively and emotionally impaired. Inability to sleep and feelings of hopelessness are also common. Severe depression may persist for 6 to 24 months. Major surgery also produces important psychological side effects, including confusion serious enough to make it hard to think clearly (especially in elderly patients). It can produce disorientation and an inability to grasp concepts and use logic.[65]

Depression and anxiety are both common effects of serious physical illness and trauma in general. People who are depressed have a hard time focusing their attention, concentrating, and remembering. They tend to overemphasize negative information; their analytic capabilities can be seriously impaired. Anxiety also degrades learning and memory and interferes with the ability to reason. These are extremely serious problems for any leader having to make crucial decisions, or any advisor upon whose counsel a leader must depend in a crisis.

FAMILIARITY

When put into a novel work situation, especially one which involves expensive, dangerous or otherwise critical systems, people tend to be very careful of what they do—for a while. But no matter how expensive or dangerous the systems might be, if things go well and all is calm for a long time, most people begin to assume that nothing will go wrong. The cutting edge of their vigilance begins to dull. Even if familiarity does not breed contempt, it does breed sloppiness.

There is no reason to expect that it is any different in dangerous technological systems. Military personnel assigned to duty that brings them directly in contact with nuclear weapons undoubtedly feel a sense of awe and danger at first. But after months of guarding them, loading them on ships or planes, and so forth, nuclear weapons are just another bomb, if not just another object.

If this seems exaggerated, consider how careful most people are when they first learn to drive. Being aware that they could get hurt or killed in a car crash tends to make beginning drivers more careful— for a while. But once they get comfortable with the act of driving, most people pay much less attention its inherent dangers. The car is just as deadly, but the act of driving has become routine.

The tendency to relax once we become familiar with a task is a common human trait, useful in most situations. When it causes vigilance to fail in dealing with critical dangerous technological systems, however, it can lead to catastrophe. There is no way to completely avoid this or any of the other fundamental problems of individual fallibility, no way to be sure that we can completely avoid recruiting workers whose reliability has been compromised by the vulnerabilities, traumas, and afflictions that are part of every human life.

Because we are, after all, social animals, programmed to interact with each other, it is not enough to consider our behavior as individuals.

When we function as part of a group, the group's behavior can be very different from the sum of our individual behaviors. It is time to consider just how dramatically that difference can affect the reliability of those who interact with dangerous technologies.

NOTES

1. Lawrence K. Altman, "State Issues Scathing Report on Error at Sloan-Kettering," *New York Times*, November 16, 1995.

2. Matthew L. Wald, "American Airlines Ruled Guilty in '95 Cali Crash," *New York Times*, September 12, 1997; Pamela Mercer, "Inquiry into Colombia Air Crash Points Strongly to Error by Pilot," *New York Times*, December 29, 1995.

3. Matthew L. Wald, "Peru Crash Is Attributed to Maintenance Error," *New York Times*, November 16, 1996.

4. "Report Criticizes Scientific Testing at FBI Crime Lab," *New York Times*, April 16, 1997; see also "FBI Practices Faulted in Oklahoma Bomb Inquiry," *New York Times*, January 31, 1997.

5. Nicole Maestri, "RPT Update 3-Bear Stearns Enters Erroneous $4 Bln Sell Order" *Reuters*, October 2, 2002 8:33PM ET.

6. See Joseph Kahn, "Gas Well Explosion and Fumes Kill 191 in China," *New York Times*, December 26, 2003; Jim Yardley, "40,000 Chinese Evacuated from Explosion 'Death Zone,'" *New York Times*, December 27, 2003; "Chinese Describe Escape from Toxic Gas," *Earthlink General News*, December 26, 2003, http://start.earthlink.net/newsarticle?cat=0&aid=D7VM86UO0_story.

7. Norimitsu Onishi, "In Japan Crash, Time Obsession May Be Culprit," *New York Times*, April 27, 2005.

8. Randal Archibold, "Utility Error Disrupts Power in Los Angeles," *New York Times*, September 13, 2005.

9. Gardiner Harris, "Report Finds a Heavy Toll from Medication Errors," *New York Times*, July 21, 2006.

10. Lindsey Gruson, "Reactor Shows Industry's People Problem," *New York Times*, April 3, 1987.

11. Larry H. Ingraham, " 'The Nam' and 'The World': Heroin Use by U.S. Army Enlisted Men Serving in Vietnam," *Psychiatry*, Vol. 37 (May 1974), 121.

12. Private communication.

13. Marianne Frankenhaeuser, "To Err Is Human: Psychological and Biological Effects of Human Functioning," in *Nuclear War by Mistake: Inevitable or Preventable?*, Report from an International Conference in Stockholm, Sweden, February 15–16, 1985 (Stockholm: Spangbergs Tryckerier AB, 1985), 45.

14. WGBH Educational Foundation, NOVA Public Television series (#1403), "Why Planes Crash" (Boston: WGBH Transcripts, 1987), 10–13.

15. James Reason, "The Psychopathology of Everyday Slips: Accidents Happen When Habit Goes Haywire," *The Sciences*, October 1984, 48.

16. Ibid.

17. Erica Goode, "The Heavy Cost of Chronic Stress: Some Can Be Benign, but Too Much Is Lethal," *New York Times*, December 17, 2002.

18. K. A. Fackelmann, "Stress Puts Squeeze on Clogged Vessels," *Science News*, November 16, 1991, 309.

19. Christopher Vaughan, "The Depression-Stress Link," *Science News*, September 3, 1988, 155.

20. B. Bower, "Stress Hormone May Speed Up Brain Aging," *Science News*, April 25, 1998; Daniel Goleman, "Severe Trauma May Damage the Brain as Well as the Psyche," *New York Times*, August 1, 1995.

21. Op. cit., Erica Goode.

22. Daniel M. Wegner, "How to Think, Say, or Do Precisely the Worse Thing for Any Occasion," *Science*, July 3, 2009.

23. Andrew E. Slaby, *Aftershock: Surviving the Delayed Effects of Trauma, Crisis and Loss* (New York: Villard Books, 1989), 33.

24. Kim Heron, "The Long Road Back," *New York Times Magazine*, March 6, 1988.

25. Op. cit. Slaby, 77.

26. Denise Grady, "War Memories May Harm Health," *New York Times*, December 16, 1997.

27. The study was reported in *Archives of General Psychiatry* (March 1991). B. Bower, "Trauma Disorder Strikes Many Young Adults," *Science News*, March 30, 1991, 198.

28. Ibid. The first three ranked disorders are phobias, severe depression, and dependency on drugs or alcohol.

29. In January 2008, the *New York Times* published a series of articles concerning violence committed by American soldiers after they returned from the wars in Iraq or Afghanistan. As described in Lizette Alvarez and Dan Frosch, "A Focus on Violence by Returning G.I.'s," *New York Times*, January 2, 2009.

30. Erica Goode, "After Combat, Victims of an Inner War," *New York Times*, August 2, 2009.

31. B. Bower, "Emotional Trauma Haunts Korean POWs," *Science News*, February 2, 1991, 68.

32. Op. cit., Slaby, xiv.

33. Ibid., 71.

34. Martin C. Moore-Ede, Frank M. Sulzman, and Charles A. Fuller, *The Clocks that Time Us: Physiology of the Circadian Timing System* (Cambridge, MA: Harvard University Press, 1982), 1.

35. In the absence of environmental time cues, the free-running sleep/wake cycle for humans is typically about 25 hours. After long periods of isolation, it can drift to as long as 30 to 50 hours. Ibid., 207 and 298–301.

36. T. H. Monk and L. C. Aplin, "Spring and Autumn Daylight Savings Time Changes: Studies in Adjustment in Sleep Timings, Mood and Efficiency" *Ergonomics*, Vol. 23 (1980): 167–178.

37. Given enough time, typically a few days to a week, an out-of-phase biological clock will reset itself, as environmental "zeitgebers" (literally "time givers" or signals, such as light and temperature variations) get the circadian rhythms back in phase with the external world. Resynchronization can be accelerated by intentionally manipulating zeitgebers. See C. F. Ehret, "New Approaches to Chronohygiene for the Shiftworker in the Nuclear Power Industry," in Reinberg, et al. ed., *Advances in the Biological Sciences, Volume 30: Night and Shift Work - Biological and Social Aspects* (Oxford: Pergamon Press, 1981), 267–268.

38. A. J. Tilley, R. T. Wilkinson, S. G. Warren, B. Watson, and M. Drud, "The Sleep and Performance of Shift Workers," *Human Factors*, Vol. 24, No. 6 (1982): 630.

39. Martin C. Moore-Ede, Charles A. Czeisler, and Gary S. Richardson, "Circadian Timekeeping in Health and Disease: Part 2. Clinical Implications of Circadian Rhythmicity," *The New England Journal of Medicine*, Vol. 309, No. 9 (September 1, 1983), 534.

40. Op. cit., Tilley, 629–631.

41. Ibid., 634.

42. Ibid., 638–639.

43. Op. cit., Moore-Ede, Czeisler, and Richardson, 531.

44. C. F. Ehret, "New Approaches to Chronohygiene for the Shiftworker in the Nuclear Power Industry," in Reinberg, et al. ed., *Advances in the Biological Sciences, Volume 30: Night and Shift Work—Biological and Social Aspects* (Oxford: Pergamon Press, 1981), 263–264.

45. Matthew L. Wald, "Sleepy Truckers Linked to Many Deaths," *New York Times*, January 19, 1995; and Matthew L. Wald, "Truckers Need More Sleep, a Study Shows," *New York Times*, September 11, 1997.

46. Op. cit., Moore-Ede, Sulzman, and Fuller, 332–333.

47. Ibid., 333–334.

48. Op. cit., Moore-Ede, Sulzman, and Fuller, 336.

49. By itself, this association does not imply either that disruptions of the biological clock necessarily trigger mental disorders or that those who suffer from mental illness always experience circadian disruptions.

50. Op. cit., Moore-Ede, Sulzman, and Fuller, 369.

51. Ibid., 372.

52. Jerrold M. Post, "The Seasons of a Leader's Life: Influences of the Life Cycle on Political Behavior," *Journal of the International Society of Political Psychology*, Vol. 2, No. 314 (Fall/Winter 1980): 42.

53. Herbert L. Abrams, "The Age Factor in the Election of 1996 and Beyond," *Stanford University School of Medicine and Center for International Security and Arms Control* (unpublished).

54. Op. cit., Jerrold Post., 44–47.

55. Ibid., 45.

56. Ibid., 46.

57. Hugh L'Etang, *The Pathology of Leadership* (London: William Heineman Medical Books Ltd., 1969), 84.

58. Bob Woodward and Carl Bernstein, *The Final Days* (New York: Simon and Shuster, 1976).

59. Ibid.,

60. Doug Thompson and Teresa Hampton, "Bush's Erratic Behavior Worries White House Aides," *Capital Hill Blue*, http://www.capitolhillblue.com; August 2, 2004.

61. Ibid.

62. Herbert Abrams, "Disabled Leaders, Cognition and Crisis Decision Making," in Derek Paul, Michael D. Intrilligator, and Paul Smoker, ed., *Accidental Nuclear War: Proceedings of the Eighteenth Pugwash Workshop on Nuclear Forces* (Toronto: Samuel Stevens Publisher, 1990), 137–138.

63. Op. cit., Herbert L. Abrams, "The Age Factor in the Election of 1996 and Beyond," 3.

64. Ibid., 139–140.

65. Ibid.

Part IV

"Solutions" That Do Not Work

8

Seeking Safety in Numbers: The Reliability of Groups

Most of the time we are at work or play, we are part of a group. We tend to think that working in groups constrains individual behavior and increases reliability by protecting us against the mistakes and foolish, crazy, or malicious acts of any one person. The nuclear military, for example, counts on groups where the possibility that an individual will become unreliable poses too great a risk. We try to render impossible the nightmarish scenarios of fiction writers—from the crazy general who manages to trigger a nuclear war to the distraught missile officer who launches nuclear weapons against an unsuspecting city—by always requiring that two or more people act together to carry out critical activities.

But we also know that observing the bad behavior of others can also encourage people to behave in ways that violate rules or social norms. From the crimes of street gangs to the reckless use of drugs, legal and illegal, peer pressure can be a powerful force in leading individuals to engage in activities that they would otherwise choose to avoid. Apart from peer pressure, there is even recent research-based evidence for a contagion effect of bad behavior: "when people observe that others violated a certain social norm or legitimate rule, they are more likely to violate other norms or rules, which causes disorder to spread."[1]

Groups are not always reliable. Despite the high skill of surgical teams, an estimated 75 percent to 80 percent of anesthesia mishaps are linked to human error, and many operating room mistakes result from interpersonal communication problems rather than technical deficiencies.[2] Even normally functioning groups, carrying out their day-to-day activities, sometimes behave in ways that create problems. Under certain conditions, group behavior can degenerate into pathological forms that actually foster risky or pernicious behavior rather than preventing it.

BUREAUCRACY, ORGANIZATIONS, AND RELIABILITY

Bureaucratic organizations link the activities of pigeonholed subgroups, each with its own specialized set of functions, into a grand organization with lines of communication, authority, and responsibility clearly specified. They often take the shape of hierarchical pyramids so that "the buck" is guaranteed to "stop here" *somewhere* within the organization for every conceivable type of decision. In theory, hierarchical bureaucracy combines the expertise of many individuals in an organizational form with a certain clarity of function and decisiveness of action. In fact, bureaucracies do not work anywhere near as well as theory would lead us to believe.

Information is the lifeblood of any management-administrative organization. It flows up to top decision makers, then down in the form of directives, to those responsible for carrying them out. The impact of directives must then be monitored, evaluated, and sent back up the organization to decision makers in a feedback loop critical to judging how well the directives are working and whether they should be changed. Without an efficient flow of accurate and relevant information, neither managers nor administrators can perform their jobs effectively. Yet bureaucracies systematically distort the flow of information in both directions.

Subordinates *must* edit the information they send up the hierarchy and present it in a manageable form. They cannot transmit all the information gathered by those on the lower levels of the pyramid or else those at the top would be overwhelmed. They could not conceivably digest it all, let alone act on it. As soon as there is editing, however, some criteria must be used to decide what to pass on and what to throw away. It is not always easy to judge which information is most important to those higher in the organization, but importance is often not even the prime criterion used. Holsti and George argue that in foreign policymaking organizations, information is filtered to fit "existing images, preconceptions, preferences, and plans." When information does not fit, it is distorted by those with a stake in a particular viewpoint to support their theory.[3]

It is, for example, entirely possible that this phenomenon, rather than incompetence at the Central Intelligence Agency (CIA) prior to the 2003 U.S.-led invasion of Iraq, that produced the incorrect assessment that Iraq possessed weapons of mass destruction threatening to the United States and its allies. It is also possible that the less-than-optimal conditions surrounding the preparation of that assessment were intentionally created by organizational superiors to insure that the news they heard would be the news they or their superiors in the administration wanted to hear. In late 2008, the recently retired Bush administration Deputy National Intelligence Director, Thomas Fingar, reportedly argued that the incorrect assessment was partly the result of the intelligence agencies being given less than two weeks to prepare their findings, whereas it often take months, if not years, "to get it right."[4]

Subordinates sometimes try to avoid superiors who might report facts they want suppressed, even when it means not reporting potentially dangerous situations.[5] Personal beliefs, rigid worldviews, and concepts of loyalty have also been shown to inhibit the communication of accurate, unbiased information to organizational higher-ups.[6] There is a "good news syndrome," a tendency for subordinates to edit out information that highlights their own errors or failures of judgment, or the errors or misjudgments of their superiors. Sometimes such "creative editing" crosses the line into blatant deception. In December 2005, Justice Department investigators concluded that officials at the Federal Bureau of Investigation (FBI) deliberately falsified documents to cover the FBI's mishandling of a terror case in Florida. They were also accused of retaliating against the agent who complained about these problems.[7]

Combined with the necessary filtering at each level of the organization, the "good news" syndrome distorts in the upward flow of information, a distortion that tends to get worse when there is more at stake. As organizational communications expert Chris Argyris put it, " . . . the literature suggests that the factors that inhibit valid feedback tend to become increasingly more operative as the decisions become more important. . . . This is a basic organizational problem . . . found not only in governmental organizations, but also in business organizations, schools, religious groups, trade unions, hospitals, and so on."[8]

The internal politics of bureaucracies can also bias information flows. Parochialism can cause subgroup leaders to consciously or unconsciously edit information so their own division looks good, believing (or rationalizing) that this is good for the organization as a whole.[9]

These distortions in the upward information flow lead to top decision makers finding themselves in an unreal world. It is hard to make good decisions when you do not really know what is going on. Worse yet, being this out of touch can lead catastrophic loss of control. That is why Arthur Schlesinger, Jr., a close confidante in the Kennedy White House, argued for the importance of what he called "passports to reality" in upper levels of government, especially in the presidency.[10]

The special presidential commission set up to investigate the explosion of the space shuttle Challenger on January 28, 1986, concluded that one of the prime causes was a breakdown in the upward flow of information at NASA. The technical problem was the failure of an O-ring seal between two segments at the rear of the right-side solid rocket booster. Challenger was launched in unusually cold weather. In cold, the O-ring was not flexible enough to seal the joint properly. But the deeper problem was the failure of the serious misgivings expressed before the launch by engineers from Morton Thiokol Inc. (builder of the booster) to reach the top management at NASA.[11]

In 1988, Secretary of Energy James D. Watkins had publicly expressed grave concern about his subordinates' performance as well as their ability to run the nation's nuclear weapons complex. In his view, these managers and supervisors were giving him unreliable information about problems at the nuclear weapons plants.[12] "When I get the briefing, I only get one side. . . . I don't have the database coming to me that I need. . . . So I am making decisions today on a crisis basis. . . . It has been a nightmare for me to try to unravel the background sufficient to make some decision. . . . It's been very confusing. . . . "[13] Eventually, Watkins said he had uncovered "serious flaws" in the procedures intended to insure that reactors in the nuclear weapons complex were safe to operate.

In the late 1980s, the Department of Energy (DoE) hired inspectors from the Nuclear Regulatory Commission (NRC) to evaluate the operation of the nuclear weapons-related reactors. One of them insisted on a full review of the (plutonium and tritium-producing) Savannah River Plant's ability to withstand earthquakes, arguing that previous DoE analyses failed to use state-of-the-art methods. The review reportedly turned up "hundreds of potential structural weaknesses" at Savannah River's P-reactor. The inspector reported that "design documentation was 'grossly inadequate'. . . . *plant managers could not put their hands on a set of complete blueprints showing the reactor as it exists now.*"[14] In August 1988, the chair of the DoE's safety group learned belatedly and indirectly that DuPont (who operated the Plant for the DoE) and local DoE officials were not keeping him informed on what action was being taken on the recommendation of his inspectors that the reactor should be shut down. By late 1988, all 14 of the reactors in the DoE's nuclear weapons complex had been closed, many of them permanently.[15]

Charged with assuring the safe operation of America's *commercial* nuclear power plants, the NRC is hardly a paragon of effective operation itself. In 1996, it was reported that "Higher level NRC managers sometimes downgrade the severity of safety problems identified by on-site inspectors without giving reasons for the change," and that "NRC inspectors who persist in pressing safety issues have been subject to harassment and intimidation by their supervisors."[16] In 1997, a GAO review found that "NRC has not taken aggressive enforcement action to force licensees to fix their longstanding safety problems on a timely basis," and argued that this laxity was so embedded in the organization that "changing NRC's culture of tolerating problems will not be easy."[17]

Without timely and accurate information, *especially* "bad news," problems that would not have been difficult to handle early on can run out of control. It is simply not wise for upper-level managers to assume that the information which flows to them on a daily basis is correct and unbiased. According to Irving Shapiro, former DuPont CEO, "Getting the right information is a substantial part of the job. . . . You get lots of information and most of it is totally unnecessary. The organization tends to want to give you the good news and not cough up the bad news. . . . But to manage well, you have to get the message across, that whatever the story is, let's get it on the table. . . ."[18]

It is not just business corporations that have a tendency to ignore or even punish the deliverers of bad news. In 2002, a nine-member expert panel at NASA warned "that safety troubles loomed for the fleet of shuttles." Shortly thereafter, NASA removed five members of the panel and two of its consultants; a sixth member was so disturbed that he quit.[19] On February 1, 2003, the space shuttle Columbia disintegrated over Texas on re-entry.

Sometimes when bad news is transmitted to them, the upper management either refuses to accept delivery or fails to take appropriate action. In April 2004, the *New York Times* reported, "President Bush was told more than a month before the attacks of September 11, 2001, that supporters of Osama Bin Laden planned an attack within the

United States with explosives and wanted to hijack airplanes."[20] On August 28, 2005, Hurricane Katrina hit the Gulf Coast of the United States hard, killing 1,800 people and doing more than $81 billion in damages.[21] Bush Administration officials said publicly that they did not know that a levee had broken and a good part of New Orleans was under water until August 30, but by early 2006 Congressional investigators learned "an eyewitness account of the flooding from a federal official reached the Homeland Security Departments headquarters . . . at 9:27 p.m. the day before and the White House itself at midnight."[22]

Other than increasing organizational transparency and removing ineffective or incompetent managers, little can be done to assure that bad news that is delivered is appropriately acted upon. But can anything be done to assure that bad news actually reaches key organizational decisionmakers? There appear to be two main schools of thought among experts in organization theory. The first is that the solution is cultural: an environment must be created that makes managers more accessible to subordinates. Probably the best way is to manage by "wandering around"—get out of the office and be seen in the corridors, the workplaces of subordinates, the cafeteria, and so forth so that informal contacts are encouraged. Most subordinates are inhibited about making an official visit to their boss's office, or even worse, to the office of their boss's boss to report a problem. That is especially true when they want to point out an error or misjudgment that either they or higher-level managers have made. Repeated informal contacts help employees feel more at ease talking to higher ups and make it less threatening to bring up bad news. Problems can then be dealt with early. It is a tremendous advantage for managers to be made aware of problems while the problems are still small.

The other main school of thought is that the answer lies in reshaping the organization, replacing the sharply rising pyramid with a much flatter structure. Hewlett-Packard, for example, divided the company into small units whose managers are therefore more visible and informally accessible to their subordinates. But the best structure and the most peripatetic managers in the world will do no good if the people who fill key management positions do not encourage open communication. Bad news will still not reach the top.

More than once, military commanders have blundered into major disasters because they have been sold an overly optimistic picture of what is actually happening on the ground. Their advice, based on that rosy view, can cause political leaders to compound earlier mistakes and multiply the disaster. During the Vietnam War, top-ranking American commanders, looking at inflated "body count" data, kept reassuring the president and his advisors that Viet Cong and North Vietnamese military forces were depleted and nearing collapse. They could see "the light at the end of the tunnel." Thinking their efforts were just about to succeed, American military and civilian leaders plunged the country deeper and deeper into the morass.

In February 1997, the Pentagon reported it could find only 36 pages of an estimated 200 pages of logs detailing detections of chemical/biological weapons it had kept during the 1991 Persian Gulf War. Since paper and computer disk copies of the logs had been placed in locked safes at two different locations in the United States after the war, the report fueled speculation by veterans that the military was either grossly incompetent or engaged in a cover up. General Norman Schwarzkopf, the commander of American forces during the war, testified before Congress that he was shocked when the Pentagon announced that thousands of troops might have been exposed to nerve gas. "We never had a single report of any symptoms at all on the part of the 541,000 Americans over there."[23]

Distortions are much more troubling in organizations in which crisis decisions with potentially catastrophic consequences must be made in a matter of minutes. Yet it is clear that distortions in the upward flow of information persist in these organizations as well. A former Air Force missileman, describing life in the hardened silos of the U.S. land-based missile forces, put it this way, "Crew members dare not tell higher command that the regulations are flouted. Noncommunication with higher command is endemic in the missile field, with the result a gap between regulations and what is really done in the capsule [missile silo]."[24]

There is not much potential for "management by wandering around" in the armed forces. By its nature, the military does not encourage casual contact between lower and higher ranks.[25] A soldier who tries to report information critical of his/her superiors is not likely to get a warm reception, as many women in the military found in the late 1990s when their complaints of sexual harassment were reportedly "met with ridicule, retaliation or indifference."[26] The system fosters obedience to orders and puts severe constraints on questioning authority.[27] Superiors run the risk of being seen as weak and indecisive if they do not "stick to their guns."

Flattening the hierarchy holds even less promise. The military must be ready to operate in a coordinated fashion from a central command to carry the day. It is consistent with military principles to break forces into small units capable of operating autonomously (in guerilla warfare, that is key), but there still must be a way of coordinating forces at critical moments. The system of rigidly enforced hierarchy with many ranks is deeply embedded in traditional, large-scale military organizations. It is not clear they could operate effectively without it. There does not appear to be any effective solution to the good news syndrome that is compatible with the basic structure and operation of military forces.

Directives

The ability to make routine the many specialized functions and rely on standard operating procedures (SOPs) is alleged to be the greatest advantage of bureaucratic organizations. Yet widespread adherence to SOPs establishes an organizational culture, style, and even ideology biased toward incremental rather than fundamental change. When directives requiring substantial departure from SOPs flow down from above, they are often reinterpreted to better fit more familiar patterns. Sometimes subordinates inform higher-level managers about these alterations. More frequently, they do not. Coalitions form among interest groups within the bureaucracy, and negotiation and compromise, rather than analysis, come to dominate policymaking. Behavior is driven by what is familiar and comfortable rather than what is best.[28]

Both bureaucratic inertia and internal politics can cause directives to be diverted, distorted, or ignored, compromising the position and flexibility of top decision makers at critical moments. Probably the most dramatic case on record involves the Cuban Missile Crisis, the confrontation that brought the world to the edge of global nuclear war. Fearing the massive U.S. advantage in strategic nuclear forces, the Soviets tried to balance the scales quickly by sneaking nuclear-armed missiles into Cuba, 90 miles from America's mainland. The United States caught them at it, threw up a naval blockade around the island, and demanded that the U.S.S.R. remove the missiles immediately. The Soviets refused, and their warships, obviously carrying additional missiles toward Cuba, continued to steam toward the blockade line. According to Robert McNamara, then Secretary of Defense, the United States did not realize at the time that Soviet forces in Cuba already had 36 nuclear warheads for the two-dozen intermediate range missiles targeted on U.S. cities. That is easily enough firepower to kill tens of millions of Americans and lay waste to a good part of the United States. The CIA had told the Administration that they did not believe there were any nuclear warheads in Cuba. This was only one of a number of ways, in McNamara's words " ... that each nation's decisions immediately before and during the crisis had been distorted by misinformation, miscalculation and misjudgment."[29]

After more than a week of crisis and confrontation, President Kennedy received a letter from Soviet Premier Khrushchev proposing that the U.S.S.R. withdraw their missiles from America's doorstep in exchange for U.S. withdrawal of its nuclear-armed missiles from Turkey, on the Soviet's doorstep. According to Robert Kennedy:

> On several occasions over the period of the past eighteen months, the President had asked the State Department to reach an agreement with Turkey for the withdrawal of Jupiter missiles in that country.... The Turks objected and the matter was permitted to drop.... He was told by the State Department that they felt it unwise to press the matter with Turkey. But the President disagreed.... The State Department representatives discussed it again with the Turks, and finding they still objected, did not pursue the matter.
>
> The President believed ... that, his wishes having been made clear they would be followed and the missiles removed. Now he learned [from Khrushchev] that the failure to follow up on this matter had permitted the same obsolete Turkish missiles to become hostages of the Soviet Union. He was angry ... our position had become extremely vulnerable, and ... it was our own fault.[30]

And so the international situation was gravely aggravated at a crucially dangerous point in human history, not by a conspiratorial plot, not by a mentally deranged or drug-addicted military officer, not even by a stress or monotony-induced failure of vigilance, but simply by the politics and inertia of bureaucracy.

The failure of subordinates to carry out management directives also played a key role in the world's worst nuclear power accident, the Chernobyl disaster of April 26, 1986.[31] The design of the RBMK reactors at Chernobyl made them particularly vulnerable to instability. The design errors were so serious, and apparently unnecessary, that most western designers who had reviewed Soviet reports on the reactors prior to the accident found them difficult to believe. The Russians were aware of these problems and knew that the reactors were dangerously unstable, especially at low power. Rather than going about the difficult and costly business of fixing them, the managers issued directives with strict rules for operation. But they apparently made little or no attempt to educate the plant operators or to strictly oversee their activities. The day before the accident, those rules were being violated. The reactor was intentionally being run below 20 percent power, a dangerously low level, and *six major safety devices had been deliberately disconnected*. After the accident, a new rule was instituted requiring that a senior person—whose primary function is to see that the rules are obeyed—be present whenever RBMK reactors were started up or shut down. According to the Soviet Minister of Atomic Energy, "this by itself would not have prevented the accident at Chernobyl, because it was the deputy chief engineer who was most responsible for breaking the rules."[32]

In the course of its investigations, the independent National Commission on Terrorist Attacks Upon the United States, also known as the "9/11 Commission," uncovered a number of breakdowns in communication at the highest levels of government, including the Pentagon and the White House during the terrorist attacks of September 11, 2001. Shortly after talking with President Bush, Vice President Cheney issued an order to the military allowing fighter pilots to shoot down passenger planes still in the air. That order was received at the Pentagon, but not transmitted by military commanders at NORAD to the fighter pilots who were then in the air over New York and Washington. According to the Commission's report, "while leaders in Washington believed the fighters circling above them had been instructed to 'take out' hostile aircraft, the only orders actually conveyed" to the fighter pilots were to try to find the planes that had been hijacked. Gaps in communications between NORAD and the Federal Aviation Administration had already prevented the fighters from being scrambled soon enough to intercept and shoot down the four hijacked planes before they had reached their targets. The last of the planes had already crashed before the White House issued the "shoot down" order. But in the words of Commission Chairman Thomas Kean, "When the president of the United States gives a shoot-down order, and the pilots who are supposed to carry it out do not get that order, then that's about as serious as it gets as far as the defense of this country goes."[33]

Hierarchical bureaucratic organizations coordinate and control the behavior of those who work within them to reliably achieve established goals. Those constraints can interfere with reliable performance by creating troubling conflicts between responsibility to the organization and personal and/or social responsibility. "Whistleblowers" are those who ultimately try to resolve this conflict by violating the organization's norms, stepping outside their bureaucratic role to expose its behavior to wider public scrutiny. Social norms of conformity are so strong that whistleblowers are often scorned and harassed for their trouble, rather than being praised for their courage. F.B.I. agent Coleen Riley testified before Congress in June 2002 "that the F.B.I.'s bureaucracy discouraged innovation, drowned investigators in paperwork and punished agents who sought to cut through the many layers of gatekeepers at the bureau's headquarters."[34]

Professionals, such as medical doctors, are placed in the same kind of conflictual situation when they work for a bureaucratic organization. One defining characteristic of a "profession" is that its practitioners are subject to a code of ethics to which their behavior is expected to conform, regardless of how or where they practice. This independent set of values may conflict with the bureaucracy's goals, creating a stressful dilemma for those who take their professional obligations seriously. The common assumption is that the professional, whether school teacher, public defender, or medical doctor, will be guided primarily by the ideals of public service and responsibility to the client embedded in ethical codes and professional training. Yet:

> studies suggest that it cannot be assumed that organizational goals will give way to professional goals. Professionals who wish to succeed may well respond to the specific immediate pressures of their employing organization before the more abstract and distant expectations of their profession. . . . The changes or modifications in professional practice which result from adaptation to organizational requirements . . . often affect the core functions of the profession.[35]

In the late 1990s, with more medical doctors in the United States working for health maintenance organizations (HMOs), some found themselves caught between their obligations to their patients and the pressures/incentives that come from the business interests of their employer.

One species of "captive" professional with particular relevance here is the military psychiatrist.[36] Psychiatrists in the military are there to assist in the control or management of behavior that the military defines as being deviant. The interest of the patient is subordinated to the interest of the organization. There are surely circumstances in which it might be dangerous to allow judgment about what is best for the patient to supersede what is necessary for safe and reliable performance of military duty, but that is not always true. There are also times when loyalty to the organizational role might be more dangerous. For example, a particular commander might believe that discharging or even changing the duty of a troubled soldier is too soft hearted and would set a bad example. The judgment of a psychiatrist trying to be a good soldier might be unduly influenced by what he or she believes the local commander wants to hear. That might not be in the best interest of either the soldier or the wider society.

Organizational Incompetence

Organizations are vital to much of the productive work that goes on in society. As social beings, we evolved to cooperate with each other and are often much more effective in achieving our common goals when we act together in groups. While groups often make us more productive and may help to overcome some of the problems of human fallibility, they certainly do not overcome them all. Rather, they add their own special dimension. Apart from the distortions in the upward flow of information and in the downward flow of directives, there are many other ways in which organizations can and do prove unreliable from time to time. They sometimes go off the rails simply because of the way they structure the interactions among the individuals that populate them and the roles those individuals are constrained to play. Perhaps the best way to illustrate the potential impact of these myriad problems on organizational behavior is with a few examples, in this instance drawn from the realm of government agencies:

- According to a report by the 9/11 Commission, in the months leading up to the terrorist attacks of September 11, 2001 officials at the Federal Aviation Administration (FAA) "reviewed dozens of intelligence reports that warned about Osama Bin Laden and Al Qaeda, some of which specifically discussed some of which specifically discussed airline hijackings and suicide operations.... "Yet the FAA failed to take tougher airport screening measures, increase the use of armed on-flight air marshals, or take other security measures that might have made a serious difference to the events that transpired on that day.[37]
- In the summer of 2003, the Columbia Accident Investigation Board issued its report about the circumstances surrounding the destruction of the space shuttle Columbia and its entire crew. It described NASA as an organization with a "broken safety culture," and found that errors and problems in the organization played a major role in the disaster. Engineers at NASA had been very concerned about the damage done to the spacecraft during liftoff and issued three requests for outside assistance to get pictures of the shuttle so that they could better assess that damage. According to the report, although a rescue mission would have been risky, one could have been mounted "if the management had recognized the severity of the problem

and acted quickly. But instead it countermanded the engineer's requests."[38]

- In June 2005, top officials in British intelligence and law enforcement sent a terrorist threat assessment report to agencies in the British government, the governments of selected foreign nations, and corporations concluding that "at present there is not a group with both the current intent and the capability to attack the United Kingdom."[39] Three weeks later, on July 7, four British suicide bombers staged a coordinated attack on the public transportation system in central London. The explosions killed 52 people and injured 770 more.[40]

- In July 2006, a federal audit of the U.S. Agency for international Development was released. Conducted by the office of the U.S. Special Inspector General for Iraq Reconstruction, the audit found that "the State Department agency in charge of $1.4 billion of reconstruction money in Iraq used an accounting shell game to hide ballooning cost overruns on its projects there and knowingly withheld information on schedule delays from Congress."

- In February 2009, the results of a two-year review of the nation's crime labs drafted by the National Academy of Sciences concluded that the forensic analyses that are so often key in a great variety of nation's criminal trials "are often handled by poorly trained technicians who then exaggerate the accuracy of their methods in court." The draft report called for a wide-ranging upgrade and standardization of forensic methods to eliminate the shoddy scientific practices they found to be all too common. Donald Kennedy, a Stanford University scientist who played a role in selecting the authors of the review, reportedly said that "the National Institute of Justice, a research arm of the Justice Department, tried to derail the forensic study by refusing to finance it and demanding to review the findings before publication."[41]

Groupthink

There are times when groups are not only unreliable, but are actually less reliable than individuals. For example, the tendency of members of a decision-making group to take greater risks than the individuals in the group would take acting alone is called "risky shift." Risky shift has been observed in dozens of controlled experiments carried out over decades.[42] One plausible explanation for this odd behavior is that the presence of others sometimes stimulates those who are part of the group to try to prove to each other how brave, tough, and daring they are.

Laboratory experiments are one thing, but there is strong evidence that this and other forms of behavior that render groups less reliable than individuals also occur in critical "real world" decision-making situations. In his landmark 1972 book, *Victims of Groupthink: A Psychological Study of Foreign-Policy Decisions and Fiascos*, psychologist Irving Janis presents a fascinating series of case studies of the deeply flawed group decision processes that produced some of the worst foreign policy disasters of the twentieth century. Janis coined the term "groupthink" to refer to "a deterioration of mental efficiency, reality testing and moral judgment that results from in-group pressures."[43] It is what happens when the members of a decision-making group become more focused on maintaining good relations with each other and achieving unanimity than on realistically analyzing the situation and critically evaluating all available alternatives. As the reputation of the FBI's famous crime lab came under fire in the late 1990s, one FBI technical expert argued that part of the problem was "an inadvertent bonding of like-minded individuals supporting each other's false conclusions."[44] Unfortunately, this state of affairs can occur even in the most critical decisions, when thousands or millions of lives are at stake.

Janis laid out eight major symptoms of groupthink:

> (1) an illusion of invulnerability . . . ; (2) collective efforts to rationalize in order to discount warnings . . . ; (3) an unquestioned belief in the group's inherent morality . . . ; (4) stereotyped views of enemy leaders as too evil to warrant genuine attempts to negotiate, or as too weak and stupid to counter whatever risky attempts are made to defeat their purposes; (5) direct pressure on any member who expresses strong arguments against any of the group's stereotypes, illusions or commitments . . . ; (6) self-censorship of deviations from the apparent group consensus . . . ; (7) a shared illusion of unanimity concerning judgments . . . ; (8) the emergence of . . . members who protect the group from adverse information that might shatter their shared complacency about the effectiveness and morality of their decisions."[45]

The power and consequences of groupthink are probably best illustrated by considering an important historical example:

On June 24, 1950, North Korean troops invaded the South, touching off the Korean War. The United States led U.N. military forces into the field to drive the Northerners back above the 38th parallel that divided communist and noncommunist Korea. Within a matter of months, the U.N. forces had achieved that objective.

During the difficult early days of the war, the Truman Administration's key decisionmaking group had developed a real sense of mutual admiration and grown insulated, self-congratulatory, and a perhaps bit intoxicated with success. In early November, the Administration decided to escalate the war, authorizing a large-scale American military action to pursue the defeated North Korean army above the 38th parallel, conquer North Korea, and unite it with the noncommunist South. They chose to ignore repeated warnings by the Chinese that they would enter the war if U.N. troops invaded North Korea.

In late November, the Chinese attacked the American invasion force. Throwing hundreds of thousands of troops into the battle, they trapped entire American units and forced the rest out of the North. The rout continued as the Chinese pushed into South Korea, nearly driving the U.N. forces out of the entire Korean peninsula. The disastrous foreign policy decision to escalate the conflict came close to snatching defeat out of the jaws of victory. What had led to that blunder?

It certainly was not lack of information. The Chinese began issuing warnings in September. In the first few days of October, they directly and explicitly warned that they would not stand idly by if the United States led U.N. troops poised on the North Korean border across the 38th parallel. The American policymaking group dismissed these repeated statements as a bluff. Then the Chinese strengthened the warning by sending forces into North Korea, which engaged South Korean and U.S. troops. Yet despite the fact that they strongly desired to avoid war with China, they still did not believe that the Chinese would enter the war in force. They were somehow able to convince each other that although the United States had not hesitated to resist aggression against an ally 8,000 miles away, the Chinese would not do the same for an ally on their border.

The State Department policy staff opposed crossing the 38th parallel, but Secretary of State Acheson made sure that the dissenters were kept far away from the policymaking in-group. In classic groupthink style, careful consideration of the alternative views of outside experts was sacrificed on the altar of group cohesion. They knew that risking war with China on the Asian mainland was risking disaster. Yet they decided to take actions that made it virtually impossible for the Chinese to stay out of the conflict.

A few days after the Chinese made good on their oft-repeated threats, General MacArthur informed President Truman that the massive Chinese assault was forcing the U.S.-led forces into a full scale retreat.[46] It took years for the U.S.-led U.N. forces to regain the territory they had lost and push the opposing forces back to the 38th parallel and out of South Korea again. The decision to invade the North had been a horrible mistake that cost millions of lives, nearly led to ignominious defeat, and, given the tenor of the times, might even have precipitated World War III. To this day, there is still no peace treaty between the two nations that share the Korean Peninsula.

The British decision to appease Hitler, the failure of vigilance at Pearl Harbor, the Kennedy Administration's Bay of Pigs fiasco (aiding Cuban exiles to invade Cuba and overthrow Fidel Castro), the long series of bad decisions in the 1960s and 1970s that marked America's disastrous Vietnam War, the foolish decisions that led the Soviet government into the morass of the War in Afghanistan during the 1980's—all of these are cases in which the groupthink played a major role. Those members of the decisionmaking in-group that knew better simply held their tongues or were silenced by group pressures. Outside experts with dissenting opinions were kept away. Treating silence as agreement and believing "if we all agree, it must be right" reinforced questionable policies favored by key decisionmakers and kept them locked into policies about which even they may have had serious doubts. In the grip of groupthink, groups can easily be less effective and less reliable than individuals.

Although not enough time has passed to thoroughly document it publicly, it is likely that the U.S. decision to invade Iraq in 2003 was also an example of groupthink in action. The modus operandi of the Bush administration's inner decision-making circle was extremely secretive. They were not only inward looking, but from all appearances utterly convinced of the rightness of their cause and assured of success. The views of dissenting outsiders, and even dissenting insiders were unwelcome. In early 2006, Paul Pillar, a high-level CIA official who oversaw government intelligence assessing the situation in the Middle East from 2000 to 2005, accused the Bush administration of distorting or ignoring prewar intelligence estimates in its determination to justify invading Iraq. He wrote, "If the entire body of official intelligence on Iraq had a policy implication, it was to avoid war—or, if war was going to be launched, to prepare for a messy aftermath. . . . What is most remarkable about prewar U.S. intelligence on Iraq is not that it got things wrong and thereby misled policymakers; it is that it played so small a role in one of the most important U.S. policy decisions in decades."[47] Almost a year earlier, even the president's commission on intelligence said, "It is hard to deny the conclusion that intelligence analysts worked in an environment that did not encourage skepticism about the conventional wisdom."[48]

Groupthink can be prevented, or at least minimized, by direct and explicit action. Specific members of the group can be assigned to criticize *every* proposal that anyone makes. Group members can be encouraged by their leader to actively debate the pros and cons of each proposal, and to search for and explain any misgivings they might have, without fear of ridicule. Outside experts of very different background and beliefs than the in-group, can be sought out and consulted, rather than excluded.

Having learned from the Bay of Pigs fiasco, the Kennedy Administration took such actions when the Cuban Missile Crisis erupted. As a result, there was much higher quality decision making during the Missile Crisis than during the Bay of Pigs, even though many of the same people were involved in both decision making in-groups.[49] Still, it is important to remember that critical errors and miscalculations were made during the Missile Crisis, which nearly led the world into the terminal disaster of general nuclear war.

Group Psychosis

Members of cohesive decision-making teams caught up in groupthink may be in denial about some elements of what is going on in the outside world, but they are still basically in touch with reality. They may be deluding themselves about their own power, wisdom, or morality, but they are not living in a world of delusion. There are, however, conditions under which the behavior of groups can become truly psychotic.

The difference, it is said, between a psychotic and a neurotic is that "the neurotic builds castles in the air, and the psychotic lives in them." Although they are disconnected from reality in important ways, psychotics may still be functional enough to avoid isolation or even detection by those around them. Because they are living in a world of delusion, they are not just unreliable; their behavior can be both bizarre and extremely dangerous.

It is easy to understand why a group of mentally deranged individuals might exhibit psychotic behavior. It is very difficult to understand—or even believe—that a group of otherwise normal people can together behave as if they were psychotic. How can such a thing happen?

Under the right conditions, a seriously deluded, but very charismatic leader can draw a group of basically normal though perhaps vulnerable people into his or her own delusional state. The most critical elements appear to be the charisma of the leader, the degree of control he or she can exert over the conditions under which the group's members live, and the extent to which the leader is able to isolate group members from outside influences, especially those of non-group friends and family. Three striking cases of group psychotic behavior in twentieth century America are briefly described below. They provide a frightening insight into the character and potential consequences of this most extreme form of group pathology.

Jonestown[50]

It was November 18, 1978, and 912 human corpses lay in the clearing in the jungle that was Jonestown, Guyana. Row after row of men, women, and children, most of them face down, their faces twisted into violent contortions by the terminal agony of cyanide poisoning, blood streaming from their noses and mouths. On the primitive stage they faced, their leader, the Reverend Jim Jones was toppled over a podium, a bullet in his head.

From an early age, Jones was fascinated by preachers and used to practice giving sermons. As a child, his was a small town life, with a stable upbringing. Yet he became adept at manipulating people and fascinated by the power that implied. A fellow minister commented, "I've never seen anyone relate to people the way he could. He would build them up, convince them that anyone as intelligent and sensitive as they were ought to do whatever it was that he wanted them to do."[51]

In 1965, Jones moved his church, called the People's Temple, from Indiana to California. He convinced many of its members that the Ku Klux Klan, the CIA, or some other external evil force would kill them if it were not for him. He encouraged them to inform on spouses or children that violated any of his rules or expressed doubts. Severe public paddlings, suicide rituals with fake poison, and sexual abuse became part and parcel of life in the Temple. Every woman close to him was required to have sex with him regularly. He used sexual activity with males in the Temple as a tool to humiliate or blackmail them, having them observed or photographed in the act.

Ignorant of this, the outside world acclaimed Jones for establishing effective drug rehabilitation programs, clinics, and nursing homes. The mayor of San Francisco appointed him chairman of the city's Housing Authority. More and more, Jones associated with high-level local and national politicians. But his growing celebrity status contained the seeds of his downfall. The press began to grow curious about the secret inner world of the People's Temple. The imminent appearance of an investigative expose in the August 1, 1977 issue of *New West* magazine prompted Jones to move the People's Temple to

Guyana, taking more than 800 members with him to build Jonestown on 27,000 acres of land he had leased.

Some of Jones's followers were young college graduates from socially progressive, upper-middle-class backgrounds; some were middle-class blacks and whites from fundamentalist religious backgrounds. More were young blacks with limited education from poor ghetto neighborhoods. The greatest percentage were elderly, also from the San Francisco ghetto.

Once ensconced in their clearing in the Guyanese jungle, Jones became even more controlling, paranoid, and psychotic. A master of mind control, he employed a variety of brainwashing techniques.[52] Life in what was originally supposed to be a peaceful, caring, non-racist utopian community became a living Hell. By November 18, Jones's followers had largely lost their ability to exercise independent judgment. They were no longer fully in control of their own minds. When he commanded them to begin the oft-rehearsed ritual of mass suicide, they were primed for mindless obedience. One by one, they came up to the vat of flavored water laced with cyanide and drank a cup of the poisonous brew. In a few minutes they began to gasp, convulse, and vomit. And soon, it was all over.

The tragedy at Jonestown seems like nothing so much as a bizarre, ghoulish nightmare. It is hard to believe that it really happened. But it did.

David Koresh and the Branch Davidian

On February 28, 1993, more than 100 armed federal agents of the Bureau of Alcohol, Tobacco and Firearms (ATF) swarmed onto the compound of a fringe religious sect known as the Branch Davidian and led by David Koresh. The agents wanted to arrest Koresh for illegal possession of firearms and search the compound for what they had been told was an impressive arsenal of illegal weapons. They encountered a barrage of high-powered gunfire that left four of them dead and 15 more wounded. The failed raid marked the beginning of an armed standoff that made headlines for months.[53]

Koresh had found his "calling" as a preacher. He was a very charismatic speaker who combined religious fervor with a talent for weaving a spell with words.[54] According to one near-convert, his style was " . . . almost schizophrenic and dissociative. . . . The words would all be there, the syntax would be correct, but when you put it together, it didn't make logical sense. . . . "[55]

Like Jim Jones, Koresh lived in a world that was an odd combination of fantasy and reality, wrapping his followers in his delusion that he was Jesus Christ, come to earth to interpret the truths contained in the biblical book of Revelation.[56] First, Koresh believed, he would become a rock star. Then, he would go to Israel and show the rabbis the biblical truths they were unable to see because of their limited wisdom and understanding, truths to which only he was privy. His presence in the Middle East would create havoc. The United States would be forced to send in troops, triggering a war that would destroy the world. He would then be crucified.[57]

While waiting for the drama to begin, Koresh lived a decidedly un-Christ-like existence. At age 24, he had married the 14-year-old daughter of a high-ranking Branch Davidian. Former Davidians allege that Koresh routinely abused even the cult's youngest followers, physically and psychologically. After a long investigation, the *Waco Tribune-Herald* reported complaints that Koresh boasted of having sex with underage girls in the cult. He had at least 15 "wives" of his own, and claimed the divine right to have sex with every male cult member's wife. Koresh claimed that the children he fathered with these women would rule the earth with him after he and the Davidian men had slain the nonbelievers.[58]

He kept his followers isolated at the group's compound and spellbound with his oratory. They believed that he was their key to entry into heaven after the soon-to-come end of the world. He controlled all the details of their daily lives, putting them through lengthy and rigorous Bible study every day, sessions sometimes lasting as long as 15 hours.[59]

Provoked by an assault on the compound by federal forces after months of armed standoff, the Branch Davidian set fire to the main building at Mt. Carmel, nearly 80 of them (including 21 children) dying together in mid-April 1993. Unlike the grotesque group suicide/murder discovered only after-the-fact in the remote jungles of Jonestown, Guyana, an awestruck public watched as flames engulfed the compound on national television. Once again a deluded, charismatic leader had led the group of people who trusted him to self-induced disaster.

Heaven's Gate[60]

In late March 1997, as the comet Hale-Bopp drew ever closer to earth, 39 men and women, dressed in loose, androgynous black clothing, swallowed a potent mixture of Phenobarbital and alcohol, and quietly lay down on their beds to die. They believed they were shedding their earthly bodies to join aliens trailing the comet in a spacecraft that had come to take them to another plane of existence, the Level Above Human.

Unlike those who died in agony in the jungle at Jonestown or were burned to death in the spare compound at Waco, the members of the cult called Heaven's Gate drifted peacefully off to their final sleep in the rooms of a large, comfortable house near San Diego. These quiet, hardworking, clean cut, celibate, and religious people had become enveloped in the ultimately deadly science fiction fantasy of their charismatic leader, Marshall Herff Applewhite.

In 1972, Applewhite was hospitalized with heart trouble. He apparently had a "near death" experience in the hospital, and a nurse there, Bonnie Nettles, told him that God had kept him alive for a purpose. "She sort of talked him into the fact that this was the purpose—to lead these people—and he took it from there."[61] Within a year, Applewhite and Nettles began traveling together, and by the summer of 1973 both were convinced that they were the "two lampstands" referred to in the biblical Book of Revelations and that God had directly revealed their "overwhelming mission" to them. They set out to recruit followers, with considerable success.

Once convinced to join the cult, members had to give away their material possessions, change their names, and break all contact with friends and family. They were not to watch television or read anything other than the "red-letter" edition of the Bible. Each member was assigned a partner and they were encouraged to always travel as a pair. Control of sexual behavior was again part of the routine, in this case in the form of complete celibacy. Camped in Wyoming in the late 1970s, cult members had to wear gloves at all times, limit their speech to "Yes," "No," and "I don't know" and otherwise communicate almost solely through written messages. They woke up and ate meals on a rigid schedule, sometimes wore hoods over their heads, and changed their work chores every 12 minutes, when there was a beep from the command tent.

Applewhite moved the cult to California in the 1990s. On their Web site, they published documents outlining their beliefs: Two thousand years ago beings from the Level Above Human sent an "Older Member" (Jesus Christ) to earth to teach people how to enter God's Kingdom, but demonically-inspired humans killed him. In the 1970s, the Level Above Human gave humanity another chance, this time sending two "Older Members" (Applewhite and Nettles) to resume the teachings.[62]

Heaven's Gate was not the subject of a pending media expose or Congressional investigation as was the Jonestown cult, nor were they surrounded by heavily armed law enforcement agents, as were Koresh and his followers. In the end, they took their own lives not because of threat, but simply so that they could ascend to the spacecraft they believed would take them to their long-sought Level Above Human.

How was a group of essentially normal people drawn so completely into the crazy world of a disintegrating psychotic like Jim Jones that they took their own lives and murdered their own children? How did a twisted fanatic like David Koresh take a group of deeply devoted believers in the teaching of a loving God to the point where they betrayed the most central principles of those teachings? How was Marshall Applewhite able to convince a group of clean-cut, well-educated adults to believe so strongly that an alien spaceship had come to take them to a better world that they all calmly and willingly committed suicide? Under what conditions can groups of basically sane people be led to behave in such insane ways?

Brainwashing

In the early 1960's, Yale psychiatrist Robert Jay Lifton published a landmark study of the brainwashing methods used by the Chinese communists during their takeover of China more than a decade earlier.[63] The essence of the process was depersonalization, achieved through repeated attacks on the individual's sense of self. The Chinese created a totally controlled, extremely stressful living environment, and after applying enough physical and psychological pressure to break the victim's will, offered leniency in return for complete co-operation. It was all quite effective. A relatively short but intense period of brainwashing overrode years of experience and belief.

The brain is an extraordinarily complex information-processing system. Just as muscles must be exercised to function properly, it appears that the brain must have a continual flow of internally or externally generated information to process. Sensory deprivation experiments have shown that halting the flow of external information can produce drastic distortions of mental function. Visual and aural hallucinations may result after as little as 20 minutes. Prolonged total deprivation may even damage brain function in ways that are difficult if not impossible to reverse. Yet the alterations of consciousness that are produced by relatively short periods of sensory deprivation can include drug-like ecstatic highs and feelings of spiritual bliss. In any case, it is clear that information is vital to normal brain function.[64]

Because people are capable of learning both facts and ways of interpreting them, experience affects the way in which the brain processes information. It is therefore logical that intense experiences can alter not only what we know, but also how we think. The brain must both take in facts and organize them around frameworks of interpretation. The key to "brainwashing" or "mind control" lies in creating intense experiences that so disrupt existing ways of thinking as to throw the brain off balance in its search for order, then to present new ways of thinking that offer relief from the mental chaos that has been created. Because chaos is so painful to the order-seeking mind, the brain will tend to lock onto the new way of thinking, even if it is quite foreign and bizarre. Since it plays off the basic way our brains work, we are all vulnerable to brainwashing. Some are much more vulnerable than others, and we are all more vulnerable during particularly stressful times than when life is flowing smoothly.

Sleep disruption, sleep deprivation, starvation, and violations of personal physical integrity (from rape to beatings to control over bodily functions such as bathroom habits) are effective mind control tools because they create enough mental disorientation and emotional chaos to throw the brain into disorder. Constant ideological harangues thrown into the mix help to further disrupt the mind's accustomed way of thinking and to provide new ways of thinking which, if accepted uncritically, promise to end the awful noise. An individual already primed for unquestioned obedience to authority by early exposure to dogmatic religion or abusive and authoritarian parenting is easier to brainwash. Perhaps that is why so many of the cults that program their followers' minds arise in the context of distorted, extremist religious faith.

Cults are not the only groups that manipulate participants to gain control over their minds. So called "mass therapies," such as Lifespring also use these or similar techniques. Founded in 1974, Lifespring presents itself as a self-help group that sells intensive seminars to help their participants come to terms with their problems and lead a more fulfilling life.[65] It uses powerful brainwashing techniques, such as high-pressure confrontation and exhausting marathon sessions. The basic course consists of 50 hours over a period of only five days. Doors to the training rooms are locked, the temperature of the room is manipulated, food is limited, trips to the bathroom are restricted, and so forth.

Lifespring searches out and highlights each participant's greatest fear. Calling that intense personal fear a "Holy Grail," the group calls on the person to confront it directly—not under the care of trained mental health professionals or even among caring friends and family, but out in the world-at-large. In the words of one recruit, " ... you have to get this Grail. If you don't get this Grail, you're not a good person, you gotta get your Grail. ... "[66] Programmed with this sort of determination, getting the Grail can be extremely dangerous. Urged to confront his paralyzing fear of heights, one man put on a business suit and jumped off a bridge high over rocks in a dangerous river. He nearly died. Another who was terribly afraid of water and had never learned to swim jumped into a river and drowned trying to swim the quarter mile to the other side.

During one incident a Lifespring trainer refused to give any assistance to a young woman who suffered a severe asthma attack, telling her that her gasping and wheezing was purely psychological. She died. Why did she not just leave? According to her Lifespring training partner, "She told me that she ... didn't want to leave; she felt very intimidated by the trainers. ... "[67] John Hanley, Lifespring's founder, admits, " ... there are people who freak out in the training. And some times they can control it and handle it ... and sometimes they just can't. ... We had a veteran who freaked out reliving his experiences in Vietnam ... he was screaming and yelling in the parking lot and he was gonna tear everything up."[68]

Brainwashing, Group Psychosis, and the Nature of Military Life

In order to mold an effective fighting force, military training must condition essentially normal people to do things that the vast majority of them would otherwise not think of doing. Militaries must condition ordinary people to put themselves in the unnatural position of killing or being killed, and to do so for relatively abstract principles such as patriotism or ideology. The many wars with which the human species has been plagued to date make it clear that they have become quite good at this conditioning. Military basic training begins with a process of depersonalization. Each recruit is made to look as much like every other recruit as possible. Heads are shaved; clothing and accessories that might carry signs of individual personality are removed and replaced by uniforms to make everyone's appearance "uniform." All conditions of the recruit's life are controlled: what food is eaten, where it is eaten and when; the bedtime and wakeup time that bracket allotted hours of sleep in barracks that are also uniformly equipped and arranged; and virtually all activities during the day. These activities are prearranged and ritualized. There are endless drills, from marching on parade grounds to inspections of clothing, living quarters, and equipment. There is constant repetition and continuing verbal and physical harassment. Punishment for even minor infractions is swift and often includes language or tasks that are intended to humiliate and degrade.

From the very beginning, members of the armed forces are conditioned to operate within a rigid, authoritarian system in which orders are to be obeyed, not debated. They are thus primed for further, perhaps pernicious manipulation by those who so completely control the conditions of their life and work. It is important to understand that this *cannot be avoided* given the mission military organizations are asked to perform. The only way to get large numbers of people to reliably do what militaries must get them to do is to establish a rigid, authoritarian system and condition the troops to follow orders.

Some American military personnel, including some in the nuclear military, have also undergone training by cults or pop-psychology mass therapies. At one point, the military itself reportedly invited Lifespring to train military personnel on several American bases in the United States and abroad. According to ABC-TV News, Lifespring held training sessions for hundreds of military personnel and their families at Vandenburg Air Force Base in California in late 1979 and again in March 1980. Among other things, Vandenburg was a Strategic Aerospace Command training base for the missile silo crews that tend the land-based missile component of the U.S. strategic nuclear forces. An Air Force captain involved in security brought these Lifespring trainings to the attention of ABC. He was concerned that " . . . we don't know if this person that we certified as being combat-ready prior to Lifespring training is the same person in terms of reliability after that Lifespring training."[69]

According to an investigative reporter who secretly attended one of the Lifespring sessions at Vandenburg:

> At several points during the training they were so vulnerable that they were . . . talking about their most intimate problems, and they were ranting and raving and screaming and cursing . . . and spitting; it was like bedlam. It was like being at a mental institution, and I think at that point they would've answered any question put to them. . . .[70]

When Harvard psychiatrist John Clark was asked if he thought Life-spring graduates in the military and other sensitive positions might be unreliable enough to be security risks, he replied, "I believe they are. . . . I don't think they can be trusted because something has happened to their mind. . . . They are loyal to Lifespring and dependent on it, as though they were dependent on a drug."[71]

Group Psychosis in the Military

Primed by military training and the obedience-oriented, authoritarian character of military life, members of the armed forces may be particularly vulnerable to being drawn into the fantasy world of a charismatic but deeply disturbed commanding officer. The bonding of military men and women to each other and their often fierce pride in their unit, so necessary to cohesiveness and effectiveness under life-threatening conditions, also predisposes them to becoming part of a group loyal to a leader who may have departed from firm contact with reality.

The ability of a charismatic leader to involve followers in acting out a script that he or she has written is greatly enhanced if they can be isolated from outside influences. An appreciation of this is obvious in the tactics of both the cults and the "mass therapies." Isolation is a normal part of life in the armed forces. Ordinarily, groups rather than individuals are isolated from the outside world for periods ranging from hours or days (in the land-based missile forces) to months (in nuclear submarines). The longer the isolation, the greater the control the commanding officer has over the environment in which subordinates live and work.

It would be relatively simple for a charismatic military commander, capable of inspiring extraordinary loyalty in troops already primed for obedience, to segregate them from the outside world. Once isolated, with control over so many elements of their lives the commander could easily create the conditions that would draw them ever more tightly into his or her world. The basic elements of such a scenario are illustrated in a 1960s vintage film, *The Bedford Incident*.[72]

The charismatic captain of an anti-submarine warfare ship, the Bedford, does not appear to be psychotic. But he is so totally wrapped up in his fierce sense of mission and determination to humiliate the enemy that his judgment is terribly distorted. Once at sea, he drives the men beyond reason and good sense. Disrupting their sleep and keeping his crew almost constantly at high levels of alert during their tour at sea, he draws them into his grand obsession. Ultimately, one of his most loyal men breaks under the strain, misunderstands a command, and launches a nuclear attack that could begin a nuclear war. Though the script is purely fiction, the story is frighteningly realistic.

Probably the most dangerous plausible place for a "Bedford Incident" like scenario in the real world is a nuclear missile submarine. The crew is largely without outside contact for months at a time, their world confined to the cramped quarters of a large metal tube sailing below the surface of the ocean. Encased day and night in a totally artificial environment from which there is no respite, the captain and ranking officers aboard have nearly complete control over the living and working conditions of the crew. The chain of command on the ship begins with the captain and is clearly specified. Obedience is expected from the crew, which has, of course, been thoroughly trained to obey orders.

There was no external physical control over the arming and launching of the nuclear-tipped missiles in American submarines until at least 1997, and there still did not appear to be any such controls on British submarines at least as recently as 2007 (see Chapter 5). It is unclear whether the missile submarines of all the other nuclear weapons states are equipped with these sophisticated controls, but it seems unlikely. Though the crews of American missile submarines are not authorized to launch without orders from the highest command authority (as is presumably true of the other nuclear states as well), it is not inconceivable that there might be a way for them to defeat the "permissive action links" and launch the nuclear weapons on board by themselves without proper authorization. That would certainly be much, much less difficult for the crews of the submarines of any other nuclear-armed states that are not fitted with these controls. And each single submarine carries enough offensive nuclear firepower to kill millions, if not to destroy an entire nation.

Transferring responsibility from individuals to groups does not make dangerous technology systems proof against the limits of safety imposed by human error or malevolence. From the banalities of bureaucracy to the arrogance of groupthink to the nightmare of group psychosis, groups not only fail to solve the problem of human fallibility, they add their own special flavor to it.

There is no way to avoid the fallibility that is an essential part of our human nature. Individually or in groups, we have accidents, we make mistakes, we miscalculate, we do things that we would be better off not doing. If we continue to create, deploy, and operate dangerous and powerful technological systems, we *will* eventually do ourselves in. That is the bottom line. After all, we are humans, not gods. The only way we can permanently prevent human-induced technological disaster is to stop relying on technologies that do not allow a very, very wide margin for the errors we cannot help but commit.

NOTES

1. Kees Keizer, Siegwart Lindenberg, and Linda Steg, "The Spreading of Disorder," *Science* (December 12, 2008), 1681.

2. Beth Azar, "Teams That Wear Blinders Are Often the Cause of Tragic Errors," *American Psychological Association Monitor* (September 1994): 23.

3. Ole R. Holsti and Alexander L. George, "The Effects of Stress on the Performance of Foreign Policy-Makers," in *Political Science Annual: An International Review*, ed. Cornelius P. Cotter, Vol. 6 (1975), 294.

4. National Journal Group, "Former Intelligence Official Faults U.S. Policy-Makers on Iraq WMD Claims," *Global Security Newswire*, December 10, 2008; http://gsn.nti.org/siteservices/print_frinedly.php?IDnw_20081210_1663.

5. Morton H. Halperin, *Bureaucratic Politics and Foreign Policy* (Washington, DC: The Brookings Institution, 1974), 235–279.

6. James Thomson, "How Could Vietnam Happen?, *The Atlantic*, 1968, 47–53; and David Halberstam, *The Best and the Brightest* (New York: Random House, 1972).

7. Eric Lichtblau, "Report Finds Cover-Up in an F.B.I. Terror Case," *New York Times*, December 4, 2005.

8. Chris Argyris, "Single Loop and Double Loop Models in Research on Decision making," *Administrative Science Quarterly*, September 1976, 366–367.

9. Op. cit., Holsti ang George, 293–295.

10. Arthur Schlesinger, Jr., *The Imperial Presidency* (Boston: Houghton-Mifflin, 1973).

11. J. Eberhardt, "Challenger Disaster: Rooted in History," *Science News*, June 14, 1986, 372.

12. Matthew L. Wald, "Energy Chief Says Top Aides Lack Skills to Run U.S. Bomb Complex," *New York Times*, June 28, 1989.

13. Ibid.

14. Eliot Marshall, "Savannah River Blues," *Science*, October 21, 1988, 364.

15. Ibid., 363–364.

16. Peter Spotts, "Nuclear Watchdog Agency Shows Signs of Losing Bite," *Christian Science Monitor*, June 6, 1996.

17. U.S. General Accounting Office, Report to Congressional Requesters, *Nuclear Regulation: Preventing Problem Plants Requires More Effective NRC Action*, May 1997: GAO/RCED-97-145.

18. Leslie Wayne, "Getting Bad News at the Top," *New York Times*, February 28, 1986.

19. William J. Broad and Carl Hulse, NASA Dismissed Advisors Who Warned About Safety," *New York Times*, February 3, 2003.

20. Eric Lichtblau and David Sanger, "August '01 Brief Is Said to Warn of Attack Plans," *New York Times*, April 10, 2004. See also David Johnston and Jim Dwyer, "Pre-9/11 Files Show Warnings Were More Dire and Persistent," *New York Times*, April 18, 2004.

21. U.S. Department of Health and Human Services, "Hurricane Katrina," http://www.hhs.gov/disasters/emergency/naturaldisasters/hurricanes/katrina/index.html (accessed September 18, 2009).

22. Eric Lipton, "White House Knew of Levee's Failure on Night of Storm," *New York Times*, February 10, 2006.

23. Philip Shenon, "Pentagon Reveals It Lost Most Logs on Chemical Arms," *New York Times*, February 28, 1997.

24. Ted Wye (a pseudonym), "Will They Fire in the Hole?," *Family*, supplement of *Air Force Magazine*, November 17, 1971.

25. Violation of strict rules against fraternization between the ranks were part of the legal charges leveled against drill instructors and other military officers during the flood of sexual harassment and rape complaints lodged by women in all branches of the armed forces during the 1990s.

26. Peter T. Kilborn, "Sex Abuse Cases Stun Pentagon, But the Problem Has Deep Roots," *New York Times*, February 10, 1997.

27. Ibid. A majority of the servicewomen responding to a 1996 Pentagon poll said that they had been sexually harassed.

28. Op. cit. Holsti and George, 295.

29. Robert S. McNamara, "One Minute to Doomsday," *New York Times*, October 14, 1992.

30. Robert F. Kennedy, *Thirteen Days: A Memoir of the Cuban Missile Crisis* (New York: Norton, 1969), 94–95.

31. Richard Wilson, "A Visit to Chernobyl," *Science*, June 26, 1987, 1636–1640.

32. Ibid., 1639.

33. Philip Shenon and Christopher Marquis, Panel Says Chaos in Administration Was Wide on 9/11," *New York Times*, June 18, 2004.

34. David Johnston and Neil A. Lewis, "Whistleblowers Recount Faults Inside the F.B.I.," *New York Times*, June 7, 2002.

35. Arlene K. Daniels, "The Captive Professional: Bureaucratic Limitations in the Practice of Military Psychiatry," *Journal of Health and Social Behavior*, Vol. 10 (1969): 256–257.

36. Ibid., 255–265.

37. Eric Lichtblau, "9/11 Report Cites Many Warnings About Hijackings," *New York Times*, February 10, 2005.

38. John Schwartz and Matthew Wald, "Report on Loss of Shuttle Focuses on Nasa Blunders and Issues Somber Warning," *New York Times*, August 27, 2003.

39. Elaine Sciolino and Don Van Natta, Jr., "June Report Led Britain to Lower Its Terror Alert," *New York Times*, July 19, 2005.

40. BBC News, "7 July Bombings: Overview," http://news.bbc.co.uk/2/shared/spl/hi/uk/05/london_blasts/what_happened/html/ (accessed September 18, 2009).

41. Solomon Moore, "Science Found Wanting in Nation's Crime Labs," *New York Times*, February 5, 2009.

42. D. J. Bem, "The Concept of Risk in the Study of Human Behavior," in *Risk Taking Behavior*, ed. R. E. Carney (Springfield, IL: Thomas, 1971).

43. Irving L. Janis, *Victims of Groupthink: A Psychological Study of Foreign-Policy Decisions and Fiascoes* (Boston: Houghton-Mifflin, 1972), 9.

44. David Johnston and Andrew C. Revkin, "Report Finds FBI Lab Slipping from Pinnacle of Crime Fighting," *New York Times* (January 29, 1997).

45. Op. cit., Irving Janis, 197–198.

46. Ibid., p. 67.

47. Paul R. Pillar, "Intelligence, Policy, and the War in Iraq," *Foreign Affairs*, March/April 2006; see also Scott Shane, "Ex-C.I.A. Official Says Iraq Data Was Distorted," *New York Times*, February 11, 2006.

48. Todd S. Purdum, "A Final Verdict on Prewar Intelligence Is Still Elusive," *New York Times*, April 1, 2005.

49. Ibid., 138–166.

50. Much of this discussion is drawn from two primary sources: "The Death of Representative Leo J. Ryan, People's Temple, and Jonestown: Understanding a Tragedy," *Hearing before the Committee on Foreign Affairs, House of Representatives* (Ninety-Sixth Congress, First Session: May 15, 1979), and "The Cult of Death," Special Report, *Newsweek*, December 4, 1978, 38–81.

51. Pete Axthelm, et al., "The Emperor Jones," *Newsweek*, December 4, 1978, 55.

52. Op. cit., Committee on Foreign Affairs, 15.

53. Tracy Everbach, "Agents Had Trained for Months for Raid," and Lee Hancock, "Question's Arise on Media's Role in Raid," *Dallas Morning News*, March 1, 1993.

54. Victoria Loe, "Howell's Revelation: Wounded Sect Leader, in a Rambling Interview, Says that He Is Christ," and "A Look at Vernon Howell," *Dallas Morning News*, March 1, 1993.

55. Ibid.

56. Ibid.

57. Ibid.

58. Mark England and Darlene McCormick, "Waco's Paper's Report Called Cult 'Dangerous,'" *Dallas Morning News*, March 1, 1993.

59. Ibid.

60. The discussion of the Heaven's Gate cult in this section is based largely on information drawn from the following articles published in the *New York Times*: Todd S. Purdum, "Tapes Left by 39 in Cult Suicide Suggest Comet Was Sign to Die," March 28, 1997; Gustav Niebuhr, "On the Furthest Fringes of Millennialism," March 28, 1997; Frank Bruni, "Leader Believed in Space Aliens and Apocalypse," March 28, 1997; Heaven's Gate, "Our Position Against Suicide" reprinted as "Looking Forward to Trip Going to the Next Level," March 28, 1997; B. Drummond Ayres, "Families Learning of 39 Cultists Who Died Willingly," March 29, 1997; Frank Bruni, "A Cult's 2-Decade Odyssey of Regimentation," March 29, 1997; Jacques Steinberg, "The Leader: From Religious Family to Reins of U.F.O. Cult," March 29, 1997; Carey Goldberg, "The Compound: Heaven's Gate Fit In With New Mexico's Offbeat Style," March 31, 1997; B. Drummond Ayres, "Cult Members Wrote a Script to Put Their Life and Times on the Big Screen," April 1, 1997; Todd S. Purdum, "Last 2 Names Released in Mass Suicide of Cult Members," April 1, 1997; James Brooke, "Former Cultists Tell of Believers Now Adrift," April 2, 1997; Barry Bearak, "Eyes on Glory: Pied Pipers of Heaven's Gate," April 28, 1997; and Todd S. Purdum, "Ex-Cultist Kills Himself, and 2nd Is Hospitalized," May 7, 1997.

61. Jacques Steinberg, "The Leader: From Religious Family to Reins of U.F.O. Cult," *New York Times*, March 29, 1997.

62. Gustav Niebuhr, "On the Furthest Fringes of Millennialism," *New York Times*, March 28, 1997.

63. Robert Jay Lifton, *Thought Reform and the Psychology of Totalism: A Study of "Brainwashing" in China* (New York: W. W. Norton, 1961).

64. This and much of the discussion on cults that follows draws heavily from Flo Conway and Jim Siegelmen, *Snapping: America's Epidemic of Sudden Personality Change* (New York: Dell Publishing, 1979).

65. On October 30, 1980, and again on November 6, 1980, ABC-TV News aired a two-part story focused on Lifespring on its "20/20" newsmagazine program. This quote is from the second part of the story (transcript, 13). The discussion that follows draws heavily from both parts of that program. Note also that as a company, Lifespring Inc. ceased to exist in 2000, though its founder John Hanley maintains a Web site at http://www.lifespringnow.com/ (accessed September 19, 2009).

66. Ibid., October 30, 1980, transcript, 15.

67. Ibid., transcript, 9 and 6.

68. Ibid., transcript, 16–17.

69. Ibid., November 6, 1980, transcript, 14–15.

70. Ibid., transcript, 16.

71. Ibid.

72. James Poe, *The Bedford Incident* (based on the novel by Mark Rascovich), produced by James B. Harris and Richard Widmark (Columbia Pictures: Bedford Productions Limited, 1965).

9

Making Technology Foolproof

More than ever before, we are dependent on a web of interconnected technical systems for our most basic needs and our most fleeting whims. Technical systems are integral to providing us with water, food, and energy, to getting us from place to place, to allowing us to communicate with each other and coordinate all the activities on which our physical and social lives depend. The more that technical systems have become central to our way of life and critical to the normal functioning of society, the greater the disruption caused when they fail.

Most of us do not really understand how any of these complex and sophisticated technical systems work, and no one understands them all. Most of the time that ignorance is not much of a problem. We do not have to know how they work to use them, and often we do not even need detailed technical knowledge to maintain or repair them. Knowing which pedal to push to speed up or slow down, when and how much to turn the steering wheel, how to back up, and so forth is all you need to know to drive a car. You do not have to understand what actually happens inside. Even those who repair cars do not need engineering or scientific knowledge of the electrical, chemical, and mechanical processes that make the car work. All they need to know is what each part does and how to adjust, repair, or replace it.

Technical expertise is also not necessary to understand why technical systems fail, and why the possibility of failure cannot be completely eliminated. It is enough to generically comprehend the inherent problems involved in developing, producing, and operating them as well as in assessing their costs, benefits, and risks.

COMPLEXITY AND RELIABILITY

The reliability of a technical system depends on the reliability of its parts and the complexity of the system. Complexity, in turn, depends upon the number of parts and how they interact with each other. All other things being equal, the greater the number of parts that must perform properly for the system to work, the less reliable the system will tend to be.[1] Multiplying the number of parts can make the whole system less reliable even if each part is made more reliable.[2] Greater interdependence among components also tends to make the system less reliable.[3]

In a system with more parts, there are more ways for something to go wrong. Given the same quality of materials, engineering, and construction, a system with more parts will fail more often. When the parts of a system are tied together more tightly, one failure tends to lead to others, dragging the system down. At first, this may only degrade performance, but if it produces cascading failures, the whole system could fail.

Sometimes achieving desired performance requires a complex design. The more complex version might also be faster, safer, cheaper, or higher precision than a simpler design. Still, increasing the number and interdependence of components rapidly overcomes even the improved reliability of each of its parts, making complicated technical systems fail more often than simpler ones.

In an internal briefing for the Air Force in 1980, military analyst Franklyn C. Spinney documented the reliability-reducing effects of complexity.[4] Using several measures, he compared the reliability of aircraft of varying complexity in both the Air Force and Navy arsenals. Data drawn from his analysis are given in Table 9.1.

All modern combat aircraft are very complex, interdependent, and high-performance technical systems. Even the most reliable of them breaks down frequently. None of the aircraft in the sample averaged more than 72 minutes of flying between failures (problems requiring maintenance), fewer than 1.6 maintenance events per flight, or had all mission essential equipment operating properly more than 70 percent of the time. Anyone whose car broke down anywhere near this often would consider that he or she had bought a world-class lemon.

Even so, differences in complexity still give rise to major differences in reliability. Table 9.1 shows a clear pattern of reduced reliability with increasing complexity, by any measure used. The contrast between the most and least reliable aircraft is striking. The highly complex F-111D

Table 9.1 Complexity and reliability in military aircraft (FY 1979)

Aircraft (AF = Air Force; N = Navy)	Relative Complexity	Average Percentage of Aircraft Not-Mission-Capable at Any Given Time	Mean Flying Hours Between Failures	Average Number of Maintenance Events Per Sortie
A-10 (AF)	Low	32.6%	1.2	1.6
A-4M (N)	Low	31.2%	0.7	2.4
A-7D (AF)	Medium	38.6%	0.9	1.9
F-4E (AF)	Medium	34.1%	0.4	3.6
A-6E (N)	High	39.5%	0.3	4.8
F-14A (N)	High	47.5%	0.3	6.0
F-111D (AF)	High	65.6%	0.2	10.2

Source: Spinney, F. C., *Defense Facts of Life* (December 5, 1980: Department of Defense unreviewed preliminary staff paper distributed in typescript by author).

averaged only 12 minutes of flying between failures, while the much simpler A-10 flew six times as long between problems. At least one piece of equipment essential to the F-111D's mission was broken nearly two-thirds of the time, while the A-10 was fully mission capable twice as often. And the F-111D averaged more than six times as many maintenance problems every time it flew.

Despite the obvious lesson of these data, the U.S. military continues to procure extraordinarily complex fighters. The twenty-first century F-35 Lightning II Joint Strike Fighter is an even more complex machine than the F-111. There is not enough experience with the F-35 yet to directly compare the two in terms of reliability, but the fact that the F-35 already has experienced a litany of serious operational problems is certainly not a good sign.[5]

Complexity-induced failures of reliability are an inherent feature of all technical systems, be they single machines or large interconnected equipment networks. The complex, interconnected telephone system fails fairly often, though most of these failures are so fleeting and trivial we scarcely notice them. Every once in a while, though, we get a spectacular illustration of just how wrong something can go when something goes wrong. At 2:25 PM on January 15, 1990, a flaw in a single AT&T computer program disrupted long distance service for nine hours. Roughly half the national and international calls made failed to connect.[6] Robert Allen, AT&T's Chairman, called it " . . . the most far-reaching service problem we've ever experienced."[7]

The program involved was part of a switching software update designed to determine routings for long distance calls. Because of the flaw, a flood of overload alarms were sent to other computers, stopping them from properly routing calls and essentially freezing many of the switches in the network.[8] Ironically, the system had been designed to prevent any single failure from incapacitating the network.[9] Furthermore, there was no sign that anything was wrong until the problem began, but once it did, it ran rapidly ran out of control. In the words of William Leach, manager of AT&T's network operations center, "It just seemed to happen. Poof, there it was."[10]

Ten years earlier, a forerunner of the Internet called "Arpanet," then an experimental military computer network, failed suddenly and unexpectedly. Its designers found that the failure of a small electronic circuit in a single computer had combined with a small software design error to instantly freeze the network.[11] In July 2004, the commission investigating the lethal collapse of part of a relatively new $900 million terminal at Charles de Gaulle Airport outside Paris noted that the complicated unconventional design of the building resulted in stresses of "great complexity" at the point where the collapse occurred.[12] After the official investigation of the space shuttle Columbia's disintegration on February 1, 2003, astronaut Mary Ellen Weber observed, "we may fix this particular problem—but I guarantee the next time astronauts get on that shuttle there will be a thousand other things that can happen."[13]

Opacity

As technical systems become more complex, they become more opaque. Those who operate ever more complex systems cannot directly see what is going on. They must depend on readings taken from gauges and instruments, which can be misleading. On August 14, 2003, a local power failure in Ohio triggered a series of events that led to the largest blackout in the history of North America. Nine months later, engineers still could not be sure how it all had happened. "Investigators, including officials from the United States and Canada, say that the eastern power grid . . . has become so large, complicated and heavily loaded in the last two decades that it is difficult to determine how a single problem can expand into an immense failure."[14] But that is exactly what happened.

When a system becomes so complex that no one—including its designers—can visualize how the system as a whole works, patchwork attempts to fix problems or enhance system performance are likely to create other hidden flaws. Just such an attempt seems to have caused the great AT&T crash of 1990. In 1976, AT&T pioneered a system called "out of band" signaling that sent information for coordinating the flow of calls on the telephone network as each call was made. Engineers who were updating the out of band system in 1988 inadvertently introduced the software flaw that caused the system to crash two years later.

If they are careful and do the job properly, those who modify very complex systems may avoid creating problems in the part of the system they are changing. However, it is virtually impossible for them to see all the subtle ways in which what they are doing will alter the performance of other parts and the overall system's characteristics and performance. If patchwork change can open a Pandora's box, why not redesign the whole system when it fails or needs updating? Fundamental redesign may be a good idea from time to time, but it is much too time consuming and expensive to do whenever a problem arises or a way of improving the system occurs to someone. In addition, frequent fundamental redesign has a much higher chance of introducing more serious problems than does patchwork change.

Although it sounds unbearably primitive, trial and error is still very important in getting complicated systems to work properly. "Bugs" are inevitable even in the most carefully designed complex systems *because* they are complex systems. Only by operating them under realistic conditions can we discover and correct unexpected (and sometimes unpredictable) problems and gain confidence in their reliability. That is why engineers build and test prototypes. How confident would you feel flying in an airliner of radically new design that had never actually been flown before?

On April 24, 1990, the NASA successfully launched the Hubble Space Telescope. Nearly a month later, Hubble produced its first blurry light image. Euphoria turned to concern, as the telescope just would not come into perfect focus, despite repeated commands from ground controllers at the Goddard Space Center outside Washington. Two months after launch, it became clear that the telescope suffered from spherical aberration, a classic problem covered in basic optics textbooks. A very slight flaw in the curvature of the Hubble's 2.4 meter primary mirror resulted in the optical system's failure to focus all incoming light at precisely the same spot. More than a decade of painstaking development had failed to prevent this crippling defect and an extensive program of testing had failed to detect it. It would be years before the Hubble could be fixed.[15]

In December 1993, a team of astronauts riding the space shuttle Endeavor into orbit finally repaired the telescope. They replaced some defective equipment and installed carefully designed corrective devices during an unprecedented series of space walks.[16] The next month, NASA jubilantly announced that although Hubble still did not meet its original design specifications, it finally could be clearly focused and was producing remarkable images of the heavens.[17] By late March 1999, the space telescope was once more in trouble. Only three of Hubble's six gyroscopes were working properly. If one more had failed, the telescope would have become too unsteady to do observations and would shut down.[18]

Sometimes even careful design, testing, *and* early experience leaves critical problems hidden until a substantial track record has been built up. The de Havilland Comet was the first commercial jet aircraft. In May 1953, a year after the inauguration of jetliner service, a Comet was destroyed on takeoff from Calcutta during a severe thunderstorm. The weather was blamed. Eight months later, a second Comet exploded soon after takeoff from Rome, Italy, in clear and calm weather. The Comets were withdrawn from commercial service for 10 weeks, then reinstated even though investigators found no flaws in the planes. In April 1954, a third Comet exploded in mid-air. One of the remaining Comets was then tested to destruction, revealing a critical problem. Repeating the normal cycle of pressurizing and depressurizing the cabin again and again caused a fatigue crack to develop in the corner of one of the windows that soon tore the plane's metal skin apart. The Comet's designers had been confident that such a problem would not develop until many more flights than the plane was capable of making during its estimated service life. They were wrong. Only the trial and error of continuing operation exposed their tragic mistake.[19]

Backup Systems and Redundant Design

Reliability can usually be improved by creating alternative routes that allow a system to keep working by going around failed components. Alternate routes not used until there is a failure are called "backup systems." For example, hospitals typically have their own backup generators to keep critical equipment and facilities operating when normal power supplies are disrupted. Emergency life vests on airliners are designed to inflate automatically when a tab is pulled, but they have a backup system—tubes through which a passenger can blow to inflate them if the automatic system fails.

Alternate routes used during normal operation can also serve as backups when failures occur. For example, calls can be routed over the telephone network from point A to point B over many possible pathways. In normal times, this allows the system load to be distributed efficiently. But it also allows calls to go through even when part of the system fails, by routing them around failed components.

"Voting" systems are another means of using redundancy to achieve reliability. "Triple modular redundancy" (TMR) has been used in the design of fault-tolerant computers. Three components (or "modules") of identical design are polled and their outputs compared. If one disagrees with the other two, the system assumes the disagreeing module is faulty, and acts on the output of the two that agree. The system can continue to operate properly even if one of the modules fails.[20]

Backup systems and redundancy can make the system as a whole *more* reliable than any of its parts.[21] Still, backup systems also fail. In 1976, the failure of both the main audio amplifier and its backup left Gerald Ford and Jimmy Carter speechless for 27 minutes before an estimated audience of 90 million viewers during the first Presidential debate in 16 years.[22] Three separate safety devices designed to prevent chemical leaks all failed during the Bhopal disaster in 1984.[23] Many back up systems failed during the nuclear power plant accident at Three Mile Island on March 28, 1979 and the much more serious accident at Chernobyl on April 26, 1986.

Adding backup systems or redundancies can also open up additional pathways to failure, offsetting some of the reliability advantages of having these backups or redundancies. Increasing the number of engines on a jet airliner makes it more likely that at least one engine will be available to fly the plane if the others fail. Because there are more engines, however, there is a higher probability that at least one engine will explode or start a fire that will destroy the aircraft.[24] One of the nine Ranger spacecraft flights intended to survey the moon before the Apollo mission failed *because* of extra systems designed into it to prevent failure! To be sure that the mission's television cameras would come on when the time came to take pictures of the moon's surface, redundant power supplies and triggering circuits were provided. A testing device was included to assure that these systems would work properly. However, the testing device short-circuited and drained all the power supplies before the spacecraft reached the moon.[25]

Common Mode Failure and Sensitivity

When several components of a system depend on the same part, the failure of that part can disable all of them at the same time. This is known as "common mode" failure. For example, if a hospital's backup generator feeds power into the same line in the hospital that usually carries electricity from the power company, a fire that destroyed that line would simultaneously cut the hospital off from the outside utility and make the backup generator useless.

At 4:25 AM on March 20, 1978, an operator replacing a burned-out light bulb on the main control panel at the Rancho Seco nuclear power plant near Sacramento dropped the bulb. This trivial event led to a common mode failure that almost triggered disaster. The dropped bulb caused a short circuit that interrupted power to key instruments in the control room, including those controlling the main feedwater system. The instrument failures also cut off information that was necessary for operators to fix the system. Equipment designed to control the reactor automatically did not take proper corrective action because it too depended on the malfunctioning instruments. Worse yet, the instruments sent false signals to the plant's master control system, which then caused a rapid surge in pressure in the reactor's core, combined with falling temperatures.[26] In an older reactor, this is very dangerous. If the Rancho Seco power plant had been operating at full power for 10 to 15 years instead of two or three, the reactor vessel might have cracked.[27] With no emergency system left to cool the core, that could have led to a core meltdown and a nightmarish nuclear power plant accident.

On January 9, 1995, construction crews at Newark Airport using an 80-foot pile driver to pound 60-foot steel beams into the ground drove one beam through a foot-thick concrete wall of a conduit six feet under ground and severed three high voltage cables serving the airport's terminal buildings. Hundreds of flights were cancelled and tens of thousands of people had to scramble to rearrange travel plans, as the airport was forced to shut down for nearly 24 hours. Local power company officials said that if one cable or even two had been knocked out, there would have still been enough power to keep the airport operating. If main power cable and auxiliary power cables ran through separate conduits rather than lying side by side in the same conduit, this expensive and disruptive common mode failure would not have occurred.[28]

Unless every part of a system has an alternative to every other part on which it depends, common mode failures can undo some or all the advantages of backup systems and redundancy. As long as there are any unique common connections, there can be common mode failures that render the system vulnerable, but duplicating every part of every system is simply not workable. In a hospital, that would involve duplicating every power line, every piece of equipment, even to the point of building a spare hospital.

High-performance technical systems are often more sensitive, as well as more complex. Greater sensitivity also predisposes systems to reliability problems. More sensitive devices are more easily overloaded, and thus more prone to failure. They are also more likely to react to irrelevant and transient stimuli. A highly sensitive smoke alarm will go off because of a bit of dust or a burned steak. False alarms make a system less reliable. They are failures in and of themselves; if they are frequent, operators will be too likely to assume that the next real alarm is also false.

Highly sensitive, complex interactive systems can also behave unpredictably, undergoing sudden explosive change after periods of apparent stability. When the Dow Jones average plummeted by 508 points at the New York Stock Exchange on October 19, 1987, losing 22 percent of its value in one day, many were quick to point an accusing finger at computerized trading practices. Programmed trading by computer had made the market less stable by triggering buying and selling of large blocks of stock in a matter of seconds to take advantage of small movements in prices. A stock market of program traders approximates the key assumptions underlying the theory of nonlinear game dynamics (including rivalry). When systems like this are modeled on a computer, strange behaviors result. The system may be calm for a while and seem to be stable, then suddenly and unpredictably go into sharp nonlinear oscillations, with both undershooting and overshooting.[29]

DESIGN ERROR

Engineering design results in a product or process that never existed before in that precise form. Mathematical verification of concepts, computer simulation, and laboratory testing of prototypes are all useful tools for uncovering design errors. Until the product or process works properly under "real world" conditions, however, it has not really been put to the test.

In 1995, after 14 years of work and six years of test flights, the B-2 bomber still had not passed most of its basic tests, despite an astronomical price tag of more than $2 billion a plane. The design may have looked fine on the drawing board, but it was not doing so well in the real world. Among other things, the B-2 was having a lot of trouble with rain. GAO auditors reported, "Air Force officials told us the B-2 radar cannot distinguish rain from other obstacles."[30] Two years later, they reported that the plane "must be sheltered or exposed only to the most benign environments—low humidity, no precipitation, moderate temperatures"—not the kind of conditions military aircraft typically encounter.[31] The B-2 was first used in combat in March 1999, during the NATO air campaign against Yugoslavia. Since they were too delicate to be based anywhere that did not have special facilities to shelter and support them, two B-2's were flown out of their home base in Missouri, refueled several times in the air so that they could drop a few bombs then hurry back to Missouri where they could be properly cared for.[32]

Designers of the space shuttle booster rockets did not equip them with sensors that could warn of trouble because they believed the boosters were, in the words of NASA's top administrator, "not susceptible to failure. . . . We designed them that way."[33] After many successful launches, the explosion of the right-side booster of the space shuttle Challenger in 1986 proved that they were very wrong.

Sometimes a design fails to perform as desired because the designers are caught in a web of conflicting or ambiguous design goals. The designer of a bridge knows that using higher grade steel or thicker concrete supports will increase the load the bridge is able to bear. Yet given projected traffic, a tight construction schedule, and a tight budget, he or she may intentionally choose a less sturdy, less expensive design that can still bear the projected load. If someday a key support gives way and the bridge collapses under much heavier traffic than had been expected, it is likely to be labeled a design error. But whose error is it: the engineer who could have chosen a stronger design, the person who underestimated future traffic, or the government officials who insisted the bridge be built quickly and at relatively low first cost?

Designers always try to anticipate and correct for what might go wrong when their design is put to the test in the real world. When designers of high-impact projects get it wrong, disaster waits in the wings. On May 12, 2008, a catastrophic earthquake struck China's Sichuan Province. Eight months later evidence accumulated that the 300-kilometer rupture of the Beichuan fault may have been triggered by forces (25 times as great as a year's natural tectonic stress) caused by the buildup of "several hundred million tons of water" behind the enormous, nearby Zipingpu Dam Apparently, the designers of the dam had seriously underestimated the danger that it would trigger such a powerful quake in this earthquake-prone region, and 80,000 people paid with their lives.[34]

Things can go wrong with complex systems in so many ways that it is sometimes difficult for even the best designers to think of them all. In June 1995, launch of the space shuttle Discovery had to be indefinitely postponed because a flock of "lovesick" male woodpeckers intent on courting pecked at least six dozen holes (some as big as four inches wide) in the insulation surrounding the external fuel tanks.[35] That would have been a hard problem to foresee. The odds are good that if one of the shuttle's designers had raised the possibility that passionate woodpeckers might someday attack the fuel tank insulation en masse, the rest of the design team would have laughed out loud.

Complexity, Interactions, and Design Error

Failure to understand or pay attention to the way components of a complex system interact can be disastrous. The Northern California earthquake of October 17, 1989, brought down a mile and a quarter stretch of the elevated Nimitz Freeway in Oakland, killing and injuring dozens of motorists. Reinforcing cables installed on columns supporting the two-tiered, elevated roadway as part of the state's earthquake proofing program may actually have made the damage worse. The quake sent rolling shock waves down the highway, and when some of the columns and cables supporting the roadbed collapsed, they pulled adjoining sections down one after another, like dominoes.[36]

On January 4, 1990, the number-three engine fell off a Northwest Airlines Boeing 727. Investigators did not find anything wrong with the design or construction of the engine, the way it was attached to the fuselage, or the fuselage itself. Instead, they reported that the loss of the engine was the result of a peculiar interaction. Water leaking from a rear lavatory was turned into ice by the cold outside air. The ice built up then broke loose, striking the engine and causing it to shear off.[37] Designers just had not anticipated that a water leak in a lavatory could threaten the integrity of the aircraft's engines.

Even the designers may find it impossible to enumerate, let alone to pay attention to, all of the ways things can go wrong in complex systems. After-the-fact investigations often determine that failure was due to a simple oversight or elementary error that causes us to shake our heads and wonder how the designers could have been so incompetent. But mistakes are much easier to find when a failure has focused our attention on one particular part of a complex design.

One of the worst structural disasters in the history of the United States occurred on the evening of July 17, 1981, when two crowded suspended walkways at the Hyatt Regency Hotel in Kansas City collapsed and fell onto the even more crowded floor of the lobby below. More than 100 people were killed and nearly 200 others were injured. When investigators finally pinpointed the problem, technical drawings focusing on a critical flaw were published. An ambiguity in a single design detail had led the builder to modify the way that the walkways were connected to the rods on which they were suspended from the ceiling. That change made them barely able to support their own weight, let alone crowds of people dancing on them. Yes, the problem should have been caught during design or construction, when it could have been easily solved. But the fatal flaw was much easier to see once the collapse forced investigators to carefully examine how the walkways were supported. At that point, the critical detail was no longer lost in the myriad other details of the building's innovative design.[38]

Unless a great deal of attention is paid to interactions, knitting together the best-designed components can still produce a seriously flawed system design. Before computers, thorough analysis of complex designs was so difficult and time consuming that good designers placed a high value on simplicity. When computers became available, designers gained confidence that they could now thoroughly evaluate the workings of even very complicated designs. This degree of confidence may have been unwarranted. After all, a computer cannot analyze a design, it can only analyze a numerical model of a design. Any significant errors made in translating an engineer's design into a computer model render the computer's analysis inaccurate; so do any flaws in the software used to analyze the numerical models.

The piping systems of nuclear power plants are so complex, it is hard to imagine designing them without computers. Yet one computer program used to analyze stresses in such piping systems was reportedly using an incorrect value for pi (the ratio of the circumference to the diameter of a circle). According to civil engineer Henry Petroski, in another " ... an incorrect sign was discovered in one of the instructions to the computer. Stresses that should have been added were subtracted ... leading it to report values that were lower than they would have been during an earthquake. Since the computer results had been employed to declare several nuclear power plants earthquake-proof, all those plants had to be rechecked. ... This took months to do."[39] Had a serious earthquake occurred in the interim, the real world might have more quickly uncovered the error—with disastrous results.

Writing the programs that model engineering designs, and writing the software the computer uses to analyze these models are themselves design processes that involve complex, interactive systems. The same cautions that apply to designing any other complex, interactive system apply here as well. A computer cannot hope to accurately appraise a design's performance in the real world unless it is given realistic specifications of system components and the way they interact. That is not an easy thing to do. Abstract numerical models are much easier to build if idealized conditions are assumed. For example, it is much easier to model the performance of an aircraft's wing in flight if it can be assumed that the leading edge is machined precisely, the materials from which the wing is made are flawless, the welds are perfect and uniform, and so on. Taking into account all of the complications that arise when these idealized assumptions are violated—and violated in irregular ways at that—makes building the model enormously more difficult, if not impossible.

The problem of spherical aberration that crippled the Hubble Space Telescope was due to an error in the curvature of its primary and/or secondary mirror of somewhere between 1/50 to 1/100 the width of a human hair.[40] Yet the error still might have been detected if the mirrors had ever been tested together. Each was extensively tested separately, but their combined performance was only evaluated by computer simulation.[41] The simulation could not have detected an error in curvature unless that error was built into the computer model. That could only have been done if the engineers knew the flaw was there. If they had known that, they would not have needed the computer to tell them about it.

No computer can provide the right answers if it is not asked the right questions. Computers are extremely fast, but in many ways very stupid. They have no "common sense." They have no "feel" for the design, no way of knowing whether anything important is being overlooked. They do what they have been told to do; they respond only to what they have been asked.

In January 1978, the roof of the Civic Center in Hartford, Connecticut collapsed under tons of snow and ice, only a few hours after thousands had attended a basketball game there. The roof was designed as a space frame, supported by a complicated arrangement of metal rods. After the collapse, it was discovered that the main cause of failure was insufficient bracing in the rods at the top of the truss structure. The bars bent under the unexpectedly heavy weight of snow and ice. The rods that bent the most finally folded, shifting the part of the roof's weight they were bearing to adjoining rods. The unusually heavy load now on those rods caused them to fold, setting up a progressive collapse of the support structure that brought down the roof. A computer simulation finally solved the problem of why and how the accident had happened, but only after investigators had directly asked the right question of a program capable of answering that question. The original designers had used a simpler computer model and had apparently not asked all the right questions. The designers had such confidence in their analysis that when workers pointed out that the new roof was sagging, they were assured that nothing was wrong.[42]

Computers are very seductive tools for designers. They take much of the tedium out of the calculations. They allow designers to reach more easily into unexplored territory. As with most things that are seductive, however, there are unseen dangers involved. Being able to explore new domains of complexity and sophistication in design means leaving behind the possibility of understanding a design well enough to "feel" when something is wrong or problematic. It is not clear whether or not this will increase the frequency of design error, but it is almost certain to increase the severity of errors that do occur. Undue confidence and opaque designs will make it difficult for designers to catch some catastrophic errors before the real world makes them obvious.

The Pressure for New Design

If we were content to use the same designs year after year in the same operating environment, design error would be much less common. "Tried and true" designs are true because they have been repeatedly tried. Using the same design over many years allows evolutionary correction of flaws that come to light, and more complete comprehension of how the design is likely to be affected by minor variations. As long as new products, structures, and systems are replicas of old and operating conditions remain the same, there is less and less scope for catastrophic error as time goes by. But that is not the world in which we live.

In our world, design change is driven by our creativity, our need for challenge, our confidence that better is possible, and our fascination with novelty. That means that the technical context within which all of the systems we design must operate is changing constantly. The social context may be changing as well, on its own or directly because of technological change. The pollution and resource depletion associated with automobile technology, for example, is affecting the design of other energy-using systems by raising the priority attached to reducing their own emissions and to increasing their fuel efficiency. The development and diffusion of technologies changes the context within which the design process subsequently takes place, which reinforces the need for new designs.

Reaching beyond existing designs brings with it a higher likelihood of error. In military systems, the pressure for new designs is intense because even a small performance advantage is believed critical in combat. It is therefore not surprising that the designers of weapons and related systems make more than their share of significant design errors. The Trident II missile was to be a submarine-launched ballistic missile (SLBM) with greater range and accuracy than Trident I. The first Trident II test-launched at sea exploded four seconds into its flight. The second test went all right, but the third test missile also exploded. The nozzles on the missile's first stage failed. They were damaged by water turbulence as the missile rode the bubble of compressed gas that propels it upward through 30 to 40 feet of water until it breaks the surface of the sea and its engines can fire. The Navy has been launching SLBMs this way for decades. Trident I had passed this test, but Trident II was much longer and nearly twice as heavy. Its designers expected that it would create more water turbulence, but miscalculated how much greater that turbulence would be.[43]

Proponents of relatively new complex technical systems are frequently overoptimistic in projecting their experience with early successes to second- and third-generation versions. Over optimism comes easily as the enthusiasm that surrounds an exciting new technology combines with a rapid rate of progress in its early stages of development. But later-generation systems are often more sophisticated and very different in scale. Being too ready to extrapolate well beyond previous experience is asking for trouble. This problem has been endemic in the trouble-plagued nuclear power industry. The first commercial plant was ordered in 1963, and only five years later orders were being taken for plants six times as large as the largest then in operation. There had only been 35 years of experience with reactors the size of Unit 2 at Three Mile Island when the partial meltdown occurred. That is very little experience for a technological system that big and complex.[44]

Thirty-five years is forever compared to the operating experience we have with many complex military systems. Design changes are so frequent and introduction of new technologies so common that extrapolation from previous experience is particularly tricky. New military systems are frequently not thoroughly tested under realistic operating conditions. It is not surprising that they often do not behave the way we expect them to at critical moments. Even when they are tried out under special test conditions, performance aberrations are sometimes overlooked in the pressure to get the new system "on-line." During flight tests beginning in 1996, one of the wings on the Navy's $70 million F-18 "Super Hornet" fighter would sometimes suddenly and unpredictably dip when the plane was doing normal combat maneuvers. Engineers and pilots struggled without success to figure out what was causing this unpredictable "wing drop," which could prove fatal in combat. Though the flaw had still not been fixed, the Pentagon authorized the purchase of the first 12 production-model F-18's in March 1997.[45]

Acts of God and Assumptions of People

Engineers build assumptions about the stresses, loads, temperatures, pressures, and so forth imposed by operating environments into the design process. Unfortunately, there is often no way of knowing all of its key characteristics precisely in advance. Consider the design of a highway bridge. It is possible to calculate gravitational forces on the bridge with great accuracy, and to know the load-bearing capabilities, tensile strength, and other characteristics of the materials used to construct the bridge (provided they are standard materials). Calculating wind stresses is less straightforward, though still not that difficult under "normal conditions." But it is much harder to calculate the strength and duration of the maximum wind stress the bridge will have to bear during the worst storms it will experience in its lifetime—or the greatest stress it will have to bear as the result of flooding or earthquake. We simply do not know enough about meteorological or geological phenomena to be able to accurately predict these occasional, idiosyncratic but critical operating conditions. For that matter, we do not always do that good a job projecting future traffic loads either.

The subtle effects of slow-acting phenomena like corrosion can also interact with the design in ways that are both difficult to foresee and potentially catastrophic. The Davis-Besse nuclear plant near Toledo was shut down in early 2002 when it was discovered that acid used in the reactor's cooling water had nearly eaten through the reactors six-inch steel lid.[46] The embrittlement of nuclear reactor vessels that results from constant exposure to radiation is also a slow-acting, subtle process that threatens structural strength.

Because making assumptions about operating conditions is risky but unavoidable, it is common practice to design for "worst case" conditions. That way, any errors made are less likely to cause the design to fail. But what is the "worst case?" Is it, for example, the worst earthquake that has ever been recorded in the area in which a bridge is being built? Or is it the worst earthquake that has ever been recorded in any geologically similar area? Or perhaps the worst earthquake that has ever been recorded anywhere?

As the assumptions escalate, so does the cost and difficulty of building the bridge. Does it make sense to pay the price for making all bridges able to withstand the same maximum earthshaking, knowing that few if any of them will ever experience such a severe test? The definition of "worst case" is therefore not purely technical. It almost inevitably involves a tradeoff between risk, performance, and cost. Even if the most extreme assumptions are made, there is still no guarantee that more severe conditions than had been thought credible (e.g., a worse earthquake than anyone had predicted) will not someday occur. During the great Kobe earthquake of 1995, at least 30,000 buildings were damaged or destroyed, 275,000 people were left homeless, and the death toll passed 4,000.[47] Japanese earthquake engineers, among the best in the world, were shocked by the extent of the damage. But ground motions in the quake were twice as large as had been expected. In the words of an American structural engineering expert, the Japanese structures "will perform well during an earthquake that behaves according to their design criteria. But . . . this] quake did not cooperate with the Japanese building codes."[48]

Engineers also try to insure proper performance by building safety factors into their designs. The safety factor is the demand on a component or system just great enough to make it fail, divided by the demand that the system is actually expected to face. If a beam able to bear a maximum load of 10,000 pounds is used to bear an actual load of 2000 pounds, the safety factor is five; if an air traffic control system able to handle no more than 60 flights an hour is used at an airport where it is expected to handle 40 flights an hour, the safety factor is 1.5. Safety factors are the allowed margin of error.

If we know how a component or system performs, and we have an established probability distribution for the demands it faces, we can calculate the factor of safety that will provide any given degree of confidence that the design will not fail. But where the component or system design is innovative and the demands it may face are unknown or subject to unpredictable variation, there is no science by which it is possible to calculate exactly what the safety factors should be. Just as in worst-case analysis, we are back to projecting, estimating, and guessing how much is enough. Unfortunately, that is the situation in constantly evolving, complex high-tech systems. When the design must face unpredictable environments *and* rivals actively trying to make it fail—as is the case in military systems—the problem of preventing failure is that much more difficult.

Peculiar confluences of circumstances can also defeat a design. Engineers examining the remains of the sections of the Nimitz Freeway destroyed in the northern California earthquake of 1989 found evidence of just such a possibility. The frequency with which the ground shook might have matched the resonant frequency of the highway. In other words, after the first jolt, the highway began to sway back and forth. By coincidence, the subsequent shocks from the earthquake may have been timed to give the highway additional shoves just as it reached the peak of its swing. Like pushing a child's swing down just after it reaches its highest point, the reinforcing motion caused the highway to sway more and more until it collapsed.[49] It would be hard for engineers to foresee this odd coincidence when the highway was being designed.

Big earthquakes are much less common in the Eastern United States than in the West, so those who design most structures built in the East are not required to and typically do not include severe earthquakes in their worst-case scenarios. Yet the most powerful earthquake ever known to hit the United States occurred in the Eastern half of the country, at New Madrid, Missouri near the Tennessee/Kentucky border in 1811. It was later estimated to have had a magnitude of about 8.7 on the Richter scale, and was one of three quakes in that area in 1811 and 1812, all stronger than magnitude 8.0.[50] The Richter scale is logarithmic (base 10). Thus, these nineteenth-century eastern quakes were more than 10 times as powerful as the magnitude 6.9 earthquake that brought down part of the Nimitz freeway and did so much damage to San Francisco in the Fall of 1989. (The New Madrid quake was nearly 100 times as strong.) They temporarily forced the Mississippi River to run backward, permanently altered its course, and were felt as far away as Washington, Boston, and Quebec.[51]

All along the Eastern seaboard there are geological structures similar to those responsible for a 7.0 earthquake that devastated Charleston, South Carolina in the late nineteenth century. Milder quakes of magnitude 5.0 are not all that rare in the East. Such a quake hit New York City in 1884.[52] A study of New York City's vulnerability to earthquake found that a magnitude 6.0 temblor centered within five miles of City Hall would do about $25 billion worth of damage. Estimates of damage in northern California from the 6.9 quake in 1989 ran from $4 billion to $10 billion.[53] So a New York City-area earthquake about one-tenth as powerful as the California quake could cause up to six times as much property damage.

The reason is partly geological: the earth's crust is older, colder, and more brittle in the Eastern United States. But it is also because designers of structures in New York do not typically include severe ground shaking in their assumptions. By contrast, designers of structures in California have been compelled to take earthquakes into account. The modern skyscrapers in San Francisco swayed during the 1989 quake, but sustained little or no damage. Virtually every structure that suffered major damage there had been built before stringent earthquake-resistance requirements were incorporated in building codes over the last 15 to 20 years.[54] The assumptions about operating environment made by designers really do make a striking difference in how products perform.

Five large nuclear reactors used to produce plutonium and tritium for American nuclear weapons were designed and built without strong steel or reinforced concrete "containments."[55] The last line of defense against the accidental release of dangerous radioactive materials, containments are built as a matter of course around civilian nuclear power reactors in the United States (and most other developed nations). The large release of radioactivity at Chernobyl where there was no containment was prevented during the Three Mile Island accident by the containment surrounding that reactor. Why were America's nuclear weapons production reactors designed and operated without this key safety feature? Ignorance may be a partial explanation. All of these reactors are "old" in terms of the nuclear business, dating back to the 1950s. The dangers of radiation were much less well understood then, and exposure to radioactive materials was often treated much too cavalierly. It is likely that the pressure to get bomb production moving, to keep ahead of the "godless Communists," also played an important role. Technical matters are not the only considerations that enter into the designer's definition of worst case.

There are many sources of error inherent in the process of designing complex technical systems. With great care, the use of fault-tolerant design strategies, and thorough testing, it is possible to keep these flaws to a minimum. But even the most talented and careful designers, backed up by the most extensive testing programs, cannot completely eliminate design errors serious enough to cause catastrophic technical failures. This is especially true where complicated, innovative, or rapidly changing technologies are involved. Many dangerous technologies are of just this kind.

MANUFACTURING AND TECHNICAL FAILURE

Even the best designs for technical systems remain only interesting ideas until they are made real by manufacturing. The process of fabricating major system components and assembling the systems themselves creates ample opportunity for error. Flawed manufacturing can translate the most perfect designs into faulty products. The more complex the system, the more sensitive and responsive it must be, the more critical its function, the easier it is to introduce potential sources of failure during manufacture. Subcomponents and subassemblies must be checked and rechecked at every step of the way. Still serious manufacturing errors persist in a wide variety of arenas.

The nuclear power industry provides innumerable examples of how easy it is for slips to occur during manufacturing. On October 5, 1966, there was a potentially devastating meltdown at the Enrico Fermi Atomic Power Plant, only 30 miles from Detroit. In August 1967, investigators discovered a piece of crushed metal at the bottom of the reactor vessel that they believed had blocked the coolant nozzles and played a key role in the accident. In 1968, it was finally determined that the piece of metal was one of five triangular pieces of zirconium installed as an afterthought by the designer for safety reasons. They did not even appear in the blueprints. This particular shield had not been properly attached.[56] Only a few weeks after the accident at Three Mile Island in March 1979, the Nuclear Regulatory Commission reported that they "had identified thirty-five nuclear power plants with 'significant differences' between the way they were designed and the way they were built."[57]

The multi-billion dollar nuclear power plant at Diablo Canyon in California sits close to an active fault. During construction in 1977, the utility hired a seismic engineering firm to calculate the stresses different parts of the plant would have to withstand in an earthquake. A little more than a week before the plant was due to open, a young engineer working for the utility that owned the plant discovered a shocking error. The utility had sent the diagrams of the wrong reactor to its seismic consultants! Guided by the faulty stress calculations that resulted, the utility reinforced parts of the plant that did not need reinforcing, while vulnerable parts were reinforced too little, if at all. More than 100 other flaws in the reactor's construction were subsequently discovered.[58]

It was not until three and a half months after the Hubble space telescope was launched in April 1990 that NASA finally determined the cause of the perplexing spherical aberration (curvature error) problem that had made it incapable of performing the full range of tasks for which it had been designed. When the Hubble mirror had arrived for final polishing at Perkin-Elmer's Danbury, Connecticut plant back in 1979, it was tested with a newly developed super accurate tester to assure that the mirror's optical properties met NASA's exacting standards. The tester showed the mirror had a small degree (one-half wavelength) of spherical aberration, well within acceptable limits for that stage of manufacture. The Perkin-Elmer team then began the final polishing process that continued until 1981, polishing out the deviation their new tester had found. The only problem was, there was an undiscovered one-millimeter error in the structure of the tester. By using it to monitor the polishing process, Perkin-Elmer had distorted, rather than perfected the mirror's surface during final polishing, creating the spherical aberration that was later to produce such headaches in the orbiting telescope.

Interestingly, the mirror had been checked with a testing device of more standard design before it was shipped to Danbury. That device had *not* shown the degree of spherical aberration that the newly developed tester had (incorrectly) detected. The company's scientists had the results of the earlier test, but were sure that the more sophisticated tester was correct. They did not bother to conduct further tests or investigate the discrepancy.[59] *Science* reported that " . . . astronomers experienced in making ground-based telescopes say they are appalled that NASA and Perkin-Elmer would rely on one single test . . . there are any number of simple and inexpensive experiments that could have seen the spherical aberration that now exists in Hubble. . . . "[60] In the excitement of meeting the kinds of technological challenges involved in designing and building complex, state-of-the-art systems, mundane matters, such as "simple and inexpensive" checks during the manufacturing process, are easy to overlook. When we have our eyes on the stars, it is all too easy to trip over our own feet.

In March 2004, a NASA official announced that some of the nations' space shuttles had been flying for 25 years with gears improperly installed in their tail rudders, a problem that under high stress could have caused the destruction of the shuttle and the loss of its entire crew.[61] In late 2004, the *Associated Press* reported "Over the last two years, federal safety officials have received 83 reports of cell phones exploding or catching fire, usually because of incompatible, faulty, or counterfeit batteries or chargers."[62] At the beginning of June 2006, the Army Corps of Engineers released a study accepting blame for the disastrous failure nearly a year earlier of the system of levees it had built to protect New Orleans against flooding. The study reportedly concluded "that the levees it built in the city were an incomplete patch-work of protection, containing flaws in design and construction and not built to handle a storm anywhere near the strength of Hurricane Katrina."[63] A report released in July 2006 concluded that improperly installed seals in the Sago Mine in West Virginia had been responsible for 12 miners failing to survive an explosion in the mine the previous January, turning it into the worst mining disaster in the United States in 40 years.[64]

COMPONENT FLAWS AND MATERIALS FAILURE

It is impossible to make high-quality, reliable products out of poor-quality, unreliable parts or seriously flawed materials. Semiconductor chips have become almost a raw material to the electronics industry. Their performance and reliability is the bedrock on which the performance and reliability of modern electronic equipment is built. In the latter part of 1984, officials at Texas Instruments (the largest chip manufacturer in the United States at the time) disclosed that millions of integrated circuits it had manufactured might not have been tested according to specification. Eighty different contractors had built the chips into more than 270 major weapons systems. The attendant publicity—and a Pentagon investigation—led officials at Signetics Corporation (the nation's sixth-largest microchip manufacturer) to audit their own chip-testing procedures. They concluded that as many as 800 different types of microcircuits they had supplied to military contractors might not have been tested properly. At least 60 different types of microchips sold to the Pentagon by IBM were also determined to have "confirmed problems."[65] In the spring of 1985, the Pentagon's inquiry found that irregularities in testing military-bound microchips were pervasive in the electronics industry. The director of the industrial productivity office at the Defense Department put it this way, "What we found was that it was common practice for the microcircuit makers to say, 'Yeah, we'll do the tests,' and then for them never to conduct them."[66]

The fact that microchips are not properly tested does not mean that they are faulty. What it does mean is that the reliability of these components, and therefore of every product that contains them, cannot be assured. If these systems fail at critical moments, the consequences can be disastrous. On June 3, 1980, the failure of a single, 46 cent microchip generated a major false warning that the United States was under land and sea-based nuclear attack by Soviet missiles. Three days later, the same faulty chip did it again.[67]

Early in the summer of 1994, Intel Corporation, by then the standard-setting computer chip manufacturer, discovered a flaw in its much touted and widely used Pentium chip.[68] The chip could cause computers in which it was embedded to give the wrong result in certain division problems that used the chip's floating-point processor.[69] Intel waited until November to publicly disclose the problem, provoking an avalanche of angry messages on the Internet from engineers and scientists disturbed with what they considered to be the company's cavalier attitude. A computer that gives the wrong results without any indication that something is wrong is no small thing. It is not just annoying or misleading, it is potentially dangerous. It does not take a great deal of imagination to see how a computer that gives the wrong results because of a flawed processor could cause a lot of damage if it were used to design, analyze, control, or operate a technologically dangerous system.

Metallurgical defects are also a major potential source of systems failure. They seem to have played a key role in the July 19, 1989 crash of United Airlines Flight 232 near Sioux City, Iowa. Apparently, a flaw in a 270-pound cast titanium alloy disk in the rear engine grew into a crack that broke the disk apart, shattering the plane's tail section. Metallurgists working for the National Transportation Safety Board found a tiny cavity in the metal, which grew into the fatal crack. The cavity was large enough to be "readily visible with the unaided eye," raising questions as to why it was not detected either at the factory or during routine maintenance.[70] It was also discovered that the manufacturer had mistakenly given two disks made at the same time the same serial number. One had failed to pass inspection, and one was destroyed. Investigators thought that the good disk might have been destroyed, and the faulty one installed in the DC-10. The company managed to convince them that that was not true.[71]

Flaws in even the simplest of components can cause very complex technical systems to fail. On April 28, 1989, ABC News reported " ... every year in this country, companies buy some $200 billion of nuts and bolts ... and put them in everything from jet planes to children's amusement rides. ... We now know that billions of bad bolts have come into this country." Peterbilt Trucks issued a recall because of a number of its customers who reported that the steering mechanism on their vehicles would suddenly stop working, causing a virtually total loss of control. Defective bolts were found. They were brittle because they had not been properly heat treated during manufacture. ABC reported, "Twice in the past year and a half, bolts holding jet engines on commercial airliners broke in flight, and the engines then fell off the airplanes. ... "[72]

Of course, bolt manufacturers also supply the U.S. military. Army documents that ABC obtained showed that defective or broken nuts and bolts were involved in 11 aircraft (mostly helicopter) accidents over the previous decade in which 16 people had died. Over 1,000 M-60 tanks were temporarily taken out of service because of defective bolts.[73] ABC News also reported that NASA had spent millions of dollars removing suspect bolts from key systems, like the space shuttle engines. There were reports that bad bolts had been supplied for the Air Force's MX missile.[74]

THE CRITICALITY OF MAINTENANCE

There is no more mundane issue in high-technology systems than maintenance. Yet without proper maintenance, the best designed and most carefully built system can slowly turn into a useless piece of high-tech junk. The "if it ain't broke, don't fix it" attitude is a prescription for endless, expensive trouble. Complex, sensitive systems often require extensive and painstaking preventive maintenance, not just after-the-fact repair.

In April 1988, the mechanics responsible for maintaining Aloha Airlines' fleet were routinely inspecting one of the airline's Boeing 737s. They failed to note that a section of the upper fuselage was starting to come loose and that the overlapping metal skins in that section were beginning to develop fatigue cracks around the rivets. On April 28, in flight, the cracks suddenly began to grow, connecting to each other and literally ripping the top off the body of the plane. A flight attendant was killed, but the pilot was able to bring a plane full of very frightened passengers down for an otherwise safe landing.[75]

In May 1998, the FAA issued an emergency order grounding dozens of older Boeing 737s because of a possible maintenance problem involving their fuel pump wiring.[76] The wiring on some 35 of the planes inspected in the first few days after the FAA found the problem showed some wear, and in nine or more aircraft the insulation was worn at least half way through.[77] Worn insulation can lead to sparks, and sparks and jet fuel are a deadly explosive combination. A decade later in April 2008, American Airlines was forced to ground its fleet of 300 MD-80 aircraft and cancel over 3,000 flights because of a mundane wiring maintenance issue: the wrapping and attachment of wires inside the aircrafts' wheel wells. The *New York Times* observed, "An aging fleet is catching up with domestic airlines, and the maintenance issues that inevitably arise are likely to worsen as the industry's jets grow older and its finances weaken."[78]

On August 9, 2004, four workers died when a section of steam pipe blew out at a nuclear power plant in Mihama, Japan. The pipe, which "had not been inspected in 28 years . . . had corroded from nearly half an inch to a thickness little greater than metal foil."[79] On August 1, 2007, the Interstate 35W bridge collapsed in downtown Minneapolis, killing 13 and injuring 145 more. Regular inspections in 17 consecutive years prior to the collapse had said that the bridge was in poor condition, yet the needed maintenance was never properly done.[80] On December 22, 2008, the holding pond at a coal-fired electric plant run by the Tennessee Valley Authority (TVA) failed, flooding 300 acres with 5.4 million cubic yards of coal ash laden with toxic chemicals and contaminating a nearby river. Decades worth of waste from the plant, which generated 1.1 thousand tons of toxics a year, had been stored in the containment. Tom Kilgore, the TVA's chief executive, said the holding ponds had leaked noticeably in 2003 and 2005 but had not been adequately repaired. Kilgore reportedly indicated that TVA "chose inexpensive patches rather than a more extensive repair of the holding ponds. . . ."[81]

Nuclear power plants are designed to "fail-safe." Any major problem, including a loss of power to the controls, is supposed to trigger an automatic shutdown or "scram" of the reactor. Industry analysts have argued that the odds of a failure of this system are no more than one in a million reactor years. Yet on February 22, 1983, the Salem-1 reactor failed in just this way, refusing to halt the fission reaction in its core when ordered to scram by a safety control system. Three days later, the "one-in-a-million" event happened again—at the same reactor.[82]

A key problem lay in a huge pair of circuit breakers, known as DB-50s (manufactured by Westinghouse) in the circuit supplying power to the mechanism that raises and lowers the core control rods. When the power is flowing and the breakers are closed, the rods can be held up out of the core to speed the fission reaction. When the automatic system orders the reactor to "scram," the DB-50s break the circuit, and gravity pulls the rods into the core, shutting down the reaction. Investigators of the incidents at Salem-1 found that a UV coil inside the DB-50s had failed. As early as 1971, Westinghouse had issued technical bulletins warning of problems with the UV coil, and in 1974 had sent out letters emphasizing the importance of properly cleaning and lubricating the coils twice a year. The utility did not heed the warnings. "Maintenance of the breakers at Salem was poor. They never got the critical attention they deserved . . . there was *no* maintenance of the UV coils between their installation in the 1970s and August 1982, when they began to fail repeatedly."[83]

At the Maine Yankee nuclear power plant high-pressure, radioactive water is pumped through metal tubes 3/4 inch in diameter and 1/20 inch thick after being heated by the reactor core. Heat conducted through the walls of the tubes turns "clean" water into "clean" steam that drives the turbine, generating electricity. Proper inspection and maintenance of these metal tubes is critical since any cracking or rupturing could allow the radioactive water from the reactor to leak into the otherwise "clean" steam and possibly escape into the environment. In early 1995, it was disclosed that about 60 percent of the plant's 17,000 tubes had severe cracks. Furthermore, the reactor had been operating in that dangerous condition for years.[84]

Maintenance cannot be an afterthought in the kind of sophisticated, complex technological systems that a modern military expects to work well even in difficult operating conditions. Given the speed of modern warfare, equipment that fails often because it is poorly maintained or because its design is so inherently complex that it cannot be properly maintained is worse than useless. It can lead to military tactics and strategies that amount to fantasies because they are built around equipment that will not perform as advertised—if it works at all—in the real world of combat.

Maintenance, lackluster and pedestrian as it may seem, requires the closest attention. From aircraft to spacecraft to highway bridges to toxic waste storage to nuclear power plants, there is persuasive evidence that improper maintenance can lead to dangerous failures of technical systems. Far from being a mere footnote in the age of high technology, it is critically important to the performance of even the most sophisticated technical systems.

The nature of technical systems themselves and the nature of their interactions with the fallible humans who design, build, and maintain them guarantee that it is not possible to eliminate all causes of failure—even potentially catastrophic failure—of complex and critical technical systems. Technology simply cannot be made foolproof. There is nothing about dangerous technological systems that makes them an exception to this rule.

Yet there are those who believe that it is possible to prevent failure of the most complex dangerous technologies by automating humans out of the system and putting computers in control. Computers are surrounded with an almost magical aura of perfection, or at least perfectibility, in the minds of many people. We sometimes think—or hope—that they can help us overcome the imperfection that is so much a part of our human nature. After all, they can bring commercial aircraft safely to the ground with remarkable efficiency and nuclear warhead to their targets with remarkable accuracy. In the next chapter, we take a closer look at these marvels of modern technology and try to understand why they are anything but a route to solving the problems of either technical failure or human error.

NOTES

1. Suppose Machine A has three parts that all must work for the machine to do its job. Assume the probability of any part's failure is independent of any other part's failure. If each part is 95 percent reliable (i.e., has only a 5 percent chance of failure), then Machine A will be 86 percent reliable (95 percent ×

95 percent × 95 percent) and it will have a 14 percent chance of failure. With five key parts, each still 95 percent reliable, Machine A would now be only 77 percent reliable (95 percent × 95 percent × 95 percent × 95 percent × 95), and its probability of failure would *increase* from 14 percent to 23 percent.

2. Suppose Machine B is much more complex than A, having 20 key parts. Assume each part must still work for the machine to work. Even if each part is only one-fifth as likely to fail as before (i.e., 99 percent reliable), the machine as a whole will still be less reliable. Machine B has an 18 percent chance of failure (82 percent reliability), as compared to the 14 percent chance of failure (86 percent reliability) of the original Machine A. Even though each part of the original machine was *less* reliable, its simpler design made the whole machine *more* reliable.

3. Suppose Machine C has three parts, each 95 percent reliable. Unlike A, if one part fails, the machine's performance degrades, but it keeps working; if two or three parts fail, the machine stops working. If the probability of failure of any part does not influence the probability of failure of any other part, there will be a 13.5 percent chance that C will have a partial failure, and less than a 1 percent chance that it will fail completely.Suppose Machine D is like C, except that the failure of one part results in a much higher strain on the remaining parts, reducing their reliability. As soon as any part fails, the reliability of the remaining parts drops from 95 percent to 75 percent because of the increased load. There will still be a 13.5 percent chance of partial failure, but the chance of complete failure will soar from less than 1 percent to 6 percent.

4. Franklin C. Spinney, *Defense Facts of Life* (Washington, DC: December 5, 1980), unpublished Department of Defense staff paper, unreviewed by the Department, and therefore not an official document.

5. According to a noted defense analyst, these include being overweight and underpowered, much less maneuverable than it should be, "too fast to see the tactical targets it is shooting at . . . [and] too delicate and flammable to withstand ground fire. . . . " Winslow Wheeler and Pierre Sprey, "Joint Strike Fighter," *Jane's Defense Weekly,* September 10. 2008.

6. Calvin Sims, "Disruption of Phone Service Is Laid to Computer Program," *New York Times,* January 17, 1990.

7. Ibid.

8. John Markoff, "Superhuman Failure: AT&T's Trouble Shows Computers Defy Understanding Even in Collapse," *New York Times,* January 17, 1990.

9. For an interesting if somewhat overlong discussion of these so-called "revenge effects" of technology, see Edward Tenner, *Why Things Bite Back* (New York: Vintage Books, 1997).

10. Ibid.

11. Ibid.

12. Craig S. Smith, "Weakened Concrete Is Cited in Collapse at Paris Airport," *New York Times,* July 7, 2004.

13. John Schwartz and Matthew Wald, "Earlier Shuttle Flight Had Gas Enter Wing on Return," *New York Times*, July 9, 2003.

14. Matthew Wald, "In Big Blackout, Hindsight Is Not 20/20," *New York Times*, May 13, 2004.

15. Waldrop, M. Mitchell, "Hubble Managers Start to Survey the Damage," *Science* (July 6, 1990), and "Astronomers Survey Hubble Damage," *Science* (July 13, 1990).

16. Cowen, Ron, "The Big Fix: NASA Attempts to Repair the Hubble Space Telescope," *Science News* (November 6, 1993), pp. 296–298.

17. Ron Cowen, "Hubble Finally Gets a Heavenly View," *Science News*, January 22, 1994, 52.

18. Ron Cowen, "Trying to Avoid Hubble Trouble," *Science News*, March 27, 1999, 203.

19. Henry Petroski, *To Engineer Is Human: The Role of Failure in Successful Design* (New York: St. Martin's Press, 1985), 176–179.

20. B. Randell, P. A. Lee, and P. C. Treleaven, "Reliability Issues in Computing System Design," Association for Computing Machinery, *Computing Surveys*, Vol. 10, No. 2 (June 1978), 135.

21. For example, suppose Machine E has two key parts, each of which is 95 percent reliable. Assume that the probability of failure of either part does not influence the probability of failure of the other. If both parts must work for the machine to work, Machine E will then be 90 percent reliable (95 percent × 95 percent)—much less reliable than either of its parts. But, if it were designed with a third 95 percent reliable part that could serve as a backup system for either of the other two, Machine E would be 99 percent reliable—much more reliable than any of its parts.

22. James T. Wooten, "Sound of Debate Is Off Air for 27 Minutes," *New York Times*, September 24, 1976.

23. Stuart Diamond, "The Bhopal Disaster: How It Happened," *New York Times*, January 28,1985.

24. For an interesting analysis of some (mainly human) reasons why redundancy is sometimes counterproductive, see Scott Sagan, "The Problem of Redundancy Problem: Why More Nuclear Security Forces May Produce Less Nuclear Security," *Risk Analysis*, Vol. 24, No. 4 (2004).

25. Charles Perrow, *Normal Accidents: Living with High Risk Technologies* (New York: Basic Books, 1984), 260.

26. Daniel Ford, "Three Mile Island: II—The Paper Trail," *The New Yorker*, April 13, 1981, 52–53.

27. *New York Times*, September 26, 1981.

28. See Robert Hanley, "Newark Airport Is Closed as Crew Cuts Power Lines," and "Blackout at Airport Shows Need for a Backup System," *New York Times*, January 10 and 11, 1995, respectively.

29. M. Mitchell Waldrop, "Computers Amplify Black Monday," *Science*, October 30, 1987, 604.

30. Tim Weiner, "B-2, After 14 Years, Is Still Failing Basic Tests," *New York Times*, July 15, 1995.

31. Philip Shenon, "B-2 Gets a Bath To Prove That It 'Does Not Melt,' " *New York Times*, September 13, 1997.

32. Christopher Clark, "B-2 Bomber Makes Combat Debut," WashingtonPost.com, March 25, 1999: 2:39 AM.

33. John N. Wilford, "NASA Considered Shuttle Boosters Immune to Failure," *New York Times*, February 3, 1986.

34. Richard A. Kerr and Richard Stone, "A Human Trigger for the Great Quake of Sichuan?," *Science*, January 16, 2009.

35. John N. Wilford, "Space Shuttle Is Grounded By Lovesick Woodpeckers," *New York Times*, June 3, 1995.

36. Katherine Bishop, "Experts Ask If California Sowed Seeds of Road Collapse in Quake," *New York Times*, October 21, 1990.

37. Eric Weiner, "Jet Lands After an Engine Drops Off," *New York Times*, January 5, 1990.

38. Op. cit., Petroski, 85–90.

39. Op.cit., Petroski, 197.

40. Op. cit., Waldrop, 1990.

41. J. Eberhart, "Solving Hubble's Double Trouble," *Science News*, July 4, 1990.

42. Op. cit., Petroski, 198–199.

43. Andrew Rosenthal, "Design Flaw Seen as Failure Cause in Trident II Tests," *New York Times*, August 17, 1989.

44. Op. cit., Perrow, 34.

45. Dale Eisman, "Super Hornet Critics Charge that Navy Hid Flaws to Get Funding," *The Virginian-Pilot*, February 2, 1998.

46. Matthew Wald, "Nuclear Plant, Closed After Corrosion, Will Reopen," *New York Times*, March 9, 2004.

47. Andrew Pollack, "Japan Quake Toll Moves Past 4,000, Highest Since 1923," *New York Times*, January 20, 1995.

48. Sandra Blakeslee, "Brute Strength Couldn't Save Kobe Structures," *New York Times*, January 25, 1995.

49. Malcolm W. Browne, "Doomed Highway May Have Pulsed to the Rhythm of the Deadly Quake," *New York Times*, October 23, 1989.

50. William K. Stevens, "Eastern Quakes: Real Risks, Few Precautions," *New York Times*, October 24, 1989.

51. Lee Hancock, "Chances Are 50-50 for Major Quake by Decade's End," *Dallas Morning News*, July 22, 1990.

52. Ibid.

53. Op. cit., William K. Stevens.

54. Paul Goldberger, "Why the Skyscrapers Just Swayed," *New York Times*, October 19, 1989.

55. David Albright, Christopher Paine, and Frank von Hippel, "The Danger of Military Reactors," *Bulletin of the Atomic Scientists*, October 1986, 44.

56. John G. Fuller, *We Almost Lost Detroit*, New York: Reader's Digest Books, 1975, 102 and 215–220.

57. Op. cit., Perrow, 36.

58. Op. cit., Perrow, 37.

59. M. Mitchell Waldrop, "Hubble: the Case of the Single-Point Failure," *Science*, August 17, 1990, 735.

60. Ibid., 736.

61. Warren E. Leary, "Shuttle Flew for Decades with Potentially Fatal Flaw," *New York Times*, March 23, 2004.

62. Associated Press, "Exploding Cell Phones a Growing Problem," *Earthlink U.S. News*, November 27, 2004; http://start.earthlink.net/newsarticle?cat=0&aid=D86I7RRG0_story.

63. John Schwartz, "Army Builders Accept Blame Over Flooding," *New York Times*, July 2, 2006.

64. Ian Urbina, "State Report Faults Seals in Mine Disaster," *New York Times*, July 20, 2006.

65. R. Jeffrey Smith, "Pentagon Hit by New Microchip Troubles," *Science*, November 23, 1984.

66. David E. Sanger, "Chip Testing Problems Abound, Pentagon Says," *New York Times*, April 16, 1985.

67. Senator Gary Hart and Senator Barry Goldwater, "Recent False Alerts from the Nation's Missile Attack Warning System," Report to the Committee on Armed Services, United States Senate, October 9, 1980.

68. "Pentium Flaw Has Some Scientists Steamed," *Wall Street Journal*, November 25, 1994.

69. For example, if X is divided by Y, then the result is multiplied by Y, the answer should again be X. If this answer is then subtracted from X, the result should be zero. The particular values of X and Y used do not matter. But performing this sequence of operations with the flawed Intel Pentium, when X = 4195835.0 and Y = 3145727.0 gives an answer of 256.

70. John Cushman, Jr., "Findings Suggest Jet's Flaw Was Visible Before Crash," *New York Times*, November 1, 1989.

71. John Cushman, Jr., "Metallurgical Mystery in Crash that Killed 112," *New York Times*, November 3, 1989.

72. ABC Television News, "20/20," Show #917, segment entitled, "Built to Break" (April 28, 1989).

73. Ibid.

74. Ibid.

75. David Nather, "Safety Issues Plague Aging Airline Fleets," *Dallas Morning News*, July 22, 1990.

76. Matthew L. Wald, "Agency Grounds Scores of 737s to Check Wiring," *New York Times*, May 11, 1998.

77. Matthew L. Wald, "Checks of 737s Show More Damaged Wiring," *New York Times*, May 12, 1998.

78. Jeff Bailey, "Aging Jet Fleets an Added Strain on U.S. Airlines," *New York Times*, April 12, 2008.

79. James Brooke, "Blown Pipe in Japan Nuclear Plant Accident Had Been Used, But Not Checked," *New York Times*, August 11, 2004.

80. Monica Davey, "Minnesota Agency Faulted on Bridge Upkeep," *New York Times*, May 22, 2008.

81. Shaila Dewan, "Tennessee Ash Flood Larger Than Initial Estimate," *New York Times*, December 27, 2008, and "At Plant in Coal Ash Spill, Toxic Deposits by the Ton," December 30, 2008; and John M. Broder, "Plant That Spilled Coal Ash Had Earlier Leak Problems," *New York Times*, January 9, 2009.

82. Eliot Marshall, "The Salem Case: A Failure of Nuclear Logic," *Science*, April 15, 1983, 280–281.

83. Ibid., 281.

84. Robert Pollard, "Showing Their Age," *Nucleus*, Union of Concerned Scientists: Summer 1995, 7. Pollard is a senior nuclear safety engineer.

10

Computers Instead of People

Are computers a possible solution to the human reliability problem? They do not drink or take drugs, they do not have family problems, and they do not become bored when repeating the same task over and over again. The truth is at certain times, for certain tasks, computers *are* much more reliable than people. Conversely, however, they can also be much less reliable. They lack common sense, good judgment, and morality. Computers can magnify human error. When the captain of an American Airlines flight to Cali, Colombia typed the wrong code into the plane's navigational computer in December 1995, the computer took the plane toward Bogota instead and flew it straight into a mountainside. It had turned a programming error into a death sentence for 159 people.[1]

Many dangerous technology systems depend heavily on computers. Computers control nuclear plants and complex toxic chemicals facilities. They also control more and more of the equipment used to manufacture weapons to increasingly exacting specifications, from sophisticated metalworking machinery to high-tech inspection and quality-control devices. They are an integral part of communications, navigation, and command and control systems critical to carrying out all aspects of military operations, from launching nuclear attack to directing forces on the battlefield to supplying them with food, fuel, and ammunition.

In some cases, computers enhance the performance of weapons systems; in others, the weapons could not perform without them. Many modern missile systems could not be controlled, launched, or accurately guided to their targets without computers. Some modern military aircraft designs are so aerodynamically unstable that they need constant, high-speed adjustment of their controls just to keep them flying—a task which would overwhelm an unassisted pilot, but is just what computers do best. For example, although it was touted as ultra-modern, the Stealth (B2) bomber was based on a design that its manu-facturer Northrop first tried in the early 1940s. Then called the "flying wing," the design was abandoned less than a decade later, when the latest jet-powered version of the plane (YB-49) proved to be extremely unstable and difficult to fly. The design was not resurrected until the 1980s when computer control technology had advanced enough to give some confidence the plane could be flown properly.

Computers follow orders without a will of their own, without raising moral questions and without exercising independent judgment. Computer programs are nothing more than sets of orders that computers follow robotically. If programs are flawed or have been covertly altered to order undesirable things, they will also follow those instructions robotically, just as if they were following proper and desirable orders.

COMPUTER HARDWARE RELIABILITY

Because computers are so often used to control other systems, their reliability takes on special significance. When control systems fail, they take the systems they are controlling with them. Failures of control can have consequences that are both serious and difficult to predict. Unlike most machines where a worn gear or wobbling wheel might only cause performance to suffer, when something goes wrong with a computer, it usually does not just become a little sluggish, it tends to freeze, crash, or even worse—do entirely the wrong thing.[2]

Hardware failures can be caused by flawed designs or by faulty components. Design flaws include errors in both the design of the computer hardware itself and the process used to manufacture it. When a manufacturing error results in a faulty component that causes trouble, the problem can be fixed by replacing it with a new "unbro-ken" copy of that part. However, that will not help when a bad design causes trouble because the same design is used in all the copies. Only redesign will fix the problem. While faulty components still can and do create havoc, hardware and software design flaws are much more often the culprits. "The computer is down" rarely means that anything physical has snapped or worn out, or even that the software is some-how "broken." The computer is probably doing exactly what the pro-grammer told it to do. Most likely, the problem is that the programmer either gave it incomplete instructions or told it to do the wrong thing.[3]

When the largest computerized airline reservation system in the United States went "down" on May 12, 1989, operations at ticket counters and 14,000 travel agencies nationwide were disrupted for 12 hours. The hardware was working just fine; the problem was a software failure. The computers had not lost any information; a software problem simply prevented them from accessing it.[4] In the summer of 2004, a computer problem grounded 150 flights on American Airlines and 100 more on U.S. Airways. An official of the computer company involved said that the likely cause was "unintentional user error."[5]

While all this may be true, it should not be assumed that mundane problems like defects in materials or faulty manufacturing will not cause devices as sophisticated and "high tech" as computers to fail or become unreliable. The complexity and sensitivity of computers makes them more vulnerable to these problems than "lower tech" equipment. As with other sensitive electronics, a bit of dirt in the wrong place, excess humidity, microscopic defects in materials, or flaws in manufacturing microminiaturized semiconductor chips can build faults into computer components or cause them to fail unexpectedly in operation. An accumulation of lint, about as mundane a problem as they get, caused both the primary and backup computer display units in the telescope pointing system aboard the space shuttle Columbia to overheat and fail in December 1990.[6] Faulty power switches were at the bottom of the failure of an electronic voting system that seriously delayed the opening of 40 percent of the polling stations in California's San Diego County in March 2004.[7]

Besides defects in the physical process of manufacturing, faulty programming may be hard-coded into computer chips during manufacture, making the computer that embeds them unreliable. In 1994, leading chip manufacturer Intel Corp. discovered a flaw in its state-of-the-art Pentium chip that created a furor among engineers and scientists. It caused errors in long division, when certain strings of digits appeared. Intel had shipped perhaps two million faulty chips. The storm the Pentium flaw caused was more a result of the company's taking months to make the problem public. The failure to promptly notify all users could have had serious consequences if the incorrect calculations were used as the basis for designs or control systems that involved critical technologies.[8] Three years later in November 1997, yet another serious design flaw was found in the Pentium processor. When a particular command was executed, the defect caused the processor to come to a complete halt.[9]

Hardware failures of some consequence can also be caused by apparently minor accidents—the proverbial cup of coffee spilled in the wrong place at the wrong time. On December 9, 1988, "An adventurous squirrel touched off a power failure ... that shut down the National Association of Security Dealers automatic quotation service [NASDAQ] for 82 minutes. ... " When power was restored, "a power surge ... disabled NASDAQ's mainframe computers and seriously damaged the electrical system at the [utility] ... making it impossible to use backup generators."[10] The problems prevented the trading of 20 million shares of stock and halted options trading at several major exchanges. This was not the first time NASDAQ's computers had serious problems. The whole system had been shut down for four hours by a hardware failure in October 1986. For that matter, it was not the first time the utility involved had to deal with rambunctious squirrels. Squirrels had been chewing on their equipment and knocking out service two or three times a day.[11]

Given the frequency of hardware/software failures and the critical nature of so many computing applications, it makes sense to try to design computer systems that can continue to operate properly despite faults. Hardware fault tolerance is simpler to achieve than software fault tolerance. Spare parts and backup systems can be combined with controlling software that brings them online when they are needed. Computer scientist Severo Ornstein reports, "I was in charge of one of the first computer systems that used programs to manipulate redundant hardware to keep going even in the face of hardware failures. ... We were able to show that the machine would continue operating despite the failure of any single component. You could reach in and pull out any card or cable or shut off any power supply and the machine would keep on functioning."[12] In 1998, researchers at the University of California and Hewlett-Packard reported that an experimental "massively parallel" computer called "Teramac," built by the company, was capable of operating effectively even though it contained "about 220,000 hardware defects, any one of which could prove fatal to a conventional computer. ... "[13] That is encouraging, but it does not mean either that this success is translatable to all other computers and applications or that no hardware flaws could prove fatal to Teramac. And there is still the problem of common mode failures.

Hardware design flaws can be very subtle and difficult to detect. They may lie hidden for a long time until some peculiar combination of inputs, directions, or applications causes them to generate a sudden, unexpected failure in what was thought to be a highly reliable, tried and true system.

The enormous information storage capacity of computer hardware has made some kinds of pedestrian, low-tech human fallibility issues highly problematic. In October 2007, the British government lost a couple of computer disks in the public mail. Unfortunately, those disks contained detailed personal information, including bank account identifiers, for 25 million Britons—40 percent of the British population. The year before in the United States, "a computer and detachable hard drive with the names, birth dates and Social Security numbers of 26.5 million veterans and military personnel was stolen from the home of an analyst."[14] Prior to the advent of modern computers, there is simply no way so much information would have been portable enough to be lost in the shipment of a single package or carried away by a single thief.

FLAWS IN SOFTWARE DESIGN

In writing software of any real complexity, it is virtually impossible to avoid programming "bugs," and equally difficult to find the ones you have not avoided. So many things can happen in complex interacting systems; there are so many possible combinations of events, it is not reasonable to suppose that designers can take all of them into account.

Complexity can breed such an explosion of possibilities that the number of possible combinations could easily overwhelm even the processing speed of state-of-the-art computers. For example, suppose you tried to design a completely general chess-playing computer program by writing the software to select moves by searching through all of the possible sequences of moves in the game. Although chess is well-defined and self-contained—in many ways much simpler than the mind boggling complexity of many ill-defined, loosely bounded real-world situations—there are some 10^{120} sequences of permitted moves. Doing this many calculations in a reasonable amount of time would even be well beyond the capabilities of a machine faster than the fastest currently available. Programmed in this way, the much simpler game of checkers would still require the computer to cope with 10^{40} (ten thousand trillion trillion trillion) combinations of moves.[15]

In most applications, it is not necessary for complex programs to work perfectly. A telephone-switching program that periodically but infrequently fails to handle a call properly may be annoying, but it is no great problem for most users. They can simply hang up and call again. Even programs designed for critical applications need not be flawless as long as whatever flaws they do have are irrelevant to the proper functioning of the system. Errors in the programming for nuclear power plants that cause trivial misspelling of words displayed in output messages fall into this category.

The problem is, there are *always* flaws and until they are found, there is no way of knowing whether they are trivial or critical. A hundred of the right kind of bugs in a complex piece of software can make little difference to the system's performance, but one critical bug buried deep within the program can cause a catastrophe. And a critical bug may not be easy to recognize even when you are looking at it. A period that should have been a comma caused the loss of the Mariner spacecraft.[16] During the 1991 Persian Gulf War, a software-driven clock whose programming resulted in a cumulative error of one millionth of a second per second caused a Patriot missile battery to fail to intercept the Iraqi Scud missile that destroyed an American barracks in Dhahran, Saudi Arabia.[17]

With the approach of the year 2000, a simple programming decision made decades earlier was about to cause an ungodly amount of trouble. It had become commonplace to write programs using a two-digit shorthand for the year, rather than the full four digits. So 1978 became 78, 1995 became 95, and so forth. The problem was that this shorthand did not allow computers to recognize any year beyond 1999. They would treat the year 2000 ("00") as the year 1900. Articles from the computer industry trade press carried "images of satellites falling from the sky, a global financial crash, nuclear meltdowns ... the collapse of the air traffic control system and a wayward ballistic missile."[18]

The decision that caused all this trouble made sense at the time. Programmers were trying to conserve memory at a time when memory was very expensive. They were also certain, given the speed of development of the computer industry that by the time the year 2000 rolled around, memory would be much cheaper and the programs they had written would be ancient history, long since retired. Memory did become much, much cheaper, but some of the "dinosaur" programs they had written were still being widely used and their decision to use two-digit years had become routine. As the year 2000 approached, up to half of the world's computer data was still interacting with programs that could only recognize the decade of a year, but not its century.

One part of this "year 2000" (Y2K) problem that received relatively little public attention was that an unknown number of the more than 70 billion computer chips built into everything from thermostats to pacemakers to nuclear submarines since 1972 were hard-coded with programming that allowed only two digits for the date. In late 1998, a report issued by the Nuclear Regulatory Commission found that the Seabrook nuclear power plant had more than 1,300 "software items" and hard-coded computer chips with Y2K problems. At least a dozen of them, including an indicator that measured the reactor's critical coolant level, had "safety implications."[19]

Solving the Y2K problem required an enormous amount of boring, pedestrian work, going through software programs line by line and methodically reprogramming them. Hard-coded computer chips could not be reprogrammed; they had to be replaced, one at a time. There was great concern that on January 1, 2000, there would be a series of computer failures that might trigger widespread chaos worldwide. But as the new millennium began, few major problems occurred. It is hard to say how much credit for this was due to all the preventative measures taken and how much was due to exaggeration of the problem, but one thing was clear: the cost was enormous. Up to $200 billion may have been spent (half in the United States) to upgrade computers and application programs to be Y2K compliant.[20]

Even the best, brightest, and most careful programmers cannot be expected to write software of any real complexity perfectly. The truth is, most programming is not written by the best, brightest, and most careful programmers, nor can it be. No matter how much effort is put into training, whenever large numbers of people are needed to do a job, it is certain that only a relatively small percentage of them will do it superbly well. This is yet another unavoidable problem.

Unlike most engineering products, when software is offered for sale it normally still contains major design flaws or "bugs." Subsequent "improved" versions may still have some of the original bugs as well as new bugs introduced when the improvements were made. David Parnas, a computer scientist with decades of experience in software engineering explains, "Software is released for use, not when it is known to be correct, but when the rate of discovering new errors slows down to one that management considers acceptable."[21] As software is tested, the rate at which new bugs are discovered approaches, but does not reach zero. Ornstein points out, "Eliminating the 'last' bug is . . . a standing joke among computer people."[22]

Why is it so difficult to find and fix all the significant errors in a complex program? Referring to what is euphemistically called "program maintenance" (changes/corrections in the program made after it is already in the hands of users), computer scientist Fred Brooks argues:

> The fundamental problem with program maintenance is that fixing a defect has a substantial (20–50%) chance of introducing another. So the whole process is two steps forward and one step back. . . . All repairs tend to destroy the structure. . . . Less and less effort is spent on fixing original design flaws; more and more is spent on fixing flaws introduced by earlier fixes.[23]

The repetitive structure found in much (but not all) computer hardware makes it possible to build hardware systems made of many copies of small hardware subsystems that can be thoroughly tested. But software design is more complex because it often does not have the repetitive structure found in the circuitry itself. The relative lack of repetitive structure contributes to making software systems generally less reliable than hardware and making programmers look less competent that they really are. Parnas argues that this " . . . is a fundamental difference that will not disappear with improved technology."[24]

How can software design flaws that have been in a complex program from the beginning remain hidden without apparently affecting the program's performance, then suddenly appear? In designing software, a programmer creates decision trees intended to include all the cases that the program might encounter. In very large and complex programs, many branches of these trees cover cases that are not all that likely to occur. As a result, much of the program will rarely if ever run in normal operation. Even if each part of every path were used as the program ran, some routes that travel these paths in a different order or combination might not run for a very long time. One day, after the program has been working just fine, a new situation causes the computer to follow a route through the program that has never been used before, exposing a problem that has been hidden there from the very beginning. So a frequently used program that has been running reliably for years suddenly and unexpectedly collapses.[25]

It is possible to design fault-tolerant software. Over the decades, programmers have developed ways of writing software that can reconfigure distributed networks of computers when part of the network crashes to work around the software (or hardware) that is down.[26] As creative and useful as such work is, it does not and cannot cure misconceptions or oversights on the part of software designers.

Even after repeated testing, exposure, and correction of errors, it is difficult to have full confidence that software does not contain some crucial hidden design flaw that will unexpectedly surface someday and take the system down. In June 1998, the billion-dollar SOHO solar space probe spun out of control because a pair of undiscovered errors in its thoroughly tested software led NASA controllers to give it an errant command.[27] When software plays a key role in switching telephone calls, controlling expensive space probes or even landing commercial airliners, that is one thing. But when it is critical to controlling technologies that can cause death and destruction on a massive scale, that is another matter entirely.

ARTIFICIAL INTELLIGENCE

Artificial intelligence (AI) as a field of computer science first emerged in the 1950s. The very name calls forth images of a super intelligence that combines the incredible power and speed of emotionless, ultimately rational, and tireless computers with the learning ability, reasoning power, intuition, and creativity of the human mind. Often, when the feasibility of some grandiose technological scheme runs up against the need for computer control systems more perfect than the human brain seems capable of creating, that magical name is invoked. In the 1980s, AI was one of the deities called upon to answer the arguments of eminent scientists critical of the technical feasibility of the Strategic Defense Initiative (SDI) ballistic missile defense system launched by the Reagan Administration. SDI's originally stated goal—to render nuclear weapons "impotent and obsolete"—required the development and construction of a vast, enormously complex system of sensors and missile attacking equipment, including perhaps 10,000 interacting computers. The system had to be coordinated quickly and accurately enough to destroy nearly every warhead launched in an all out attack against the United States. If ordinary computer techniques could not do it, artificial intelligence could find a way to make it all work.

From its very inception, AI has been a field of dreams. The most spectacular dream was to create a computer program that would possess the full range of human problem-solving abilities. Reflecting this lofty goal, one of the seminal AI programs, written in 1957, was called the General Problem Solver. Within 20 years, the whole idea of creating a general problem-solver was seen as extraordinarily naive.[28]

AI programming became more task specific, and some successes were achieved. Some viable "expert systems" began to emerge as early as the 1980s. Once again the name, which could have come out of an advertising agency, promises much more than the programs have been able to deliver.

"Expert systems" are created by interviewing human experts in a particular area and trying to incorporate their knowledge into a computer program intended to carry out a particular task.[29] The most successful expert systems are programs that have built-in "rules of thumb" ("heuristics") gleaned from analyzing the procedures followed by people considered to be "experts" in whatever specific task (say medical diagnosis) is being mimicked. In essence, they are *workable* ways of achieving *good* results *most* of the time. They do not *always* result in good choices, let alone in finding and making the best choice. Expert systems are therefore *not* flawless computer systems of superhuman intelligence and expertise. They are considered to be doing well if they perform anywhere near as well as the human experts they imitate.

Why has the dream faded? Artificial intelligence seems to be caught in an inherent tradeoff. AI programmers can take the "logic programming" approach in which the computer breaks up all the possibilities into complex cases and sub-cases with rules specifying how to sort through them in deciding what to do. This is thorough, but it can result in the need to consider enormous numbers of possibilities. Or to avoid a combinatorial morass, programmers can use heuristic decision rules to dramatically narrow the range of possibilities that must be considered. "Deep Blue" is the remarkable IBM chess-playing parallel processing computer system that ultimately defeated world chess champion Garry Kasparov in May 1997. It is programmed to consider possible moves for any arrangement of the pieces. But since there are many more board positions in a game of 40 moves (10^{120}) than the estimated number of atoms in the universe, even a system as powerful and fast as Deep Blue must make use of heuristics. It can only partially preview possible moves until there are just a few pieces still on the board.[30] Whichever approach they take, programmers cannot be sure that the software they write will find the best solution in complex, real-world decision situations.

We usually assume that if we just keep at it, we will eventually solve the problem. But it may also be that as we press toward the fringes of knowledge, we will increasingly encounter situations like this where what we are running up against is not just the boundaries of what we know, but the boundaries of what is knowable.

One of the key problems with artificial intelligence may go right to the heart of the fundamental assumption that underlies it—the ability of computers to mimic the functioning of the human mind.[31] In computing, the real world is represented in an essentially linear and rigid way. Computers move with lightening speed through the branches of complex decision trees, following a predetermined set of rules. The human brain appears to work in a much more flexible and very nonlinear way. Rather than repeatedly and "mindlessly" searching through the same network of decision trees for a solution, it is capable of stepping back and looking at a particularly baffling problem from an entirely differently perspective. Sometimes that different angle is all that is needed to find a solution. The brain takes leaps, doubles back, and is capable of reordering and reprogramming its own thinking processes. It has proven very difficult, if not impossible, for AI researchers to come anywhere near duplicating this process.[32] As David Parnas once put it, "Artificial intelligence has the same relation to intelligence as artificial flowers have to flowers. From a distance they may appear much alike, but when closely examined they are quite different. . . . "[33]

Artificial intelligence may indeed have great promise, but it is not magic. There is no reason to believe that it will somehow allow us to circumvent or overcome the problems of technical (and human) fallibility that are built into the world of computers.

COMPUTER VIRUSES, WORMS, AND OTHER FORMS OF "MALWARE"

Computers can also become unreliable because they have succumbed to malicious software ("malware"). Probably the best-known malware is the computer virus. A virus is a program that can "infect" other programs by inserting a version of itself. The virus then becomes part of the infected program. Whenever it is run, the computer acts upon the set of instructions that make up the virus in exactly the same way that it acts on all the other instructions in the program. Programs accessed by an infected program can also become infected. A virus that copies itself into the "operating system" (the program that supervises the execution of other programs) will have access to and therefore can infect every program run on that system.

Whether a virus is benign or malicious depends on what its instructions tell the computer to do. Viruses can be helpful. For example, a compression virus could be written to look for uninfected "executables," compress them, and attach a copy of itself to them. When an infected program ran, it would decompress itself and execute normally. Such a virus could be extremely useful and cost effective when space was at a premium.[34] A "compiler" is a program that translates the kind of higher-level computer language used by most programmers (e.g., FORTRAN, Pascal, C, and PL/1) into binary "machine language" that computer hardware can understand. A compiler that compiles a new version of itself is another useful virus—it is altering another program (the original compiler) to include a version of itself (the modified compiler).[35]

Some viruses are neither helpful nor harmful in and of themselves. A playful programmer might, for example, create a virus which causes the programs it infects to display a mysterious message. After the message appears, the virus self-destructs, leaving the program uninfected and its users bewildered.

Malicious viruses can do a lot of damage. They can corrupt data files by manipulating them in some destructive way or by erasing them entirely. They can destroy infected programs or change them in ways that make them unreliable. Having a program "killed" by a virus is bad enough, but it can be worse if the infected program appears to be operating normally when it is not. A failed program is more obvious, more easily removed and replaced. A program that is subtly corrupted may continue to be relied upon, like a steam pressure gauge that has become stuck at a normal reading. By the time the malfunction is discovered, a disaster may be unavoidable. Delayed action viruses are particularly difficult to control. Because they do not affect the behavior of infected programs right away, those programs continue to be used, and the virus has a chance to infect many more programs before anyone realizes anything is wrong.

Malicious viruses can also be designed to overload a single computer system or interconnected network. When the infected program is run, computers are given a rapidly expanding volume of useless work or the communications network is jammed with a proliferation of meaningless "high priority" messages. While this kind of virus may do little or no long-term damage, it can temporarily take the system down completely, in a so-called "denial-of-service" attack. In 1996, some Internet servers were suddenly bombarded by requests for service coming from randomly generated false addresses at the rate of more than 100 per second. For a time service failed as the attacks jammed the system.[36] In January 2003 "Slammer," the fastest spreading computer virus to that date, infected hundreds of thousands of computers within three hours. It took cash machines offline, caused delays of scheduled airline flights, and generally wreaked "havoc on businesses worldwide."[37]

In the early days, each computer was a standalone device, so an out-of-control, self-reproducing program could at worst incapacitate one machine. But as computer networks became more common, the possibility of such a program spreading and causing widespread damage grew. At the same time, computers became more central to all phases of life, from health care to industry to education to communications to the military. The threat posed by malicious viruses grew much more serious.

With the incorporation of networked computers into more aspects of our lives, the problems generated by infectious malware are virtually guaranteed to spread. "Smartphones" are an interesting example. As mobile phones grew in popularity, they became a mainstay of our business and personal communications. When email and Internet access capabilities were added to these devices, they became vulnerable to malware infection. Researchers have argued that cell phones using Multimedia Messaging Service messaging protocol that allows rapid sharing of media and programs between cell phones around the world are particularly vulnerable to the epidemic spread of computer viruses. In 2009, *Science* magazine reported the conclusion of P. Wang and his colleagues that "the increasing market share of smartphones ... will soon reach a ... point, beyond which mobile [phone] viruses could become far more damaging and widespread than current computer viruses. ... "[38]

Real Virus Attacks

Computer viruses have become a very real threat to computer security. As early as 1988, *Time* magazine reported "a swarm of infectious programs ... descended on U.S. computer users ... an estimated 250,000 computers, from the smallest laptop machines to the most powerful workstations, have been hit. ... "[39] There were almost a dozen major virus or virus-like attacks reported in the late 1980s. One particularly interesting virus attacked Arpanet, a nationwide Defense computer communications system linking more than 6,000 military, corporate, and university computers around the United States—a forerunner of today's Internet.[40]

Early in November 1988, Arpanet was invaded by a rapidly multiplying program that spread throughout the network. It made hundreds of copies of itself in every machine it reached, clogging them to the point of paralysis. Hundreds of locations reported attacks, including the Naval Research Laboratory, NASA's Ames Research Center, Lawrence Livermore (nuclear weapons) Laboratories, and the Naval Ocean Systems Command.[41] Designed by Robert Morris, a 23-year-old graduate student in computer science at Cornell, the program entered the network through a secret "backdoor" Morris discovered that had been left open years earlier by the designer of Arpanet's email program.[42] The attack, launched by Morris sitting at his computer in Ithaca, New York, concealed its point of origin.[43]

Although Morris apparently meant no harm, he made a single programming error that allowed his infiltration of Arpanet to turn into the country's most serious computer virus attack to date. The program was designed to detect copies of itself in a computer so as to avoid infecting a machine more than once. This would have caused it to move slowly from machine to machine and forestalled the explosion of copies that jammed the system. But Morris thought someone might discover the signal that prevented the program from infecting a computer and use it to stop the invasion. So he instructed each copy to randomly choose one of 15 numbers when it encountered that signal, and if the number was positive, to infect the computer anyway. One of the numbers was supposed to be positive, so the probability of overriding the "stay away" signal would only be 1 in 15 (7%). But a programming error made 14 of the 15 numbers positive, raising the probability of override to more than 93 percent.[44] The program kept re-infecting already infected machines with additional copies of itself, spinning out of control.[45]

In the words of Chuck Cole, deputy computer security manager at Lawrence Livermore (nuclear weapons) Laboratories, " ... a relatively benign software program can virtually bring our computing community to its knees and keep it there for some time."[46] In the late 1980s, scientists at Livermore were warned by government security that some of the Lab's 450 computers were infected by a virus set to become active that day. Many users stopped doing their work and frantically made backup copies of everything. The warning turned out to be a hoax. An attack does not even have to be real to be disruptive.[47]

Viruses, virus-like attacks, and virus hoaxes continue to proliferate. In the late 1990s, anti-virus sleuths reportedly were finding about six new viruses a day.[48] By late 2008, the problem had gotten exponentially worse: the *New York Times* reported "Security researchers at SRI

International are now collecting over 10,000 unique samples of malware daily from around the globe."[49]

By March 2009, it was reported that a "vast electronic spying operation" had infiltrated nearly 1,300 computers in 103 countries and "stolen documents from hundreds of government and private offices around the world."[50] In August 2009, a concerted effort by top security experts from universities, industry, and government had still not been able to eradicate or trace the origins of a malware program called Conficker that emerged nine months earlier. Conficker exploited flaws in Windows software to turn more than five million government, business, and household computers in more than 200 countries into "zombies," machines that could be controlled remotely by the program's authors. In effect, they had created a virtual computer that had "power that dwarfs that of the world's largest data centers."[51]

Defending Against Viruses and Other Malware

There are basically three strategies for defending computer systems against attack: preventing exposure, "immunization," and after-the-fact cure. In the abstract, it seems that immunization would be best since it avoids the restrictions that might be required to prevent exposure and the problems associated with infection. Can computer systems be immunized against virus attack?

Viruses are programs just like any other normal user program that runs on the system. They differ only in what they instruct the computer to do, not in what they are. There is no general structural way to stop them. As Cohen points out, "It has been proven that in any system that allows information to be shared, interpreted, and retransmitted, a virus can spread throughout the system."[52] Software "vaccines" must first detect a virus program, then destroy or disable it. A vaccine program may try to detect viruses in control software or other legitimate programs by checking to see whether the amount of space those programs ordinarily occupy has changed. Or it may check whether the program has been rewritten at an unauthorized date. It might carefully watch key locations within control programs that are convenient places for viruses to hide, stop the computer, and alert the operator if anything tries to change that part of the operating system.[53]

It is also possible to physically "hard-code" or "burn" part or even all of a computer's master control programming onto a computer chip so that no software running on the computer could modify them.[54] But hard-coding the operating system onto computer chips is generally impractical. Most operating systems are complex. Complexity implies both a high probability of bugs and the need to modify the operating system over time. Hard-coding greatly complicates debugging as well as updating and testing the operating system. Rather than simply reprogramming software, new computer chips must be designed, manufactured, shipped, and physically installed.[55]

Viruses or virus-like programs can be designed to avoid signaling their presence by enlarging the programs they infect or changing "write dates." They can hide in atypical places and do not need to infect the computer's master control system. There are so many possible ways to write a virus that no one anti-viral program could be effective against all of them, just as no single vaccine is effective against all the viruses that might infect the human body. The advent of computer viruses has launched a kind of "arms race" between those writing viruses and those trying to stop them. Any program written to block the current strain of viruses can be circumvented by the next type of virus, written specially to defeat that defense. Raymond Glath, a virus expert whose company produces anti-viral software, has said, "No anti-virus product can offer a 100 percent guarantee that a virus cannot slip past it in some way."[56]

In late 2005, Israeli researchers unveiled a different theoretical inoculation approach for dealing with viruses being propagated through a computer network. The network would be seeded in advance with computers "armed with software that can trap and identify new viruses" and generate programs to block the virus. When a new virus is detected, the system would launch "an internal counter-epidemic of self-propagating, protective messages" that would protect any as yet uncontaminated computer from being infected with the virus.[57]

Looking at viral detection from a theoretical point of view, Fred Cohen concludes, " . . . a program that precisely discerns a virus from any other program by examining its appearance is infeasible . . . in general it is impossible to detect viruses. . . . "[58] He argues, "I can prove mathematically that it's impossible to write a [general] program that can find out in a finite amount of time whether another program is infected."[59] Cohen goes on to say, " . . . any particular virus can be detected by a particular detection scheme. . . . Similarly, any particular detection scheme can be circumvented by a particular virus. . . . "[60] Even if a computer were protected by a highly effective anti-viral program, a serious attack might still occupy so much of the machine's resources that it could not be used for its normal purposes during the attack.[61] Thus, the virus would have taken the computer out of action anyway, at least for a time. Immunization may seem a superior approach in the abstract, but it has many problems in practice.

Curing an infected computer can require a painstaking process of search and destroy that must work its way through all levels of the system and through all backup data and program files. This can be very time consuming. Creating separately stored backup copies of all programs and data does not assure that a computer purged of a virus can be easily returned to its original state. If the infection occurred before some of the backups were made (likely in the case of delayed action viruses), backups can be a source of re-infection rather than defense.[62]

It is possible to imagine a sophisticated, evolving virus capable of detecting protective or curative programs and modifying itself to circumvent them. Another possibility is the so-called "retro-virus," a program that communicates with the copies of itself with which it has infected other programs. If it detects that these clones have been wiped out, it assumes they have been destroyed by an attempt to disinfect the system, and shuts itself off. Then after lying dormant for weeks or months, it awakens and re-infects the system. The user, thinking the system has been cleared of the infection, discovers that the same virus has mysteriously reappeared.[63] A shutdown can buy time to search for a virus and work out a method of destroying it. By forcing a shutdown, however, the virus has succeeded in disabling the system, at least temporarily.

If immunization is problematic and cure may still temporarily cripple the system, what about preventing exposure in the first place? Protective "firewalls" can be built by using additional layers of passwords or encryption.[64] The system can also be physically isolated. Or physical barriers can be created. A decade ago, engineers at Sandia Laboratories attached a tiny electromechanical combination lock to a computer chip to prevent anyone who does not have the combination from using the system.[65]

More complex passwords and more involved identification measures do make it more difficult for an outsider to introduce a virus. Making it harder to break into a system can also make it harder to use, however, and such an approach does nothing to prevent insiders from causing trouble. According to an informal survey of corporate data security officers conducted for the FBI in the late 1990s, insiders launched more corporate computer attacks than outsiders.[66]

Any technical system designed to permit some people access while denying it to others contains within it the seeds of its own defeat. If it is possible to figure out how it works, it is possible to figure out how to get around it. Computer security systems are no exception. After the Arpanet virus attack, a former top information security chief at the Pentagon put it bluntly, "No one has ever built a system that can't be broken."[67] It does not always take an expert with long years of experience to find a way in. In 1995, a 22-year-old researcher, Paul Kocher, discovered a weakness in the so-called "public key" data security technologies used by most electronic banking, shopping, and cash systems.[68] A few years later, Kocher and his colleagues developed an ingenious scheme for breaking the coding system used in "smart cards," the credit-card like devices with special electronic encryption and data processing circuitry built into them that can be used for everything from cash cards to identification.[69]

That same year, University of California researchers cracked the world's most widely used cellular phone encryption scheme, designed to prevent digital phones from being "cloned" by unauthorized users.[70] In early 2005, a team of computer scientists at Johns Hopkins University cracked the codes of "immobilizer" systems that include tiny RFID chips imbedded in car keys to prevent car theft.[71] The following month, "three codebreakers demonstrated a way to break the Secure Hash Algorithm (SHA-1), a . . . standard cryptographic function crucial to many electronic transactions, including digital signature schemes and password verification."[72]

For many computer programmers, figuring out how to break into a computer security system designed to keep them out is just the sort of intellectual challenge that got them interested in programming in the first place. The fact that this game playing can often be done by remote control with relatively little chance of getting caught makes it that much more appealing. According to one security expert, "There are enough hackers out there that someone will manage a serious compromise of security."[73] Sometimes human error gives them a hand, even in very dangerous technological systems. In the 1990s the Federal Advisory Committee on Nuclear Failsafe and Risk Reduction discovered "an electronic backdoor to the naval communications network used to transmit launch orders to U.S. Trident missile submarines. Unauthorized individuals . . . could have hacked into the network . . . and sent a nuclear launch order . . . to the subs."[74]

The only sure way to protect a computer system against infection (or hackers) is to block *all* incoming flows of information. That means no user could enter any new data or load any new programs, making the system close to useless. If the system were physically isolated to prevent any unauthorized users from even getting into the rooms where the access terminals are, someone might still be able to break in, just as burglars do on a daily basis. As long as anyone is, or ever was, on the inside, that person could intentionally or accidentally infect or otherwise interfere with the proper operation of the system. Physical isolation prevents information sharing and user interaction, one of the great advantages of modern computer systems. Isolation may be workable in some unusual, highly specialized applications, but it is *not* a realistic solution in general.

Unfortunately, in the electronic as well as the physical world, perfect security is an illusion.

COMPUTER FALLIBILITY AND DANGEROUS TECHNOLOGIES

In the minimal time that dangerous technologies allow for decision making in a crisis, even a short-lived computer malfunction can cause disaster. The breakdown of even a single, seemingly unimportant, inexpensive piece of computer hardware—at the "right" time in the "right" place—can threaten global security. When that infamous 46-cent computer chip failed at NORAD on June 3, 1980, it triggered a false warning of attack by Soviet submarine-launched and land-based missiles. According to NORAD officials, the circuit board containing the chip had been installed five years earlier and was periodically tested.[75] Yet neither these periodic checks nor anything in the circuit board's behavior in *five years* of operation brought the problem to light until the false warning that put America's nuclear military on high alert.

This dramatic incident illustrates two points made earlier: that critical flaws can remain hidden for extended periods of time and that computer problems can and do produce spectacular failures, not merely degraded performance. As it turns out, it is also an example of the deeper, more structural issue of flawed design. The communications computer used only very superficial methods for checking errors. Much more effective methods of detecting errors were commonly used in the commercial computer systems of the day, methods that would likely have prevented this dangerous false warning. Former Defense Communications Agency chief, General Hillsman, said of the false alert, "Everybody would have told you it was technically impossible."[76]

Failures of military software are legion. They have led to enormous problems in everything from the performance of particular weapons to the mobilization of the Armed Forces as a whole.

During the 1982 war between Britain and Argentina over the Falklands/Malvinas Islands, the British destroyer HMS Sheffield was sunk by a French-built Exocet missile. The ship's radar reportedly detected the incoming missile, but its computers had been programmed to identify Exocet as a "friendly" missile (because it was French). No alarm was sounded and no attempt was made to take any defensive action. Four of the six Exocets fired by the Argentines during that war hit their targets, destroying two ships and seriously damaging a third.[77]

As for general mobilization, in late 1980 the U.S. military ran a simulation called "Proud Spirit" to test the nation's readiness to mobilize for war. During the exercise, "A major failure of the computerized Worldwide Military Command and Control System... left military commanders without essential information about the readiness of their units for 12 hours at the height of the 'crisis.'"[78] With the demands of the major mobilization being simulated, the computer became so overloaded, it temporarily shunted updating readiness information into a buffer memory. Later, the buffer simply refused to discharge it.[79]

Saboteurs, spies, or disgruntled employees might intentionally manipulate software to sabotage computers controlling dangerous technologies. They could also be compromised inadvertently by playful programmers. It could even happen accidentally, as when a "keyboard error" at Rockwell International stripped away software security for *six months*, allowing unauthorized personnel to make changes in the raw flight software code for the space shuttle.[80]

Intentional software manipulation could be used to steal data (i.e., to spy). After all, computers store and manipulate information. Whether they are industrial, political, or military spies, information is what spies are looking for. If they can access information by hacking into a computer, they are saved the messy and dangerous business of making their way past armed guards, breaking into locked rooms and rifling filing cabinets. They are also likely to gain quicker access to much more information. If they can do it anonymously from a remote location, they are exposed to even less risk. By early 2005, federal and state law enforcement officials were already saying that the increasingly popular Wi-Fi wireless networks, many of which were unsecured, had made it far easier for sophisticated cyber-criminals to make themselves untraceable on the Internet.[81]

Virus-like invasions and other programming attacks on computer systems have frequently been used to steal passwords and other access codes for later use. Access codes could not only allow an unauthorized user to break in to computer files, but to do so in someone else's name. The 1988 Arpanet virus stole passwords and directed messages to a remote monitoring computer to hide the location of the perpetrator.

How vulnerable are the computers in dangerous technological systems to this kind of break-in? In late 1988, the *New York Times* reported that a few years earlier, without any fanfare, an unspecified number of military computer experts had decided to try to get past the safeguards protecting some vital, highly classified Federal government computers to see if it could be done. "One expert familiar with those efforts ... said that they found those safeguards to be 'like swiss cheese' to enterprising electronic intruders. ... They went after the big ones and found it incredibly easy to get inside.' "[82] *Newsweek* reported on a study conducted in 1985 by the office of Robert Brotzman, director of the Department of Defense Computer Security Center at Fort Meade. According to that report, the study "determined that only 30 out of 17,000 DOD computers surveyed met minimum standards for protection. Brotzman's conclusion: 'We don't have anything that isn't vulnerable to attack from a retarded 16-year-old.' "[83]

In 1996, GAO issued a report to Congress of its investigation into Pentagon information security. DoD has over 2.1 million computers as well as 10,000 local and 100 long distance networks. According to the study, Defense Information Systems Agency data indicated that there may have been as many as *250,000 attacks on DoD computers the preceding year, 65 percent of them successful.* GAO reported:

> Attackers have seized control of entire Defense systems, many of which support critical functions, such as weapons research and development, logistics and finance. Attackers have also stolen, modified and destroyed data and software. ... They have installed unwanted files and 'back doors' which circumvent normal system protection and allow attackers unauthorized access in the future. They have shut down and crashed entire systems and networks, denying service ... to help meet critical missions. ... The potential for catastrophic damage is great.[84]

DoD did not dispute the report's findings.[85]

But the mid-1990s, let alone the mid-1980s, was a long time ago in the rapidly advancing world of computer technology. Surely these problems must have been solved by now. Actually, the situation may have gotten worse. Governments and their militaries are increasingly directing resources and attention to a new international arms race in offensive and defensive cyberweapons. The United States has already spent billions of dollars on this effort. By 2009, the Pentagon had commissioned military contractors to develop a top-secret Internet simulator to use as a platform for developing cyberweapons that might be able to shut down a country's power plants, telecommunications systems, commercial aviation, financial markets, and so forth—or to frustrate another nation's attempt to sow such chaos in the United States by electronic means. When Russia and Estonia confronted each other in April 2007, a group of cyber attackers commandeered a slew of computers around the world, which they used to attack Estonia's banking system. When Russia invaded Georgia in August 2008, cyber attacks denied the Georgians access to the Internet by locking up their servers. In April 2009, the *New York Times* reported, "Thousands of daily attacks on federal and private computer systems in the United States—many from China and Russia, some malicious and some testing chinks in the patchwork of American firewalls—have prompted the Obama administration to review American strategy."[86]

Spying creates problems. Aside from copying advanced designs, an opponent that better understands how a weapons system works can more easily find ways to defeat it. A terrorist group that can better understand the details of how a nuclear power plant or other dangerous technological system is laid out is more likely to find its weak points. Yet this may not be the biggest problem.

It is not difficult to imagine that a virus could be used to sabotage a key data processing or communications system during a terrorist attack or a developing international political or military crisis. Planted in advance and designed to "hide," the virus could lie dormant until the critical moment, then be triggered by an external command or even by the system itself as it passed some measurable level of crisis activity. Once triggered, the virus would rapidly overload the system, jamming it long enough to render it useless. The Arpanet attack demonstrates that large interconnected networks of computers can be crippled by programs whose design caused them to multiply out of control. It reportedly disabled many of the computer systems at the Naval Ocean Systems Command in San Diego for more than 19 hours by overloading them.[87] Even false alarms or hoaxes, such as that at the Livermore Laboratories in 1988, can result in lengthy disruptions of key computer systems.

Juxtaposed against the quick reaction times required in the event of nuclear attack (30 minute *maximum* warning of an ICBM launch), this kind of jamming may be virtually as effective as destroying the system. Virus-provoked overloading of crucial computer links could also be used in a political or conventional military crisis or terrorist action to temporarily "blind" or "cripple" the opponent. Pre-planted viruses could interfere with computers that are key to electronic information gathering or communication, making it difficult for military or anti-terrorist forces to see what is happening in an area of critical confrontation or to effectively deploy and control their own people. By the time the electronic smokescreen could be cleared, the attacker may have succeeded or at least gained an important advantage.

Malicious viruses can also be designed to destroy critical data or programs, requiring extensive and time-consuming repair.[88] They could interfere directly with critical programming that controls dangerous technological systems. If the programming that launches, targets, or otherwise operates individual weapons or weapons systems could be infected by such a virus, it would not be necessary to destroy the weapons. Their programming or data could be so scrambled as to render them unusable. Computers that control critical functions at nuclear plants, toxic chemical facilities, and the like could also be subjected to electronic attack. Even if they are not accessible from any external network, disgruntled insiders or agents who had infiltrated the system could launch such an attack.

It is possible to design fairly sophisticated electronic security procedures that will make it difficult for outsiders to connect, or insiders to gain access to parts of the system they are not authorized to reach. But it has been shown repeatedly, even in the case of military-linked computer systems, that sufficiently creative and/or persistent programmers can find a way to break through these software barriers. In July 1985, the prosecutor involved in the case of seven teenage hackers arrested in New Jersey revealed that they had "penetrated 'direct lines into . . . sensitive sections of the armed forces.' "[89] In 1987, the so-called "Hannover hacker" broke into the 20,000 computer scientific research and unclassified U.S. military system Internet from his apartment in West Germany. The hacker "showed uncommon persistence . . . and gained access to more than 30" different computers ranging from "the Naval Coastal Systems Command in Panama City, Florida to Buckner Army Base in Okinawa."[90] There are no grounds for believing that all of the loopholes have been—or even can be—closed.

The networks of satellite-based computers that have become so central to communication, attack warning, espionage, and the like cannot be kept under lock and key. They must be accessible from remote locations. Computers that service complex worldwide logistics must be connected to many other computers in many other places in order to carry out their mission. Their usefulness lies in their ability to receive and process information from many different entry points. They cannot practically be completely physically isolated.

It might be possible to physically isolate certain key, smaller, special-purpose dangerous technology computer systems. There is not much information publicly available about the military computers that include nuclear war plans and control nuclear weapons, but they may fit into this category. Will enough guards, walls, locks, and other security measures offer virtually perfect protection to such critical systems? The answer is no. Someone, somewhere, sometime must write the programming imbedded in all computers. Others must periodically access the programming to modify, update, or test it. As long as that is true—and it is always true—those with authorized access will be able to make unauthorized changes. No matter how much technology advances, systems of external protection can never protect perfectly against the actions of insiders.

Deliberate attack by viruses and other forms of malware is a clear threat to the integrity and efficiency of the computer systems on which so many modern systems and technologies depend, dangerous and otherwise. In the end, however, it may not be this kind of malevolence, so much as ordinary human error and the limits imposed by our inherent imperfectability that pose the greatest threat of catastrophic computer failure. We can and should do whatever we are able to do to frustrate deliberate attempts at sabotage, but we should not delude ourselves into thinking that these efforts will eliminate even that problem. In the end, it may well be the faulty component, the misplaced comma, the simple error of logic, the inability to foresee every possibility, and the less-than-brilliant job of manufacturing or programming that are even more likely to do us in.

NOTES

1. Associated Press, "Pilot's Wrong Keystroke Led to Crash, Airline Says," *New York Times*, August 24, 1996.

2. See, for example, S. M. Ornstein, "Deadly Bloopers," unpublished paper, June 16, 1986, 5.

3. Ibid., 7.

4. A. Salpukas, "Computer Chaos for Air Travelers," *New York Times*, May 13, 1989.

5. Associated Press, "Computer Failure Grounds and Delays Flights on 2 Airlines," *New York Times*, August 2, 2004.

6. William J. Broad, "New Computer Failure Imperils U.S. Shuttle Astronomy Mission," *New York Times*, December 9, 1990.

7. Associated Press, "Firm Blames Equipment for E-Vote Glitch," NYTimes.com, April 15, 2004.

8. D. Clark, "Some Scientists Are Angry Over Flaw in Pentium Chip, and Intel's Response," *Wall Street Journal*, November 25, 1994.

9. Communication to the author from an Intel employee in the fall of 1997.

10. K. N. Gilpin, "Stray Rodent Halts NASDAQ Computers," *New York Times*, December 10, 1988.

11. Ibid.

12. Op. cit., Ornstein, 12.

13. James R. Heath, Philip J. Keukes, Gregory S. Snider, and R. Stanley Williams, "A Defect-Tolerant Computer Architecture: Opportunities for Nanotechnology," *Science*, June 12, 1998, 1716.

14. Eric Pfanner, "Data Leak in Britain Affects 25 Million," *New York Times*, November 22, 2007.

15. M. M. Waldrop, "The Necessity of Knowledge," *Science*, March 23, 1984, 1280.

16. Op. cit., Ornstein, 10.

17. I. Peterson, "Phone Glitches and Other Computer Faults," *Science News*, July 6, 1991.

18. John M. Broder and Laurence Zuckerman, "Computers Are the Future But Remain Unready for It," *New York Times*, April 7, 1997.

19. Ivar Peterson, "Year-2000 Chip Danger Looms Large," *Science News*, January 2, 1999, 4.

20. "Y2K Bug," *Encyclopedia Britannica* (http://www.britannica.com/EBchecked/topic/382740/Y2K-bug (accessed October 15, 2009).

21. D. L. Parnas, "Software Aspects of Strategic Defense Systems," *American Scientist*, September–October 1985, 433 and 436.

22. Op. cit., Ornstein, 10.

23. Brooks, 121.

24. Op. cit., Parnas, 433.

25. Op.cit., Ornstein, p.11.

26. See, for example, "Software for Reliable Networks," *Scientific American*, May 1996.

27. James Glanz, "Chain of Errors Hauled Probe into Spin," *Science News*, July 24, 1998, 449. SOHO is an acronym for Solar and Heliospheric Observatory.

28. Op. cit., Waldrop, March 23, 1984, 1280.

29. John McCarthy, "What Is Artificial Intelligence?: Applications of AI" Stanford University Computer Science Department, http://www-formal .stanford.edu/jmc/whatisai/node3.html; revised November 12, 2007 (accessed October 15, 2009).

30. Ivars Peterson, "The Soul of a Chess Machine," *Science News*, March 30, 1996, 201.

31. The best and most comprehensive exposition I have seen of the distinction between what computers do and what the human mind does when it thinks is Theodore Roszak's, *The Cult of Information: A Neo-Luddite Treatise on High-Tech, Artificial Intelligence, and the True Art of Thinking* (Berkeley: University of California Press, 1994).

32. In 1996, a computer program "designed to reason" rather than solve a particular problem developed "a major mathematical proof that would have been called creative if a human thought of it." This is impressive, but very far from proving that the human mind can be effectively duplicated by electronic means. See Gina Kolata, "With Major Math Proof, Brute Computers Show Flash of Reasoning Power," *New York Times*, December 10, 1996.

33. Op. cit., Parnas, 438.

34. F. Cohen, "Computer Viruses: Theory and Experiments," *Computers & Security*, February 1987, 24.

35. Ibid., 29.

36. John Markoff, "A New Method of Internet Sabotage Is Spreading," *New York Times*, September 19, 1996.

37. Brian Krebs, "A Short History of Computer Viruses and Attacks," *washingtonpost.com*, February 14, 2003; http://www.washingtonpost.com/ wp-dyn/articles/A50636-2002Jun26_3.html (accessed October 15, 2009).

38. Shlomo Havlin, "Phone Infections," *Science*, May 22, 2009.

39. P. Elmer-De-Witt,"Invasion of the Data Snatchers," *Time*, September 26, 1988, 62.

40. Higher security military communications were split off from Arpanet into a second network called Milnet in 1983.

41. J. Markoff, " 'Virus' in Military Computers Disrupts Systems Nationwide," *New York Times*, November 4, 1988.

42. J. F. Shoch and J. A. Hupp, "The 'Worm' Programs—Early Experience with a Distributed Computation," *Communications of the ACM*, Vol. 25, No. 3 (March 1982): 172.

43. J. Markoff, "Invasion of Computer: 'Back Door' Left Ajar," *New York Times*, November 4, 1988.

44. C. Holden, "The Worm's Aftermath," *Science*, November 25, 1988, 1121–1122.

45. J. Markoff, "Invasion of computer: 'Back Door' Left Ajar," *New York Times*, November 4, 1988.

46. J. Markoff, " 'Virus' in Military Computers Disrupts System Nation-wide," *New York Times*, November 4, 1988.

47. Op. cit., Elmer-De-Witt, 65. During the mid to late 1990s, hoax virus messages were repeatedly spread widely over the Internet.

48. Steve Lohr, "A Virus Got You Down? Who You Gonna Call?," *New York Times*, August 12, 1996.

49. John Markoff, "Thieves Winning Online War, Maybe Even in Your Computer," *New York Times*, December 6, 2008.

50. John Markoff, "Vast Spy System Loots Computers in 103 Countries," *New York Times*, March 29, 2009.

51. John Markoff, "Defying Experts, Rogue Computer Code Still Lurks," *New York Times*, August 27, 2009.

52. F. Cohen, "On the Implications of Computer Viruses and Methods of Defense," *Computers & Security*, April 1988, 167.

53. Op. cit., Elmer-De-Witt, 67.

54. F. G. F. Davis and R. E. Gantenbein, "Recovering from a Virus Attack," *The Journal of Systems and Software*, No. 7 (1987), 255–256.

55. Ibid., 256.

56. R. M. Glath, "A Practical Guide to Curing a Virus," *Texas Computing*, November 11, 1988, 14. See also F. G. F. Davis and R. E. Gantenbein, "Recovering from a Virus Attack," *The Journal of Systems and Software*, No. 7 (1987); and S. Gibson, "Tech Talk: What Were Simple Viruses May Fast Become a Plague," *Infoworld*, May 2, 1988.

57. P. Weiss, "Network Innoculation," *Science News*, December 3, 2005.

58. Op. cit., Cohen, February 1987, 28.

59. E. Marshall, "The Scourge of Computer Viruses," *Science*, April 8, 1988, 134.

60. Op. cit., Cohen, February 1987, 30.

61. Op. cit., Davis and Gantenbein, 254.

62. Op. cit., Cohen, April 1988, 168.

63. S. Gibson, "Tech Talk: Effective and Inexpensive Methods for Controlling Software Viruses," *Infoworld*, May 9, 1988, 51.

64. For an interesting nontechnical discussion of how this type of security works, see "Special Report: Computer Security and the Internet," *Scientific American*, October 1998, 95–115.

65. P. Weiss, "Lock-on-a-Chip May Close Hackers Out," *Science News*, November 14, 1998, 309.

66. Peter H. Lewis, "Threat to Corporate Computers Is Often the Enemy Within," *New York Times*, March 2, 1998.

67. *U.S. News & World Report*, "Could a Virus Infect Military Computers?," November 14, 1988, 13.

68. John Markoff, "Secure Digital Transactions Just Got a Little Less Secure," *New York Times*, November 11, 1995.

69. Ivar Peterson, "Power Cracking of Cash Card Codes," *Science News*, June 20, 1998, 388; see also Peter Wayner, "Code Breaker Cracks Smart Cards' Digital Safe," *New York Times*, June 22, 1998.

70. John Markoff, "Researchers Crack Code in Cell Phones," *New York Times*, April 14, 1998.

71. P. Weiss, "Outsmarting the Electronic Gatekeeper," *Science News*, February 5, 2005; John Schwartz, "Graduate Cryptographers Unlock Code of 'Thiefproof' Car Key," *New York Times*, January 29, 2005.

72. Charles Seife, "Flaw Found in Data Protection Method," *Science*, March 4, 2005.

73. *U.S. News & World Report*, " 'Hackers' Score a New Pentagon Hit," July 29, 1985, 7.

74. Bruce G. Blair, "Primed and Ready," *Bulletin of the Atomic Scientists*, January/February 2007, 36.

75. D. Ford, *The Button: America's Nuclear Warning System—Does It Work?* New York: Simon and Shuster, 1985, 79.

76. Ibid.

77. L. J. Dumas, "The Economics of Warfare," in *Future War: Armed Conflict in the Next Decade*, ed. F. Barnaby (London: Michael Joseph, 1984), 133–134.

78. *New York Times*, "Simulated Mobilization for War Produces Major Problems," December 20, 1980.

79. Ibid.

80. R. Doherty, "Shuttle Security Lapse," *Electronic Engineering Times*, June 6, 1988.

81. Seth Schliesel, "Growth of Wireless Internet Opens New Path for Thieves," *New York Times*, March 19, 2005.

82. M. Wines, "Some Point Out Chinks in US Computer Armor," *New York Times*, November 6, 1988, 30.

83. R. Sandza, "Spying Through Computers: The Defense Department's Computer System Is Vulnerable," *Newsweek*, June 10, 1985, 39.

84. General Accounting Office, *Information Security: Computer Attacks at Department of Defense Pose Increasing Risks*, Report to Congressional Requesters (Washington, DC: May 1996), 2–4.

85. Philip Shenon, "Defense Dept. Computers Face a Hacker Threat, Report Says," *New York Times*, May 23, 1996.

86. David E. Sanger, John Markoff, and Thom Shanker, "U.S. Steps Up Effort on Digital Defenses," *New York Times*, April 28, 2009.

87. J. Markoff, " 'Virus' in Military Computers Disrupts Systems Nationwide," *New York Times*, November 4, 1988.

88. The "Brain," "Lehigh," "Flu-Shot 4," "Hebrew University," and "Ft. Worth" viruses all destroyed data in the systems they infected. For descriptions of these viruses, see P. Elmer-DeWitt, "Invasion of the Data Snatchers," *Time*, September 26, 1988; and H. J. Highland, "Computer Viruses: A Post Mortem," *Computers and Security*, April 1988.

89. *U.S. News & World Report*, " 'Hackers' Score a New Pentagon Hit," July 29, 1985, 7.

90. P. Elmer-De-Witt, "A Bold Raid on Computer Security," *Time*, May 2, 1988, 58.

Part V

Finding Solutions That Do Work

11

Judging Risk

We have seen that technological systems fail for a great many reasons, rooted in the nature of technology itself and in the fallibility of its human creators. We have also seen that when those failures involve dangerous technologies, the possibility of catastrophe is all too real. If we are to avoid stumbling into technology-induced disaster, we cannot leave the choice of which technological paths we follow and which we forego to chance. We all have a stake in making that choice a conscious and deliberate one in which we all participate. In order to make intelligent decisions in a world that combines determinate and random elements, we must first understand the limits imposed by chance itself and by the difficulty of estimating how likely the events that concern us might be.

RISK AND UNCERTAINTY

Although the terms are often used interchangeably, technically "risk" and "uncertainty" are not the same thing. When there is "risk," we do not know exactly what will happen, but we do know all possible outcomes as well as the likelihood that any particular outcome will occur.[1] "Uncertainty" refers to a situation in which we do not know all possible outcomes and/or we do not know the probability of every outcome. We are lacking some information that we have in a case of risk: uncertainty equals risk minus information. That missing information is important to our ability to predict what will happen.

Consider trying to predict the likelihood of randomly drawing a spade from a standard deck of 52 playing cards. This is risk: we know there are 13 spades; we know there are 52 cards. Therefore, we know that the probability of drawing a spade is one out of four (13/52), or 0.25. If this random drawing were repeated 10,000 times (replacing the card we drew each time before drawing again), we can predict with considerable accuracy that we will come up with a spade about 2,500 (0.25 × 10,000) of those times. Now suppose we have a deck made up of 52 playing cards that are the randomly assembled odds and ends of other decks. We have absolutely no idea how many cards of each suit there are. This is a situation of uncertainty: we still know all the possible outcomes (any card that we draw will either be a heart, diamond, club, or spade), but we do not know the probability of any of them. There is no way to predict the likelihood of randomly drawing a spade from such a deck, no way to accurately predict about how many times a spade will come up in 10,000 random drawings.

The uncertainty would be greater still if an unknown number of the 52 cards was not from an ordinary deck, but instead had an arbitrarily chosen word, symbol, or number printed on it. Now it would not only be impossible to predict the likelihood of randomly drawing a spade, it would be impossible to make any useful or sensible prediction as to what would be printed on the face of the card that was chosen. We neither know all the possible outcomes, nor the probability of any of them.

Even in situations of risk, it is impossible to reliably predict what will happen on any particular occurrence. Either "heads" or "tails" will come up whenever a fair coin is tossed; each is equally likely. If the coin is tossed 6,000 times, we can predict with real confidence that "heads" will turn up about 3,000 times, but we have no idea which of the 6,000 tosses will come up heads. When there is risk, repetition and aggregation increase predictability. This principle is essential to writing life insurance. Actuarial tables based on past experience allow insurers to estimate every policyholder's probability of living through the year or dying. All possible outcomes are known, and there are reasonably accurate estimates of their likelihood. By insuring a large number of people, companies are able to predict how many claims they will have to pay in any given year and set premiums accordingly. Though no one knows which particular policyholders will die, aggregation allows accurate prediction of the overall number of deaths that will occur.

The complex and immensely useful mathematics of probability and statistics provide vital tools for analysis, prediction, and decision making in situations of risk. But their usefulness in situations of uncertainty falls as the degree of uncertainty increases. Greater uncertainty means less information, and information is the raw material of analysis and decision making. Because there are much better tools available, analysts and decision makers would rather deal with risk than uncertainty. It is easier to figure out what to do. If this bias encourages careful and systematic data collection, it is a good thing. But if, as is so often the case, it leads analysts and decision makers to assume or to convince themselves that they are dealing with risk when they are actually dealing with uncertainty, it can be misleading and even downright dangerous.

If it were possible, the great majority of us would just as soon eliminate risk entirely in most situations. We would like to guarantee the outcome we prefer. Of course, we would rather face risk with some possibility of positive outcomes than the certainty of disaster, but unpredictability means lack of control, and lack of control can be very anxiety producing. Not knowing what will happen makes it hard to know what to do or how to adjust our actions to achieve our goals. Though we all like the thrill of taking a chance from time to time, we still want to "win." That may partially account for the popularity of amusement park rides like roller coasters that create the illusion of taking the ultimate life or death risk while virtually guaranteeing that we will survive the experience.

All this adds to the psychological bias toward wanting to believe that there is a higher degree of predictability in situations than actually exists. Events that are risky may be assumed to be deterministic. Cases of uncertainty may be assumed to be cases of risk. And since unpredictability is more anxiety producing when there is more at stake, it is likely that this bias gets worse in more critical situations.

Then there is the problem of ego. For an analyst or decision maker to admit that a situation is one of uncertainty rather than risk amounts to admitting that he/she is unable to see all the relevant outcomes and/or evaluate their probabilities. It is an admission of ignorance or incapability. This is yet another source of psychological pressure to misrepresent what is actually uncertainty as risk instead. Those we call "experts" are even more likely to fall into this trap. It is a real boost to the ego to be called an "expert." Experts are not supposed to be ordinary people who just know more than most of us about some things; they are supposed to know everything about everything within their area of expertise. It is therefore difficult for someone called an "expert" to admit to ignorance or incapability.

When experts are hired as consultants, clients want to know what will happen, or at least what could happen and how likely it is.

They are not interested in being told the situation is uncertain and therefore unpredictable. That does not do wonders for the reputation or the future flow of consulting work. The desire to remove the anxiety that unpredictability produces is often stronger than the desire to know the truth.

When experts in risk assessment fail, it is usually not because they have incorrectly calculated the probability of outcomes, but because they have not foreseen all possible outcomes. They were dealing with uncertainty when they believed—or at least pretended—that they were dealing with risk.

Unpredictability interacts with dangerous technological systems in two important and partly contradictory ways:

1. Many dangerous technology systems are built as an attempt to protect against unpredictability (e.g., building nuclear power plants to assure a predictable energy supply; building up nuclear arsenals in an effort to increase security by deterring enemy attacks).
2. Unpredictability is built into the design, manufacture, operation, and maintenance of these technologies (because of the problems of fallibility).

WHEN THE "IMPOSSIBLE" HAPPENS

When faced with the possibility of disaster, it is common to seek refuge in the belief that disaster is very unlikely. Once we are convinced (or convince ourselves) that it is only a remote possibility, we often go one step farther and translate that into "it will not happen." Since there is no point in wasting time worrying about things that will never happen, when we think about them at all, we think about them in passing, shudder, and then move on.

It is certainly not healthy to become fixated on every remote disaster that might conceivably occur. That is the essence of paranoia. Paranoid thinking is not illogical, but it does not distinguish between possibility and probability. Psychoanalyst Erich Fromm put it this way, "If someone will not touch a doorknob because he might catch a dangerous bacillus, we call this person neurotic or his behavior irrational. But we cannot tell him that what he fears is not *possible*. Even full-fledged paranoia demonstrates that one can be insane and yet that one's capacity for logical thinking is not impaired."[2]

There is a big difference between paranoid fixation and a proper appreciation for the critical distinction between what is highly unlikely and what is completely impossible. We simply cannot afford to ignore events whose consequences are catastrophic, even if their likelihood is very small. Along with its many benefits, the enormous growth of our technological capacity to affect the world in which we live has created the prospect of unparalleled, and perhaps terminal, disaster. General nuclear war and global ecological catastrophe are among the most extreme examples. They may not be very likely, but as long as they are still possible, we can ill afford to ignore them. Events that are extremely unlikely do happen from time to time.

There is also an important difference between the probability that an event will occur during any one trial of the chance experience and the probability that it will occur *at least once* in many, many repetitions. Even if the probability of an event occurring on any *single* repetition is extremely low, *as long as it is not zero* the probability that it will happen at least once rises to one (certainty) with continued repetition. Given enough time and enough opportunity, anything that is possible will happen. This is the mathematical principle behind "believe-it-or- not" statements like "a group of monkeys, choosing typewriter keys at random, will eventually type out all the works of Shakespeare."

Our ability to accurately estimate the true probability of a very unlikely event is also questionable, especially if the event is the result of a complex process. As the last two chapters make clear, it is extremely difficult, if not impossible, to foresee all the ways things can happen in complicated systems. Unless it is possible to see all the paths to a given outcome, it is hard to be sure that our estimates of its probability are reliable. These problems are worse for very unlikely events because there is much less experience with them.

Rare Events

Even exceptionally unlikely events do occur. The great power failure that left New York City without electricity for an entire summer day in 1977 began with a series of lightning strikes of electric supply lines that according to the power company's president "just never happens."[3] After-the-fact accounts pointed out," . . . double lightning hits were a rare event," and went on to outline the extraordinarily unlikely series of subsequent failures of grounding systems, circuit breakers, timing devices, and other technical and human components.[4] This highly improbable blackout led to widespread arson and looting, with property damage estimated at $350 million to $1 billion.[5]

Natural phenomena are certainly not the only source of rare events. On October 31, 1968, an Australian fighter jet shot itself down when bullets fired from the plane's own cannon ricocheted off the ground.[6] In late spring 1989, the Oak Hill Country Club near Pittsford, New York was playing host to the U.S. Open golf tournament. On June 16, four golfers had holes-in-one on the sixth hole within two hours! In 88 previous Opens, only 17 holes-in-one had ever occurred. *Golf Digest* reportedly estimated the odds of any four golfers having holes-in-one at the same hole on the same day at 332,000 to 1 (0.000003). But as the *Times* put it, " . . . not even a Las Vegas odds-maker could come up with credible odds on four golfers using the same club to make a hole-in-one at the same hole in the United States Open within two hours."[7] Remarkably, 12 years later in July 2001, something very much like this incident happened again. Four golfers at the Glen Ridge Country Club in New Jersey all got holes-in-one at the same course on the same day, although these were on different holes and not during a professional tournament.[8]

On February 25, 1988, the *New York Times* reported a story drawn from a British psychiatric journal, yet bizarre enough to sound like a headline story from the *National Inquirer*. Five years earlier, a mentally ill teenager suffering from a disabling obsessive-compulsive disorder that caused him to wash his hands constantly, decided to take his own life. He went into the basement of his house, put a .22 caliber rifle in his mouth, and pulled the trigger. The bullet lodged in the left frontal lobe of his brain, and, except for a few fragments, was later removed by surgeons. It destroyed only the section of the brain responsible for his compulsive hand washing, without causing any other brain damage. Five years later, there was still no sign of the obsessive- compulsive behavior or any other brain related problem.[9]

Seventeen years later on January 12, 2005, 23-year-old Colorado resident Patrick Lawler went to a dentist complaining about a minor but persistent toothache he had had for nearly a week. A dental X-ray revealed the problem—he had a four-inch nail through the roof of his mouth that had barely missed his right eye and embedded itself an inch and a half into his brain. Lawler knew that a nail gun he had been using six days earlier had accidentally backfired, but did not realize that it had shot a nail through his mouth and into his brain. Doctors were able to remove it during a four-hour surgery, with apparently no ill effects.[10]

In October 1985, a New Jersey woman won the top prize in that state's lottery. Four months later, she did it again! Her February winnings added $1.5 million to the $3.9 million she had won before. The odds of winning the first prize in the October lottery were about 1 in 3.2 million; the odds of winning February's first prize were about 1 in 5.2 million. After consulting a statistician, state lottery officials announced that the odds of one person winning the top prize in the same state lottery twice in a lifetime were roughly 1 in 17.3 trillion (0.00000000000058).[11]

Neither the world of high technology in general, nor the world of the military in particular is immune to such events. In August 1985, the Air Force attempted to launch a Titan 34D rocket, its main vehicle for putting heavy satellites into orbit. The Titan exploded.[12] Then, on January 28, 1986, after 24 successful space shuttle program launches in a row, NASA's attempt to launch the space shuttle Challenger ended in a tragic explosion, killing its crew of seven. Two and a half months later, the Air Force's next attempt to launch a Titan 34D ended in another explosion.[13] A week later on April 25, a Nike-Orion research rocket misfired as NASA tried to launch it at White Sands Proving Grounds in New Mexico. It was the first failure in 55 consecutive missions of that rocket.[14] At the end of the next week, a Delta rocket, considered one of the most dependable available, misfired and had to be destroyed in flight by NASA controllers.[15] *Within only nine months, the United States experienced the failure of five highly reliable launch vehicles*. This remarkable series of events grounded most of America's space launch fleet, temporarily leaving the United States with sharply diminished capability to put payloads into orbit.

On February 19, 1994, six hospital emergency room workers at Riverside General Hospital in California mysteriously collapsed after drawing blood from a 31-year-old woman whose life they were trying unsuccessfully to save. Working with samples of the woman's blood and the air from her sealed coffin, scientists at Lawrence Livermore Laboratories concluded that the workers may have been poisoned when an extremely unusual set of chemical reactions in the woman's body and the syringe used to draw her blood produced a close relative of nerve gas! They theorized that she had taken in the solvent DMSO (dimethyl sulfoxide), possibly in an analgesic salve, which was then converted in her body—through just the right combination of temperature, the oxygen she was being given, and blood chemicals—into dimethyl sulfone. When her blood was drawn, the dimethyl sulfone became dimethyl sulfate gas, a highly poisonous chemical warfare agent.[16]

How Likely Are Rare Events?

As strange as it may seem, the likelihood of an event is greatly affected by exactly how the event is defined. Suppose we want to estimate the probability of getting "heads, tails, heads, tails" when we flip a fair coin four times. If the event is defined as "getting two heads and two tails" in four flips of a coin, its probability is 6/16 or 0.375. If the same event is instead described as "getting alternating heads and tails," its probability is only 2/16 or 0.125. And if it is described as "getting heads on the first toss, tails on the second, heads on the third and tails on the fourth," its probability drops to 1/16 or 0.0625. Now let us reconsider the case of the New Jersey woman who won top prize in that state's lottery twice within four months. The Rutgers University statistician consulted by state lottery officials calculated the odds at 1 in 17.3 trillion. That is correct if the event is defined as buying one ticket in each of two New Jersey state lotteries, and winning the first prize each time. But suppose the event is defined as someone, somewhere in the United States winning the top lottery prize twice in four months? Stephen Samuels and George McCabe of the Department of Statistics at Purdue University calculated that probability at only about 1 in 30.[17]

How could it be so high? Remember, millions of people buy lottery tickets each month, often buying many tickets in a number of different lotteries. (The woman who won the New Jersey lottery twice had bought multiple tickets each month.) Samuels and McCabe then calculated the odds if seven years, rather than four months, were allotted for the double win; the odds grew to better than 50-50 (1 in 2).[18] If the time is extended indefinitely, a double lottery winner becomes a virtual certainty.

In the *Journal of the American Statistical Association*, mathematicians Persi Diaconis and Frederick Mosteller argued for what they call " ... the law of truly large numbers. ... With a large enough sample, any outrageous thing is likely to happen." Elaborating further, " ... truly rare events, say events that occur only once in a million ... are bound to be plentiful in a population of 250 million people. If a coincidence occurs to one person in a million each day, then we expect 250 occurrences in a day and close to 100,000 such occurrences a year. ... Going from a year to a lifetime and from the population of the United States to that of the world ... we can be absolutely sure that we will see incredibly remarkable events."[19]

Many events that intuitively seem very unlikely are actually much more likely than they seem. Consider what mathematicians call the "birthday problem." Suppose 23 people, assembled randomly with respect to their date of birth, are brought into a room. What are the odds that at least two of them have the same birthday? With only 23 people, and 365 days (actually 366) to choose from, it seems very unlikely that there would be any matches. Actually, the odds are 50-50; it is just as likely there will be a match as there will not be. If the birthdays only have to be within one day of each other, the chances are 50-50 with as few as 14 people in the room. If the definition of a match includes any pair of birthdays within one week of each other, the odds are fifty-fifty with as few as seven people in the room.[20]

It is reasonable to draw three crucial conclusions from all of this:

1. Events that are extremely unlikely do occur from time to time.
2. The probability of at least one occurrence of any event, no matter how unlikely, rises toward certainty given enough time and opportunity (as long as it is not completely impossible).
3. Events that we believe to be very unlikely may not actually be anywhere near as unlikely as we think.

The last of these raises an important question. Just how accurately are we able to estimate risks in the real world?

RISK ASSESSMENT

In a letter to *Science*, nuclear engineer Karl Z. Morgan wrote, "I often recall my argument with friends in the Atomic Energy Commission shortly before the United States launched a space rocket which, together with its 17,000 curies of plutonium-238, was incinerated ... over the Indian Ocean in April 1964 [on the very first launch]. My friends had tried unsuccessfully to convince me before the launch that the probability of something like this happening was of the order of 10^{-7} [one in ten million]; if so we were very unlucky."[21]

On May 25, 1979 one of the engines on an American Airlines DC-10 tore loose as the plane took off from O'Hare airport in Chicago. Normally, the loss of an engine, even in such a dramatic fashion, does not make the plane impossible to fly. When the engine tore off, however, it severed control cables and hydraulic lines that made it very difficult for the pilot to control the plane and at the same time disabled critical warning indicators. Without normal control and without proper warning, the plane crashed, taking 273 victims with it. There had been a similar DC-10 accident in Pakistan in 1977. On January 31, 1981, a DC-10 at Washington's Dulles airport also had an engine failure combined with slat (and flap) control damage that lead it to crash during takeoff. Nine months later, another DC-10 blew an engine while taking off from Miami, and the debris severed the slat control cables. Because it happened early enough in the flight, the crew was able to successfully abort the takeoff. Various engineering studies at McDonnell Douglas had estimated the probability of the combined loss of engine power and slat control damage during takeoff at less than one in a billion. Yet that exact combination of events had occurred four times in only five years.[22]

How could we be so wrong? Risk assessment is a difficult and complex process that runs up against the limits of our technical knowledge and our personal humility. That is especially true when dealing with complicated systems and trying to assess the probabilities of events that occur only infrequently.

Risk assessment has something in common with computer programming. In programming, as in risk assessment, it is necessary to imagine all the ways things can go wrong. But after programmers write a piece of complex software, they can run the program again and again to debug it. Risk assessors cannot keep running their "program" to catch their errors. They must correctly see all the ways thing can go wrong *in advance*. For complex systems, especially those that are technologically advanced, this is a monumental and sometimes impossible task. The problem is that much greater when people are an integral part of the system's operation. We are much better at predicting the behavior of technical components than that of human beings.

Formulating Problems and Making Assumptions

While there is a scientific core to risk assessment, it is as much art as science. Many assumptions must be made, and it is difficult for even the most unbiased and disinterested analysts to avoid the distortions that the assumptions may introduce. They may not even know whether on balance the resulting bias overstates or understates the risks being analyzed. Risk assessors are not always as explicit about their assumptions as they should be. Apart from any deliberate attempt to mislead (which is always a possibility), it may be necessary to assume so many things to make a complicated risk assessment tractable that it is too tedious and time consuming to state them all. But the underlying assumptions are critical to the ultimate results.

The impact these assumptions can have on the risk assessor's ultimate conclusions is strikingly illustrated by the global financial crisis that began in 2007 and has continued into 2010. For years financial institutions have offered very lucrative pay to attract statistical risk analysts away from other areas to work on creating complex models aimed at accurately assessing the financial risks to which these institutions were exposed as a result of their operations. Yet, despite the dangers made obvious by the collapse of the huge hedge fund Long Term Capital a decade earlier, these models apparently did not give the institutions accurate guidance.[23] Their models worked well enough under normal conditions, but did not fare so well in predicting major systemic crisis.[24] This contributed significantly to putting even some very large banks in grave danger of failing.

Part of the problem was a handful of critical assumptions that underlay these risk assessments, including the following:

1. Market prices of the financial instruments that banks traded day-to-day were reasonably predictable and would behave as they did in the past.
2. Banks would be easily able to sell sub-prime mortgages and other "problematic" assets if they needed to get their hands on cash.
3. Risks that banks ran in trading in the financial markets were isolated from the risks they ran of not getting paid back by those to whom they had made loans.
4. "Bad news" in part of the financial market would not stimulate a broader contagion that would make it hard for banks to borrow money when they needed it.[25]

None of these assumptions were correct. One of the most basic assumptions of risk assessment is that the situation being analyzed is actually one of risk and not uncertainty. Creating a comprehensive list of outcomes in complex systems is often beyond our capabilities. In many real-world situations, our unwillingness to admit how little we actually know about what will happen can lead us into serious trouble.

One way to deal with the problem of uncertainty in risk assessment is to specify all the known outcomes, and then create a residual category called "all other outcomes." If reasonable estimates of probabilities can be worked out for all the outcomes that can be listed, the probability of the "all other outcomes" category would simply be one minus the sum of those probabilities. The probability attached to this residual category is an estimate of the degree of uncertainty in the situation, of the limits of our knowledge. As long as the residual probability is small enough, the risk assessment should still be useful.

Unfortunately, this common strategy does not work nearly as well as one would think. The probabilities estimated for specific outcomes tend to be significantly affected by how complete the listing of outcomes and paths to those outcomes is. Making the list less complete should enlarge the probability of the "all other outcomes." But it often also changes the probabilities that assessors attach to those outcomes that have not been altered. That should not happen.

Consider a common tool of risk assessors, the "fault tree." Creating a fault tree begins by specifying some kind of system failure, then constructing branches that represent different paths to that failure. Once the tree is created, the probabilities of all the events along all the paths represented by the branches are estimated. From these, the probabilities of each path to failure, and thus the probability of the failure itself can be estimated. The problem is that the probabilities risk assessors attach to the main branches of the tree are influenced by how many branches (paths to failure) they have foreseen.

Consulting authoritative sources, Baruch Fischhoff, Paul Slovic, and Sarah Lichtenstein constructed a fault tree for the failure of a car to start.[26] The full tree included seven main branches. They then created several versions of the tree that differed from the original in which paths and outcomes were listed and which were left out. First they gave a group of laypeople the complete and various "pruned" fault trees. Each person was asked to estimate the probability that a car would fail to start for each of the reasons listed as branches of the version of the tree that they were given. If the completeness of the fault tree did not distort the subjects' risk estimates, leaving out a branch should just transfer the probability of that branch to the branch labeled "all other problems." In other words, the probability assigned to "all other problems" by those given *incomplete* fault trees should equal the probability assigned to "all other problems" by those given *complete* trees plus the probabilities attached to the branches left out.

Those given incomplete tree diagrams did assign higher probability to "all other problems," but it was not high enough to account for what had actually been left out: "Pruned tree subjects clearly failed to appreciate what had been left out. For pruned tree Group 1, 'other' should have increased by a factor of six ... to reflect the proportion of failures ... left out]. Instead, 'other' was only doubled, whereas the importance of the three systems that were mentioned was substantially increased."[27] Although risk estimates for the branches that were the same in pruned and complete trees should have been identical, they were higher when assessors were looking at incomplete trees.

The experiment was then repeated, with subjects specifically told " . . . we'd like you to consider . . . [the fault tree's] completeness. That is, what proportion of the possible reasons for a car's not starting are left out, to be included in the category *all other problems*?"[28] While saying this did increase the probability that pruned tree subjects assigned to "all other problems," it was still much less than it should have been. In another experiment, the subjects were all given complete trees, but with different branches split or fused in different ways. This too affected the assessment of risks: "In every case, a set of problems was perceived as more important when it was presented as two branches than when presented as one. . . . "[29]

When the first version of the pruned/complete fault tree experiment was redone with a group of experienced auto mechanics, those given the complete diagram assigned an average probability of 0.06 to the residual branch. The probabilities they assigned to the branches that were later left out in the pruned version totaled 0.38. Those shown the pruned tree should therefore have assigned a probability of about 0.44 (0.06 + 0.38) to the "all other problems" category, but they actually assigned an average probability of only 0.215. The expert's risk assessment suffered the same basic distortion as the laypersons assessment had.[30]

These studies strongly suggest the troubling possibility that the way a risk assessment is set up may strongly influence its results. Risk analysis appears to be less solid and objective than its formal structure and mathematical nature leads most of us to believe. Since a proper understanding of risk is critical to many important public policy decisions, the ease with which the results of risk assessments can be manipulated is disturbing.

The history of risk assessment in civilian nuclear industry is a good case in point. The federal government has heavily subsidized nuclear power since its inception. Early on, nuclear power was pushed especially hard, with relatively little attention to possible public health effects.[31] The mid-1950s saw the first systematic attempt on the part of the Atomic Energy Commission (AEC) to assess the risks of a major nuclear accident. Known as WASH-740, this study used "worst case" analysis, concluding that a large radioactive release could kill 3,400 people, injure 43,000, and cause $7 billion in property damage.[32] Policymakers were not pleased with the results and the AEC did not put much effort into publicizing them. Because neither nuclear industry nor the insurance companies were willing to assume a financial risk of that magnitude, the federal government rushed to the rescue with the Price-Anderson Act of 1957, which sharply limited private sector liability for nuclear power accidents. Had it not done so, it is quite possible that nuclear power would have died on the vine.

In 1964, the AEC conducted a follow-up to WASH-740. The worst-case results were even more alarming this time, with 45,000 deaths projected as the possible consequence of a major nuclear plant accident. The Commission did not even publish that study.[33] Then in 1972, the AEC set up a group headed by Norman Rasmussen to conduct what became known as the Reactor Safety Study, or WASH-1400, which structured the problem differently. Rasmussen's team was told to calculate the probabilities, as well as the consequences, of a broad array of possible nuclear reactor accidents. By focusing attention on average risk rather than worst case, the AEC hoped the results would be less alarming. The team used fault tree analysis. This time, the results came out "right," supporting the idea that nuclear power was safe. Among its famous conclusions, "a person has as much chance of dying from an atomic power accident as of being struck by a meteor."[34] Despite serious objections to the structure and method of this study from inside and outside the AEC, it released all 14 volumes in the summer of 1974. Five years later, the NRC (AEC's successor), finally repudiated the study, though not with the fanfare that accompanied its earlier release.[35]

The Dimensions of Risk and Its Consequences

Although not technically correct, in common usage "risk" is loosely used for any chance situation that involves a dangerous or otherwise undesirable outcome. There is a considerable literature, most of it based in psychology, relating to how people perceive risk in this looser sense, and how their perceptions affect their judgment of the magnitude and acceptability of any particular risk.

Slovic, Fischhoff, and Lichtenstein list nine key dimensions of risk in the form of questions:

1. Is the risk one that people expose themselves to voluntarily (e.g., breaking a leg while skiing), or is it imposed on them (e.g., getting cancer from polluted air)?
2. Does the risk have immediate consequences (e.g., getting killed in a high-speed auto accident) or are they delayed (e.g., becoming infected with the AIDS virus)?
3. Are the people exposed aware of the risk?
4. Are the risks understood by relevant scientists or other experts?

5. Is it possible to take action to avoid or mitigate the consequences of the risk (e.g., building levies to protect low-lying areas against flooding)?
6. Is the risk a new one or one that people have known about for a long time?
7. Is the risk chronic, claiming only one or a few victims at a time (e.g., cirrhosis of the liver in alcoholics) or is it acute and catastrophic, claiming many victims at the same time (e.g., a disastrous tsunami)?
8. Are people able to think calmly about the risk (e.g., cancer caused by smoking tobacco) or do they find it terrifying (e.g., being eaten by a shark)?
9. When the risk is realized, how likely is it that the consequences will be fatal?[36]

We can summarize these dimensions as *voluntariness, immediacy, personal knowledge, scientific knowledge, control, familiarity, catastrophic potential, dread,* and *severity.*

In psychological experiments, an interesting difference emerged between experts and laypeople asked to judge the riskiness of various causes of death: "When experts judge risk, their responses correlate highly with technical estimates of annual fatalities. Laypeople can assess annual fatalities if they are asked to (and produce estimates somewhat like the technical estimates). However, their judgments of 'risk' are related more to other hazard characteristics (e.g., catastrophic potential, threat to future generations) and as a result tend to differ from their own (and experts') estimates of annual fatalities."[37] In other words, the experts in these experiments rely on a narrow, stripped down, quantitative concept of average risk, while the laypeople have a much broader, richer, and more qualitative view of risk.

These differences in concept result in sharp disagreements as to the riskiness of hazards that rate very differently on the nine risk dimensions. When asked to rank 30 activities and technologies in order of risk, well-understood, chronic, voluntary risks with little dread (like smoking) were ranked similarly by experts and laypeople. But those that had considerable catastrophic potential and dread (like nuclear power) were ranked very differently. Smoking was on average ranked third or fourth by laypeople and second by experts; nuclear power was on average ranked first by most laypeople and twentieth by experts.[38]

Furthermore, among laypeople, race and gender seem to matter as well. In studies done in the early 1990s and reproduced almost a decade later, white men were found to put more trust in experts and to rate a variety of risks lower than either white women or people of color.[39]

Surely experts have more information than laypeople. But if the information they have is primarily on average risk, it may mislead at least as much as it informs. Information, especially "hard" quantitative data, can be seductive as well as informative. It can direct attention away from qualitative issues because they are more vague and subjective than quantitative measures. Yet as "soft" and difficult to integrate into "objective" calculations as they may be, these other dimensions of risk are much too important to ignore.

Experts often believe if the lay public would only take the time to look at the data, they would agree with the experts' point of view. But these disagreements about risk do not simply come from stubbornness or lack of data. They are the result of fundamentally different ideas about what risk means and how it should be judged. These differences in concept will not evaporate in the presence of more data. Of course, stubbornness and preconception can be a real problem—for experts as well as laypeople. Research shows that when people strongly hold to a particular point of view, they tend to interpret the usefulness and reliability of new evidence on the basis of whether or not it is consistent with their belief. If it is, it is relevant and useful; if not, it is obviously wrong or at least not representative.[40] When Plous gave supporters and opponents of a particular technology identical descriptions of a non-catastrophic disaster involving that technology in three studies at Wesleyan University, " . . . supporters focused on the fact that the safeguards worked, while opponents focused on the fact that the breakdown occurred in the first place." Supporters were therefore reassured by the account while opponents were more convinced than ever that the technology posed a real threat.[41]

There are no scientific grounds for dismissing either the more quantitative expert or the more qualitative lay view of risk. Decision theorists have actually developed strategies for decision making under risk and uncertainty that offer general support for both approaches. The "expected loss" strategy argues that each possible hazardous outcome of an activity or technology should be weighted by the chance that it would occur, then combined in a weighted average estimate of the danger involved. Different risks can then be rated and compared on the basis of this average. The possibility of catastrophe is not ignored, but if it is judged extremely unlikely it will have a very small weight and thus little effect on the average. Experts are essentially using this approach when they judge riskiness by average annual fatality rates. On the other hand, decision theorists have also developed the "minimax" strategy. This approach focuses attention on the worst possible outcome associated with each activity or technology whose riskiness is being judged. The strategy then calls for ranking the activities or technologies in terms of their worst cases. The riskiest is the one whose worst outcome is the most damaging of all the worst outcomes, the next most risky is the one whose worst outcome is the next most damaging and so on.

Assessing the risk of alternative activities or technologies with "minimax" is more conservative than using "expected loss." It minimizes the damage even if you are very unlucky and the worst possible outcome occurs. The minimax approach seems more closely related to how laypeople evaluate risk. Though it is possible to argue that the "expected loss" strategy is superior to the "minimax" strategy, it is just as possible to argue the opposite. There is no objective way of deciding which approach is better. That depends on which characteristics of risk are considered more important.

Overconfidence

Even those who are experts in system design, operation, or maintenance tend to overlook some possible pathways to failure. It is easy enough to understand how this could happen, but troubling because it leads to systematic *underestimates* of the chances of failure. If we believe our own risk estimates, we are likely to be overconfident in the system's reliability. When dealing with dangerous technological systems, overconfidence is hazardous to our health.

There is little evidence that we do, or that we can, completely overcome this problem even when the stakes are extremely high. Among the many criticisms leveled at the Rasmussen nuclear reactor safety study was that it did not consider improper maintenance, failure to obey operating rules when shutting the reactor down, and sabotage as possible causes of system failure. The study did consider human error.[42] However, it neither enumerated nor evaluated all of the important ways in which human error could lead to potentially disastrous reactor failure. After the NRC withdrew its endorsement, analysis of actual operating experience showed that the report's overconfidence was pervasive: "all values from operating experience fell outside the 90 percent confidence bands."[43]

Fischhoff argues that pathways to failure involving human error or intentional misbehavior are the most likely to be overlooked.[44] It is extremely difficult to think of all the ways people might cause problems. Who would expect that a technician would decide to use a lit candle to look for air leaks and so begin a fire that would disable the emergency core cooling system at the Brown's Ferry nuclear power plant? How could it be foreseen that the operators at Three Mile Island would repeatedly misdiagnose the reactor's problems and so take inappropriate actions that made matters worse? A second common cause of omissions is the failure to anticipate ways that the technological, societal, or natural environment in which the system operates may change. Suppose unusually heavy rainfall in a normally dry area causes flooding because of inadequate drainage systems, crippling the area's transportation system. Transportation designers would have had to foresee both the possibility of excessive rainfall and the fact that designers of the drainage system would fail to provide for that possibility, in order for flooding to have been included as a pathway to failure of the transportation system.

A third category is failure to foresee all the interactions within the system that affect the way the system as a whole functions. The National Academy of Sciences study of the effects of nuclear war, for example, concluded that survivors could still grow food because food plants could tolerate the additional ultraviolet light that would penetrate the earth's partially destroyed ozone layer.[45] However, they failed to consider that the increased ultraviolet (among other things) might make it virtually impossible for people to work in the fields to grow and harvest the crops. Yet another related cause of omissions is overlooking the chances of common mode failure (see Chapter 9).

Finally, omissions arise from overconfidence in our scientific and technical capabilities. Most risk analyses assume that the system has been correctly designed and that it will not fail if all of its components work properly. But as we saw in Chapters 9 and 10, this degree of confidence is not always warranted. When Challenger exploded in 1986, NASA's estimates of the odds of a space shuttle failing catastrophically ranged as high as 1 in 100,000 flights. By the time the space shuttle Columbia broke up in 2003, NASA's official estimate was only 1 in 250. Even that may have been overconfident, since NASA actually lost two space shuttles in only 113 flights.[46]

Psychological studies of risk perception and assessment throw some light on just how overconfident we can be about what we know. In one study, participants given paired lists of lethal events were asked to indicate which cause of death in each pair was more frequent. Then they were asked to give the odds that each of their answers was correct. Many people expressed high confidence in the accuracy of their answers: odds of 100:1 were given 25 percent of the time. Yet about one out of every eight answers given so confidently were wrong (less than one in a hundred should have been wrong at those odds).[47] In another related experiment, when subjects gave odds of 1,000:1, they would have been correct to give odds of only 7:1; at 100,000:1, they should have given odds of just 9:1. There were even a large number of answers for which people gave odds of 1,000,000:1 that they were right—yet they were wrong more than 6 percent of the time (true odds of 16:1). Subjects gave odds of 1,000:1 or greater about 25 percent of the time,[48] and the subjects were so sure of their answers that they were willing to bet money on them. Extreme overconfidence was "widely distributed over subjects and items."[49]

Two disturbing conclusions can be drawn from this literature. First, overconfidence is not only widespread; it also gets worse as the task at hand becomes more difficult. One would hope and expect that people confronted with more trying tasks would be more tentative about the correctness of their conclusions. Yet, "overconfidence is most extreme with tasks of great difficulty."[50] Second, and even more troubling, experts seem as likely to be overconfident as laypeople, especially when they must go beyond the bounds of available hard data. For example, seven internationally known geotechnical engineers were asked to predict how high an embankment would cause a clay foundation to fail, and to give confidence limits around this estimate that would have a 50 percent chance of enclosing the correct value. None of them estimated the failure height correctly. Worse, the proper height did not even fall within *any* of the confidence intervals they specified.[51]

Some psychologists argue that the "heuristics" (rules of thumb) that people adopt when dealing with difficult problems may lead them astray. When asked to give a range of estimate for a particular risk, people often use a heuristic called "anchoring and adjustment." They think first of a specific number ("anchor"), then adjust up and down from that number to get confidence limits that seem right to them. It is easy to influence the choice of that anchor by the way the question is asked. Yet once the anchor is chosen, it has such a powerful impact that they tend to give upper and lower bounds that are much too close together.[52]

It has also been argued that people work through complex decision problems one piece at a time. To make the process less stressful, they ignore the uncertainty in their solutions to earlier parts of the problem when they are working through later parts. Because the solutions of the later parts are built on the solutions of the earlier parts, ignoring the tentativeness of their earlier conclusions makes them overconfident in the final conclusions.[53] Overconfidence can also be a reaction to the anxiety that uncertainty creates: "One way to reduce the anxiety generated by confronting uncertainty is to deny that uncertainty ... people faced with natural hazards ... often view their world as either perfectly safe or predictable enough to preclude worry. Thus, some flood victims interviewed ... flatly denied that floods could ever recur in their area."[54] Since anxiety is greater when the problem is more complex or consequential, this might explain why people are often more overconfident in such situations. But this does not bode well for our ability to assess the risks posed by dangerous technologies, where the problems we confront are almost always both complex and highly consequential.

Former military chemical warfare instructor Peter Tauber tells this story about appearing on an educational television panel with noted scientist Sam Cohen, the "Father of the Neutron Bomb":

> With a flourish I brandished U.S. Army Technical Manual. ... "Military Chemistry and Chemical Agents" ... [which] had quite a number of ... " correction pages"—e.g. page 26, 1.4, "replace word before 'mix' ('then') with 'DO NOT EVER' ... For all those pages of corrections they still never got the formula for the ... nerve agent compound ... Tabun ... quite right. ... I merely pointed these out to stress the folly of betting our lives on systems and personnel prone to consequential error. ... Dr. Cohen gestured with imperious amusement. ... If only we knew what he knew, he assured us. ... Dr. Cohen snatched a pack of cigarettes, shook one free and without missing a beat, continued to lecture us on the perfectibility of complex systems, as he proceeded ... to light the filter end."[55]

A CALCULUS OF CATASTROPHE

In addition to all the problems we have discussed, our ways of evaluating risks that involve dangerous technologies are also flawed because we have yet to develop a clear way of thinking about low probability/high social consequence events. As a result, we confuse very unlikely events that have enormous social consequences with very unlikely events that are harmful to those involved but have little effect on society.

Beginning with the famous "meteor" statement in the Rasmussen report, in the long debate over nuclear power, it has been common to see statements like "the probability of getting killed as a result of a nuclear core meltdown in a power reactor is about the same as the chance of dying from a snake bite in mid-town Manhattan." In one sense, the comparison seems valid. Both possibilities are very unlikely and after all, dead is dead. But the underlying events are not at all comparable. A snakebite might kill the person bitten, but it has little effect on the wider society. But a core meltdown that breaches a reactor's containment and spews deadly radioactivity all over the landscape is an event of enormous long-term social consequence, with the potential for taking hundreds of thousands of lives and doing billions of dollars of property damage. The huge difference in their social impact puts these two kinds of events into entirely different categories. Even if their probabilities were mathematically identical, they are simply not comparable.

We also do not know how to properly aggregate technological risks that have low probability but high social consequence. We look at each risk separately. If its probability is low enough, we call it "acceptable," and often unconsciously translate that into "it will not happen." Then we develop and use the technology involved. But looking at these risks at a point in time and in isolation from each other is foolish. They are neither static nor independent. They cumulate in three different ways, all of which make catastrophe more likely. First, they grow over time. No matter how unlikely an event is (unless it is impossible), the probability that it will happen at least once increases to certainty as time goes by. Second, the larger the number of dangerous technologies we create and the more widespread their use, the more low probability/high consequence events there are. The more low probability/high consequence events there are, the higher the probability that at least one of them will happen. As we add new toxic chemicals to the environment, the probability of ecological disaster increases; as we add to the size and variety of arsenals of weapons of mass destruction, the probability of catastrophic war or weapons accident increases, and so on. In other words, the more ways there are for technology to generate a catastrophe, the more likely it is that at least one of them will do us in.

Finally, there is the very real possibility of synergies. As the actual and potential damage we do by one route increases, it may increase the probability of disaster by another route. A future Chernobyl, spewing dangerous radioactivity across national borders, might just be the spark that could ignite an already tense international situation and cause a war in which weapons of mass destruction are used. Or chronic ecological damage done by one set of toxics could magnify the impact of a different set of toxics to the point of precipitating a major environmental disaster. With greater numbers of increasingly complex dangerous technologies in use, it becomes impossible to even imagine all the ways in which synergies like these could raise the probability of technology-induced disaster to the point of inevitability.

It would be very helpful to have a "calculus of catastrophe"—a method that focused our attention on low probability/high social consequence events and allowed us to clearly evaluate the risk associated with each of them over time and in combination with each other. We could then monitor the total social risk and take care not to do anything that pushed it above the level we consider acceptable. Such a calculus of catastrophe would be particularly useful in making social decisions about risks involving dangerous technologies. But here too we must avoid the illusion of a simple technical fix. No matter how clever, no method of calculation is going to solve the problem of differing judgments as to what dimensions of risk are most important, or save us from the hard work of dealing with the complex social, economic, and political problems of making decisions on technological risk. No method of calculation can completely come to terms with the inherent unpredictability that human fallibility unavoidably introduces.

It may seem strange, even foolish, to be so concerned about the possibility of highly improbable technological disasters, especially those that have never happened. Yet *everything that has ever happened, happened once for the first time.* Before then, it too had never happened. For some types of catastrophic events, we cannot afford even a first time.

In this and in the previous chapters, we have carefully considered the many ways in which disaster might be triggered by the clash of our innate fallibility and the dangerous technologies with which we have filled our world. We have looked at a wide range of often terrifying problems. Now it is time to look for solutions. In the next chapter, we consider what we can do to prevent our technological society from succumbing to one or another of the self-induced catastrophes that lie waiting for us along the path we have been traveling.

NOTES

1. The likelihood of any event is mathematically measured by its "probability," expressed as a number ranging from zero (the event is impossible) to one (the event is certain). This can be thought of as reflecting the fraction of

trials of many repeated performances of the chance experience that would result in that particular outcome.

2. E. Fromm, "Paranoia and Policy," *New York Times*, December 11, 1975.

3. V. K. McElheny, "Improbable Strikes by Lightning Tripped Its System, Con Ed Says," *New York Times*, July 15, 1977. See also W. Sullivan, "Con Ed Delay Cited as a Possible Cause," *New York Times*, July 16, 1977.

4. "Investigators Agree New York Blackout of 1977 Could Have been Avoided," *Science*, September 15, 1978, 994–996.

5. R. D. McFadden, "Disaster Status Given New York and Westchester to Speed Loans; Services Resume After Blackout," *New York Times*, July 16, 1977.

6. *International Herald Tribune*, October 31, 1968.

7. D. Anderson, "Four! A Hole-In-One Record at Open," *New York Times*, June 17, 1989.

8. Susan Stamberg, "Weekend Edition, with Scott Simon," *National Public Radio*, July 14, 2001; Hour 1), transcript 18–22.

9. "Doctor's Say Gun Wound to Brain Eliminated a Man's Mental Illness," *New York Times*, February 25, 1988.

10. National Desk, "Cause of a Man's Toothache? A 4-Inch Nail," *New York Times*, January 17, 2005.

11. R. D. McFadden, "Odds-Defying Jersey Woman Hits Lottery Jackpot 2nd Time," *New York Times*, February 14, 1986.

12. P. M. Boffey, "Fifth US Rocket Failure Is Acknowledged by NASA," *New York Times*, May 10, 1986.

13. A. F. Manfredi and C. DiMaggio, "Effect of U.S. Space Launch Vehicle Problems on the SDI Decision Timeline" (Washington, DC: Congressional Research Service, 86-783 SPR, July 17, 1986), 1.

14. Op. cit., P. M. Boffey.

15. Op. cit., Manfredi and Di Maggio, 1.

16. Associated Press, "Hospital Fumes That Hurt 6 Are Tied to Nerve Gas," *New York Times*, November 5, 1994.

17. P. Diaconis and F. Mosteller, "Methods for Studying Coincidences," *Journal of the American Statistical Association*, December 1989, 859.

18. Ibid.

19. Ibid.

20. Ibid., 858, Diaconis and Mosteller give a formula for calculating the approximate number of people (N) that need to be in the room to have a fifty-fifty chance that at least two of them will have one of c categories of event (birthday, anniversary, graduation date, etc.) within k days of each other: $N = 1.2 [c/(2k +1)]^{1/2}$

21. K. Z. Morgan, "Radiation-Induced Health Effects," letter to *Science*, January 28, 1977, 346.

22. C. Perrow, *Normal Accidents* (New York: Basic Books, 1984), 137–139. See also National Transportation Safety Board, document AAR-82-3, April 6, 1982, 14–17.

23. Rob Jameson, "The Blunders that Led to the Banking Crisis," *New Scientist*, September 25, 2008; http://www.newscientist.com/article/mg19926754.200-the-blunders-that-led-to-the-banking-crisis.html (accessed October 30, 2009).

24. There were other more important underlying causes for the financial crisis than inaccurate risk assessment.

25. Op. cit., Rob Jameson.

26. B. Fischhoff, P. Slovic, and S. Lichtenstein, "Fault Trees: Sensitivity of Estimated Failure Probabilities to Problem Representation,"*Journal of Experimental Psychology: Human Perception and Performance* Vol. 4, No. 2 (1978), 330–344.

27. Ibid., 334.

28. Ibid., 335.

29. Ibid., 340.

30. Ibid., 342.

31. L. Clarke, "The Origins of Nuclear Power: A Case of Institutional Conflict," *Social Problems* Vol. 32 (1985), 474–487.

32. L. Clarke, "Politics and Bias in Risk Assessment," Vol. 25, No. 2 (1988), 158–159.

33. Ibid., 159. See also J. Primack, "Nuclear Reactor Safety: An Introduction to the Issues," *Bulletin of the Atomic Scientists*, September 1975, 16.

34. U.S. Atomic Energy Commission, *Reactor Safety Study: An Assessment of Accident Risks in United States Commercial Nuclear Power Plants* (Washington, DC: U.S. Nuclear Regulatory Commission, 1975); D. F. Ford, "A Reporter at Large: The Cult of the Atom-Part II," *New Yorker*, November 1, 1982, 58.

35. Op. cit., Clarke (1988), 159.

36. P. Slovic, B. Fischhoff, and S. Lichtenstein, "Facts and Fears: Understanding Perceived Risk," in *Societal Risk Assessment: How Safe Is Enough?*, ed. R. C. Schwing and W. A. Albers (New York: Plenum Press, 1980), 195.

37. P. Slovic, "Perceptions of Risk," *Science*, April 17, 1987, 283.

38. Ibid., 281.

39. Ruth Bennett, "Risky Business: The Science of Decision Making Grapples with Sex, Race, and Power," *Science News*, September 16, 2000.

40. P. Slovic, "Perceptions of Risk," *Science*, April 17, 1987, 281.

41. S. Plous, "Biases in the Assimilation of Technological Breakdowns: Do Accidents Make Us Safer," *Journal of Applied Social Psychology* 21, 13 (1991), 1058.

42. R. K. Weatherwax, "Virtues and Limitations of Risk Analysis," *Bulletin of the Atomic Scientists*, September 1975, 31.

43. Kristin Schrader-Frechette, "Scientific Method, Anti-Foundationalism and Public Decision-Making," *Risk: Issues in Health and Safety* 1(1) (1990), 29.

44. B. Fischhoff, "Cost Benefit Analysis and the Art of Motorcycle Maintenance," *Policy Sciences* No. 8 (1977), 181–183.

45. National Research Council, National Academy of Sciences, *Long-Term Worldwide Effects of Multiple Nuclear-Weapons Detonations*, Washington, DC: National Academy of Sciences, 1975.

46. Charles Seife, "Columbia Disaster Underscores the Risky Nature of Risk Analysis," *Science*, February 14, 2003, 1001.

47. Op. cit., Slovic, Fischhoff, and Lichtenstein, 185.

48. Ibid., 558.

49. Ibid., 554 and 561.

50. S. Lichtenstein, B. Fischhoff, and L. D. Phillips, "Calibration of Probabilities: the State of the Art to 1980," in *Judgment Under Uncertainty: Heuristics and Biases*, ed. D. Kahneman, P. Slovic, and A. Tversky (Cambridge, MA: Cambridge University Press, 1982), 315–318.

51. M. Hynes and E. Vanmarcke, "Reliability of Embankment Performance Predictions," *Proceedings of the ASCE Engineering Mechanics Division Specialty Conference* (Waterloo, Ontario: University of Waterloo Press, 1976), 31–33.

52. Op. cit., Lichtenstein, Fischhoff, and Phillips, 333. See also S. Plous, "Thinking the Unthinkable: The Effects of Anchoring on Likelihood Estimates of Nuclear War," *Journal of Applied Social Psychology* 19, 1 (1989), 67–91.

53. Ibid.

54. Op. cit., Slovic, Fischhoff, and Lichtenstein, 188.

55. Peter Tauber, "Have Banana Peel, Will Travel: Erraticism 101," *The Sciences* (New York Academy of Sciences, May 2000).

12

Escaping the Trap: Four Steps for Preventing Disaster

The clash between our growing technological power and our unavoidable fallibility has laid us open to grave danger. We have not only gained the knowledge required to trigger any of a variety of global catastrophes, we have constructed the means. There is no longer any corner of the planet so remote that it is safe from the havoc we can now cause, intentionally or by accident.

There is no reason to believe that we will ever forget how to do that which we have learned how to do. And there is every reason to believe that our frailty, our potential for malevolent behavior, and our tendency to make mistakes are a permanent part of what makes us human. Are we then doomed to a world of increasing disaster, a world that may someday even see our species die by its own hand?

Ultimately, none of the dangerous technologies we have developed is really beyond our control. We can change them, limit how we use them, or even eliminate them entirely. No external force compelled us to create the dangers we now face, and no external force will prevent us from getting rid of them. Yet the technologies that have put us in danger were not developed by accident. They were brought into being to serve purposes important enough to make us want to explore the ideas that lie behind them and then build the devices that gave those ideas physical form. It is possible to prevent disaster without giving up all the benefits that led us to create dangerous technologies and cause us to continue relying on them. We can find ways to enjoy the advantages of modern technology without exposing ourselves to self-induced catastrophe.

There is no technological fix, simple or complicated, for the problems created by the deadly combination of human fallibility and dangerous technology. But there is a solution. It lies at the intersection of technological capability, social wisdom, and political will. There are four steps we must take. The first begins with recognizing that there are some technologies whose potential for catastrophe is so great that fallible human beings have no business dealing with them at all. No matter what short-term benefits they seem to provide, in the long run they are simply too dangerous.

STEP 1: ABOLISH WEAPONS OF MASS DESTRUCTION

Given the inevitability of eventual technical failure or human-induced disaster, abolition is the only sensible, long-term viable way of dealing with technologies capable of destroying human society and driving our species to extinction. Nuclear, biological, and chemical weapons of mass destruction are the clearest present day examples of such technologies. But is abolition really practical? Can the genie be put back into the bottle?

The Canberra Commission

In 1996, the government-sponsored Canberra Commission presented a comprehensive plan for abolishing nuclear weapons to the United Nations. It was not the first time nuclear abolition was proposed. But this time, the proposal came from a group with very unusual credentials, from four of the five nuclear weapons states. The Commission included General George Lee Butler, former commander-in-chief of the U.S. Strategic Command, the officer in charge of all American strategic nuclear weapons from 1991 to 1994; Michel Rocard, Prime Minister of France from 1988 to 1991; Field Marshal Michael Carver, former chief of the British Defense Staff; Roald Sagdeev, former science advisor to the President of the Soviet Union; and Robert MacNamara, former U.S. Secretary of Defense and a key figure in developing the idea of security through "mutually assured destruction" that drove much of the nuclear arms race. The lengthy report argues, "The proposition that large numbers of nuclear weapons can be retained in perpetuity and never used—accidentally or by decision—defies credibility. The only complete defense is the elimination of nuclear weapons and assurance that they will never be produced again."[1]

The Commission goes farther, arguing that nuclear weapons have little military utility and do not even provide security, their main *raison d'etre*:

> Possession of nuclear weapons has not prevented wars ... which directly or indirectly involve the major powers. They were deemed unsuitable for use even when those powers suffered humiliating military setbacks [as did America in Vietnam and the Soviets in Afghanistan]. ... Thus, the only apparent military utility that remains for nuclear weapons is in deterring their use by others. That utility ... would disappear completely if nuclear weapons were eliminated ...
>
> The possession of nuclear weapons by any state is a constant stimulus to other states to acquire them. The world faces threats of nuclear proliferation and nuclear terrorism. ... For these reasons, a central reality is that nuclear weapons diminish the security of all states.[2]

The year before the Canberra Commission, U.S. Air Force General Charles A. Horner, head of the North American Aerospace Defense Command (NORAD) had become the first active-duty officer to publicly call for the abolition of nuclear weapons, saying "I want to get rid of them all. ... Think of the high moral ground we secure by having none. ... It's kind of hard for us to say ... 'You are terrible people, you're developing a nuclear weapon' when the United States has thousands of them."[3] Then on December 8, 1996, 60 retired generals and admirals from the all of the declared nuclear-armed nations (the United States, Russia, England, France, and China) signed a joint statement at the United Nations endorsing the idea that nuclear weapons can and should be completely eliminated.[4]

Very much the insider, with decades of experience in the U.S. nuclear military, General Butler continued to forcefully argue the case for abolition. In an article based on his speech to the National Press Club in early 1998, Butler declared:

> "I was engaged in the labyrinthine conjecture of the strategist, the exacting routines of the target planner, and the demanding skills of the air crew and the missileer. I was a party to their history ... witnessed heroic sacrifice and the catastrophic failure of both men and machines. ... Ultimately, I came to these judgments:
>
> - ... the risks and consequences of nuclear war have never been properly weighed by those who brandish them [nuclear arms]. ...
> - The likely consequences of nuclear war have no political, military or moral justification.
> - The threat to use nuclear weapons is indefensible.

" . . . We cannot at once keep sacred the miracle of existence and hold sacrosanct the capacity to destroy it. . . . It is time to reassert . . . the voice of reason. . . . "[5]

The movement in support of nuclear abolition has continued to grow, even among some of the most prominent Cold Warriors. On January 4, 2007, the *Wall Street Journal* published an editorial "A World Free of Nuclear Weapons," jointly authored by Henry Kissinger (former U.S. Secretary of State and Assistant to the President for National Security), George Schultz (former U.S. Secretary of State), William Perry (former U.S. Secretary of Defense), and Sam Nunn (former chairman of the Senate Armed Services Committee). The concluding paragraph of the editorial began, "We endorse setting the goal of a world free of nuclear weapons and working energetically on the actions required to achieve that goal. . . . " December 2008 saw the launch of a new organization, Global Zero, focused on the goal of eliminating nuclear weapons within 25 years and supported by former U.S. President Jimmy Carter, former Soviet Premier Mikhail Gorbachev, former U.S. Defense Secretary Frank Carlucci, former Pakistani Foreign Minister Shaharyar Khan, retired chief of the Indian Air Force S.P. Tyagi, former British Foreign Secretary Malcolm Rifkind, and a number of others with similar credentials.[6]

This idea has also been gaining currency among sitting national leaders. On January 21, 2008, Gordon Brown, Prime Minister of Great Britain, was quoted as saying, "we will be at the forefront of the international campaign to accelerate disarmament among possessor states and to ultimately achieve a world that is free from nuclear weapons."[7] A few months later, President of France Nicolas Sarcozy, while not endorsing abolition, broke ground by saying he would cut France's nuclear arsenal to fewer than 300 warheads, "half the maximum number of warheads we had during the Cold War."[8] On September 24, 2009, presiding over the Summit on Nuclear Nonproliferation and Nuclear Disarmament of the U.N. Security Council, U.S. President Barack Obama presented UN Resolution 1887 of 2009 for a vote. All of the members of the Security Council (including France, China, Russia, the United States, and Britain) voted in favor of the resolution about which President Obama said, "The historic resolution we just adopted enshrines our shared commitment to the goal of a world without nuclear weapons. And it brings Security Council agreement on a broad framework for action to reduce nuclear dangers as we work toward that goal."[9]

Although most people are not aware of it, the five major nuclear-armed states (the United States, Britain, Russia, China, and France) have had a formal treaty commitment for many years to do exactly what the generals, the admirals, the Canberra Commission, and by now so many others have urged. Along with 182 other nations, they signed the Treaty on the Nonproliferation of Nuclear Weapons, Article VI of which reads, "Each of the Parties to the Treaty undertakes to pursue negotiations in good faith on effective measures relating to cessation of the nuclear arms race at an early date and to nuclear disarmament.... "[10] For the United States, Russia, and Britain, that commitment is very longstanding. They signed the Treaty more than 40 years ago (1968). It is long past time for them to keep their promise.

Inspection and Verification

The elimination of nuclear, chemical, and biological weapons of mass destruction is not only practical; it is critical to our security and our survival. Even so, abolition will not, and should not be achieved overnight. A careful process of arms reduction requires time to verify that each nation has done what it has agreed to do at each step of the process. It also takes time for everyone to gain confidence that the process is working. Care must be taken to create a world open enough to inspection that any meaningful violation of the treaty will be quickly detected.

For more than 10 years after the defeat of Iraq in the 1990 to 1991 Persian Gulf War, U.N. inspectors combed the country for evidence of clandestine programs aimed at developing biological, chemical, and nuclear weapons of mass destruction. As a condition of ending that war, Iraq had agreed to allow free access. Nevertheless, the Iraqi government continually interfered with and harassed the inspectors. From spying on them to find out about "surprise" inspections to blocking access to so-called "presidential sites," Iraq tried every trick in the book to frustrate the inspection process.[11] Yet despite all of Iraq's obstructionism, unarmed U.N. inspectors were able to find and destroy more weapons of mass destruction in Iraq than were destroyed by military action during the first Persian Gulf War.[12] And the conclusion of the IAEA inspectors in the months prior to the 2003 U.S.-led invasion that Iraq's nuclear weapons program and stockpiles of weapons of mass destruction had long since been eliminated turned out to be exactly correct. Despite a great deal of effort by U.S. forces in the absence of an obstructionist Iraqi government in the years after the 2003 occupation, no such weapons were ever found.

Inspectors from the International Atomic Energy Agency (IAEA) first tried "environmental monitoring" in Iraq in the early 1990s. The technique uses advanced technologies to test for the presence of materials likely to be emitted by forbidden activities. It has made substantial progress over the years. In 2003, the Deputy Director General of the IAEA said, "You can sample dust from a truck that has passed by a factory. You can see a single atom. It's really marvelous."[13] Though IAEA used this approach only to search for nuclear materials, there is no reason why it could not also be used to detect at least some suspicious chemical or biological weapons activities. Environmental monitoring can certainly provide evidence of illicit activities at declared sites and may be able to detect covert activities at facilities that have never been declared. According to the U.S. Office of Technology Assessment:

> In the month between the end of the [1990 to 1991] war and the start of inspections, Iraq removed much of the most incriminating equipment . . . and concocted stories to explain the remainder. Inspectors took samples of materials within and near facilities, and swipes of dust that had collected on the surfaces of equipment. These were analyzed at various laboratories . . . [and] played a key part in demolishing Iraq's cover stories and exposing its nuclear weapon program. . . . *No industrial process can prevent minute traces of materials from escaping.* Even the most sophisticated filtration systems can only reduce, not eliminate, releases. . . . [14] (emphasis added)

Environmental monitoring can also be used to detect secret nuclear weapons tests carried out in violation of the Comprehensive Test Ban Treaty. An on-site inspection team armed with tubes for boring into the suspected site of the test could find evidence of telltale radioactive trace gases weeks to more than a year after even a "well-contained" clandestine explosion.[15] Although it would be unwise to depend solely on environmental monitoring (or any other single technique) to verify compliance with abolition agreements, it should be extremely useful as part of an inspection and verification regime.

Even without intrusive inspection, it is much more feasible today than it once was to monitor clandestine attempts to test nuclear weapons. By 2009, "The scientific and technical community . . . [had] developed a well-honed ability to monitor militarily significant nuclear test explosions anywhere in the world, above ground or below, and to distinguish them from mine collapses, earthquakes and other natural or nonnuclear phenomena." For example, the relatively small North Korean underground nuclear test (in 2006) was "promptly detected and identified."[16]

Nevertheless, effective inspection is greatly facilitated by both open access and fierce determination on the part of well-trained inspectors, fully equipped and strongly supported by the parties to the disarmament agreement. Ironically, the same problems of error and fallibility that call us to abolish weapons of mass destruction also make it obvious that no system of inspection can be guaranteed to work perfectly all the time. There is simply no way to eliminate risk from the world. Yet the risks involved in carefully and deliberately eliminating weapons of mass destruction are far less threatening than the certainty of eventual catastrophe we face if we do not.

What if some nation some day does manage to slip through a crack in the system and secretly build a nuclear weapon without being detected? It may seem obvious that that would give them a decisive political advantage in a world without weapons of mass destruction. But it is not at all obvious. The rogue government would have to make it known that it had violated the treaty and built such a weapon in order to have any chance of political gain. There is no way to achieve political advantage from a weapon no one knows about. That revelation would instantly make it a pariah, subject to immediate political isolation, economic boycott, or even conventional military attack by every other nation in the world. There is nothing that country could do with one or two nuclear weapons to prevent this from happening. It could not even be sure to devastate the nation it had built the bomb to threaten, let alone prevail against such overwhelming odds.

After all, in the aftermath of the horrors of World War II, the United States was the only nation that had nuclear weapons or the knowledge required to build them. Yet America's nuclear monopoly did not allow it to bend every other nation to its will. If a powerful United States could not even dictate terms to a world of heavily damaged, war-weary nations (the Soviet Union being the most striking example), it is difficult to see how a rogue nation with one or two nuclear weapons would be more able to dictate terms to a strong, healthy, intact world at peace. The leaders of that nation might be able to cause a lot of pain, but there is nothing they could actually gain by having one or two nuclear weapons that would be worth the enormous risks involved in building them.

Inspection and verification systems do not have to be perfect for the abolition of nuclear, chemical, and biological weapons of mass destruction to succeed. They only need to be good enough to insure that any attempt to build a significant arsenal will be caught and exposed.[17] Perfection is simply not required. It is not easy to design, and even more difficult to implement systems that good, but it is certainly possible. Some interesting, creative, and practical ideas have already been suggested.[18]

Apart from the problems of developing a disarmament process supported by effective systems of verification and inspection, it is not a simple matter to physically eliminate the world's stockpiles of weapons of mass destruction. There are serious technical issues involved in safely destroying the weapons.

Destroying Chemical and Biological Weapons

The Chemical Weapons Convention (CWC) entered into force in 1997. By signing, Russia, the United States, and 182 other nations agreed to destroy all their chemical weapons and weapons production facilities.[19] Chemical weapons can be destroyed by a number of processes, including incineration and chemical neutralization.[20] In 1996, a panel of the National Research Council endorsed neutralization as the technology best able to protect human health and the environment while effectively destroying the chemical weapons by the Army's deadline.[21]

Nevertheless, the United States chose to rely mainly on incineration, having built its first prototype chemical weapons incinerator ("JACADS") on Johnston Atoll in the Pacific years earlier.[22] A number of other incineration plants were built in the continental United States over the years. But not until 2009 did the Pentagon propose to build the first plants to destroy chemical weapons by neutralization.[23] By that time, the United States had already destroyed 60 percent of its more than 30,000 metric ton stockpile; although it was projecting that it would not finish destroying the arsenal until 2021, nine years after its deadline.[24] Russia, which had managed to destroy only about 30 percent of its more than 40,000 metric ton arsenal by 2009, was also making steady progress though it too would not meet its 2012 deadline.[25]

At the end of World War II, the allies captured nearly 300,000 tons of chemical weapons manufactured by Nazi Germany. More than two-thirds of them were dumped into the oceans by the victorious British, French, and American forces. They are still there. The United States continued to dispose of chemical weapons in the oceans until 1970.[26] The Chemical Weapons Convention specifically excludes weapons dumped into the sea before 1985. Some scientists believe it is less dangerous to leave them alone than to try to recover and destroy them, especially if they are in deep water. Others argue that they are a ticking time bomb that must be defused, pointing out that some mustard gas bombs have already washed up on beaches.[27] This is yet another dimension of the chemical weapons disposal problem we cannot afford to ignore, but we must think long and hard before deciding what to do.

Though there is no public information available on the size or content of any stockpiles of biological weapons that might exist, there has been considerable biological warfare research. In early 1998, Dr. Kanatjan Alibekov ("Ken Alibek"), a medical doctor and former colonel in the Soviet military, began to speak publicly about the size and scope of the Soviet/Russian biological weapons development effort.[28] Alibekov, who left the former Soviet Union for the United States in 1992, had been second-in-command of a branch of that program. He claimed that, by the early 1980s, the Soviets had prepared many tons of anthrax and plague bacteria and smallpox virus that could be mounted for delivery by intercontinental ballistic missile with only a few days notice. Although the offensive program was officially cancelled by Soviet President Gorbachev in 1990 and again by Russian President Yeltsin in 1992, research continues on defense against biological attack (developing vaccines, protective masks, etc.). Alibekov believed that some offensive research—including work on "genetically altered antibiotic-resistant strains of plague, anthrax, tularemia and glanders" and work on the Marburg virus—might have been continued under the cover of the ongoing defensive program.[29]

The U.S. offensive biological weapons program was officially cancelled in the 1970s, but as in Russia, "defensive" research continues that could provide cover for otherwise forbidden offensive research. After all, it is hard to develop a good defense, say an effective vaccine, without knowing what form of which organism might be used in the attack and knowing what organism might be used requires a degree of offensive research.

It should be possible to destroy any biological toxin that has been accumulated as a result of this research (and any clandestine weapons programs) by processes similar to those used to destroy toxic chemicals. But even greater care must be taken to destroy the virulent living organisms that have been studied, especially any that are genetically modified or newly created and do not exist in nature. Biotoxins cannot reproduce, but living organisms can. Not even one of these genetically engineered germs can be allowed to escape into the environment alive.

Dismantling Nuclear Weapons

Nuclear weapons cannot be incinerated or neutralized the way chemical and biological weapons can. They must instead be carefully taken apart. Since the Pantex Plant in Amarillo, Texas was the final assembly point for all American nuclear weapons during the long Cold War, it logically became the point of disassembly when arms reductions began. More than 11,000 of the 21,500 nuclear warheads in the U.S. arsenal at the end of the Cold War were disassembled during the 1990s.[30] By January 2007, there were still nearly 10,000 nuclear warheads (5,700 of which were considered active/operational) in the U.S. stockpile, though plans had been announced in 2004 to cut the stockpile to about 6,000 by 2012.[31]

The Cooperative Threat Reduction Program—set up to help previously hostile countries reduce their arsenals of weapons of mass destruction—had already assisted these countries (chiefly Russia) in dismantling more than 7,500 nuclear warheads by the end of 2008 (89% of its 2012 goal).[32] We have made some real progress in reducing nuclear arsenals, but there is still a long way to go before we rid ourselves of these terrifying weapons and therefore of the inherent insecurity from which a world with nuclear weapons must suffer.

By 1998, roughly 10,000 plutonium "pits" (radioactive cores) from dismantled nuclear weapons were in storage at Pantex, a facility with a less-than-pristine safeguards history (see Chapter 3). That amounts to about 30,000 to 40,000 kilograms of plutonium stored as pits in one location. According to the GAO, those pits were stored in containers "that both the Department [of Energy] and the Defense Nuclear Facilities Safety Board believe are not suitable for extended storage. . . ."[33] Recognizing that it was too dangerous to store this much plutonium metal in this form, the Department of Energy (DoE) announced in 1997 that it had decided to follow a two-track approach to more permanent disposal.[34] Some of the plutonium would be "vitrified" (mixed with other highly radioactive waste then solidified as a glass) and stored in steel canisters as nuclear waste. The rest was to be converted into oxide form, mixed with oxides of uranium, and fabricated into mixed oxide (MOX) reactor fuel to be "burned" in civilian nuclear power reactors.

Weapons-grade uranium from dismantled nuclear weapons can also be made less dangerous by mixing with natural uranium and making the "blended down" uranium into standard, low-enriched reactor fuel. By 2009, some 10 percent of the electricity being generated in the United States came from dismantled nuclear weapons: 45 percent of the fuel in American nuclear power reactors was blended down uranium from dismantled Russian nuclear warheads; another 5 percent from dismantled U.S. weapons.[35]

On the surface, for plutonium the MOX option seems more promising. "Burning" plutonium in reactors sounds like a more complete way to destroy it. Using it to generate electric power also sounds more cost effective than just throwing it away, but a closer look reveals serious technical and economic problems with MOX. For one thing, plutonium does not actually "burn" in reactors the way oil or coal burns in a furnace. Running MOX through reactors does convert weapons grade plutonium (Pu 239) both into plutonium isotopes less well suited for nuclear weapons (Pu 238, 240 and 242) and into other elements, but it also turns some uranium into plutonium.[36] In the end, there is still plenty of plutonium to be disposed of, now contaminated with newly created highly radioactive waste. The plutonium could still be separated and used for weapons, although it would require a fairly sophisticated processing plant.[37] More troubling still, the MOX approach may encourage international commerce in plutonium-based fuels, raising the risk of accidents, theft, nuclear terrorism, and nuclear weapons proliferation.

The MOX option is also not cheap. It requires new production facilities as well as changes in existing facilities. According to calculations by William Weida (based on DoE data released in 1996), the total life cycle cost of fuel fabrication, capital investment, and operations involved in using MOX fuel in commercial nuclear power reactors is between $2.5 billion and $11.6 billion.[38] The cost of safely storing the extra nuclear waste generated by the MOX approach must also be added. By late 2006, the estimated cost of the ongoing effort to build a MOX plant at Savannah River, Georgia had more than quadrupled, from $1 billion to $4.7 billion. At the same time, critics calculated that the cost of building an equivalent vitrification plant would be billions of dollars less.[39]

Yet another problem with the MOX option is that the metallic element gallium is alloyed with the plutonium used in American nuclear weapons. In high concentrations, gallium interferes with fabrication of MOX fuel. It also chemically attacks the zirconium metal used in the tubes that hold reactor fuel, causing potentially serious deterioration problems after the fuel is used. Current processes for extracting gallium are expensive and generate large quantities of liquid radioactive waste. Gallium is not known to pose any particular problem for vitrification.[40]

The DoE had opened a vitrification plant at Savannah River in March 1996. Since the plant is designed to deal with liquid nuclear waste, using it to vitrify metallic plutonium pits would require first processing them into some form that could be mixed with the waste. The facility mixes radioactive waste with molten glass, pours the mixture into stainless steel cylinders, and solidifies it into glass "logs" 10 feet long and 2 feet in diameter. Operations at the plant are highly automated, since this process only renders the radioactive material easier to handle and harder to extract, not less deadly. (A worker standing by one of the glass-filled steel cylinders would absorb a lethal dose in minutes.) Like the MOX option, this approach is not cheap. The $2.4 billion plant produced only one glass log per day, making the cost over the projected life of the plant about $1.4 million per log. At that rate, it would take 25 years even to vitrify all of the 36 million gallons of liquid waste already stored in 51 underground tanks at the site.[41]

Vitrification is far from ideal, but at the current state of knowledge, it seems to make more sense than MOX. Whatever approach we use, it is important to keep in mind that the end product will still be dangerous radioactive material that has to be safely stored, kept isolated from the environment, and protected for very long periods of time. Even so, it will pose much less danger to us than the danger posed by continuing to store pure weapons-grade plutonium metal, let alone by continuing to maintain large stockpiles of intact nuclear weapons.

We humans are a contentious and quarrelsome lot. It will be a long time, if ever, before we learn to treat each other with the care and respect with which we would all like to be treated. As cantankerous and conflictual as we may be, there is no reason why we cannot learn to manage and resolve our conflicts without maintaining huge arsenals of weapons that threaten our very existence. In this increasingly interconnected and technologically sophisticated world of fallible human beings, these weapons have become too dangerous for us to continue to rely on them for our security. The very idea that weapons of mass destruction can provide security is an idea whose time has gone. And nothing should be less powerful than an idea whose time has gone.

STEP 2: CHOOSE NEW, MORE EFFECTIVE SECURITY STRATEGIES

Abolishing weapons of mass destruction will not by itself make us secure. In a troubled world of nations and people that still have a great deal to learn about getting along with each other, we need some other way of protecting ourselves. But how?

The Canberra Commission's answer is for the major powers to rely on large and powerful conventional military forces instead. While that might be an improvement, there are other more effective and much less expensive ways of achieving security. Among the alternative security strategies worthy of a closer look are non-offensive defense, the more efficacious use of economic sanctions, and using properly structured economic relations to create positive incentives to keep the peace.

Non-Offensive Defense

One form of "non-offensive defense" is a military approach known as the "porcupine" strategy. Its advocates argue that a nation is most secure when, like a porcupine, it has a very strong defense against anyone who tries to attack and yet has no capability to launch an attack against anyone else. Because it has no offensive military capability, there is no reason for other nations to worry that it might attack them. Therefore, there is no reason for other nations to build up their own military forces to counter the threat of attack, and no reason to think about launching a preemptive strike to prevent it. At the same time, powerful defensive forces make it clear to any potential aggressor that attacking would be pointless, since the devastating losses they would suffer would overwhelm any possible gains.[42] It is also possible to conceive of nonmilitary or mixed versions of non-offensive defense.[43]

Non-offensive defense may be an easier strategy to sell politically to nations that do not have "great power" status, but even the largest and most powerful nations would do well to thoroughly explore the advantages it offers. Unlike more traditional offensive military approaches to security, the more nations decide to adopt it, the more secure all of them become.

Economic Sanctions

A different but compatible approach is to use nonmilitary forms of force against countries that threaten the peace. Most prominent among them are economic sanctions (e.g., trade embargoes, financial boycotts, and the freezing of assets abroad). The strongest sanction is a total ban on exports to and imports from the target country. Economic sanctions have a reputation for being weak and nearly always ineffective, but they actually work much better than most people think.

In 1990, Gary Hufbauer, Jeffrey Schott, and Kimberly Elliott published a comprehensive analysis of the use of economic sanctions in international relations since World War I.[44] They analyzed 116 cases, most of which involved the sanctions applied by the United States. Using a consistent set of criteria to score each case, they judged that economic sanctions were successful about a third of the time. That may not seem terribly impressive, but no security strategy, military or nonmilitary, works all the time. Even if they cannot reliably do the whole job, sanctions clearly do work well enough to play a serious role in providing security.

If we can better understand why sanctions succeed when they do and why they fail when they do, we can surely make them work better. To that end Hufbauer, Schott, and Elliott tried to analyze the conditions under which economic sanctions have historically succeeded. Their conclusions were as follows:

1. The higher the cost sanctions imposed on the target country, the more successful they were. On average, sanctions that cost the target country about 2.5 percent of its Gross National Product (GNP) succeeded, while sanctions that cost 1 percent or less failed.
2. Sanctions do not work quickly. Overall, an average of almost three years was required for success.
3. The success rate was higher when the target country was much smaller economically than those applying the sanctions.
4. Applying sanctions quickly and in full force increases the chances that they will succeed.[45] The world economy has become considerably more interconnected since they did their analysis, making it even less likely that economic sanctions can be successfully applied by any one nation acting alone. There are too many alternative trade partners and too many ways of circumventing sanctions with the help of nations that do not support them. At the same time, increased international interdependence has made broadly imposed sanctions more likely to work.[46]

The current process for imposing multilateral economic sanctions against countries threatening the peace can and should be strengthened. One way to do this would be to create a special body within the United Nations to deal exclusively with sanctions and peacekeeping. With a broader membership than the Security Council, this body would decide on what kind of sanctions, if any, should be imposed when accepted norms of international behavior are violated. Certain acts could be defined in advance as so unacceptable as to trigger powerful sanctions automatically. For example, it could be agreed that any nation that has invaded or attacked the territory of another nation would be subject to an immediate and total trade embargo by all U.N. members. The embargo would be lifted only when the attacks ceased, its military was entirely withdrawn, and an international force was installed to monitor compliance.[47]

Economic force can often be a useful and effective substitute for military force. But make no mistake, it is still a strategy of force and it will still cause a lot of pain, often among those most vulnerable. The greatest contribution economic relationships can make to global security lies not as a form of punishment, but as a means for strengthening positive incentives to keep the peace. It allows us to get out of the trap of always reacting to crises and move instead toward a more proactive strategy of preventing conflicts from degenerating into crisis and war.

Economic Peacekeeping

More than 30 years ago, Kenneth Boulding set forth his "chalk theory" of war and peace.[48] A piece of chalk breaks when the strain applied to it is greater than its strength (ability to resist that strain). Similarly, war breaks out when the strain applied to the international system exceeds the ability of that system to withstand the strain. Establishing stable peace requires that strains be reduced and strength be increased. So to realize the potential contribution of the international economic system to global security, we must look for and implement a combination of strain-reducing and strength-enhancing strategies. The principles underlying economic peacekeeping, approaches to implementing those principles, and the kinds of institutional supports that would make them more effective (along with other related issues) are discussed in detail in my book, *The Peacekeeping Economy: Using Economic Relationships to Build a More Peaceful, Prosperous and Secure World.*[49]

Strategies that incorporate the four basic principles of economic peacekeeping very briefly summarized below should help to build the kind of international economic system that will provide nations strong positive incentives to keep the peace:

1. *Establish Balanced, Mutually Beneficial Relationships.* Mutually beneficial, balanced economic relationships (as opposed to unbalanced, exploitative economic relationships) create interdependence that binds the

parties together out of mutual self-interest. Everyone has strong incentives to resolve any conflicts that arise to avoid losing or even diminishing the benefits that these relationships provide. The European Union is a working, real-world example of how effective this approach can be. The people of its 25 member nations have fought many wars with each other over the centuries. But though they still have many disagreements—some of them quite serious—they are now bound together in a web of balanced economic relationships so mutually beneficial that they no longer even think in terms of attacking each other. They debate, they shout, they argue, but they do not shoot at each other.

2. *Seek Independence in Critical Goods.* Economic independence reduces vulnerability and therefore increases feelings of security. But economic independence runs counter to the idea of tying nations together in a web of mutually beneficial relationships embodied in the first principle. If independence is emphasized where vulnerability is most frightening and interdependence is emphasized everywhere else, a balance can be struck that resolves this dilemma. Since vulnerability is greatest where critical goods are involved (such as staple foods, water, and basic energy supplies), they should be excepted from the general approach of maximizing interdependence implied by Principle 1.

3. *Emphasize Development.* There have been more than 150 wars that took the lives of over 23 million people since the end of World War II. More than 90 percent of them were fought in the less developed countries.[50] Poverty and frustration can be a fertile breeding ground for conflict. Without sustained improvement in the material conditions of life of the vast majority of the world's population living in the Third World, there is little hope for a just and lasting peace. It is also much easier to establish balanced, mutually beneficial relationships among nations at a higher and more equal economic level. They simply have more to offer each other.

4. *Minimize Ecological Stress.* Competition for depletable resources has been a source of conflict over the millennia of human history. Acute environmental disasters that cross borders, such as the nuclear power accident at Chernobyl and the oil fires of the 1990 and 1991 Persian Gulf War, also generate conflict. So do chronic international ecological problems such as acid rain and global warming. Developing renewable energy resources, conserving depletable resources by recycling, and a whole host of other environmentally sensible strategies are thus not only important for ecological reasons, they also help reduce the strain on our ability to keep the peace.

Dealing with the Terrorist Threat

Neither weapons of mass destruction nor conventional military forces nor economic sanctions can protect us from acts of terrorism perpetrated by subnational groups. The threat posed by terrorists motivated by the frustration borne of grinding poverty and/or the deliberate social and political marginalization of those who cannot make themselves heard will not be effectively countered by any strategy of force. Even the most brutal, unfree police state cannot rid itself of terrorism. Its "anti-terrorist" campaigns will only replace "private sector" terrorism with government terrorism. The terrorism that grows from these roots can be countered by economic and political development that provides better economic opportunities and more civilized, nonviolent avenues for those with serious, legitimate economic, social, and political grievances to be heard and to seek remedy for their problems.[51]

The terrorism that arises from unbridled hatred, ethnic prejudice, sadism, and other forms of mental illness and social pathology will not yield to this treatment. It must instead be fought with high-quality intelligence, solid police work, international cooperation, and the full force of legitimate legal authority. Terrorists driven by hate and paranoia are also the most likely to one day see dangerous technologies as weapons of choice or targets of opportunity. Abolishing nuclear, chemical, and biological weapons and eliminating stockpiles of weapons-grade materials will take us a long way toward overcoming this real and potentially catastrophic threat to our security. However, the threat will never be completely eliminated until we have found a way to eradicate terrorism itself.

It is important that terrorism, whatever its source, be taken seriously and opposed at every turn, but it is even more important that we keep the threat in perspective. We cannot afford to accept serious compromise of our civil liberties simply because it is clothed in the rhetoric of anti-terrorism. If we do, the terrorists will have done incalculable damage to us. And we will one day wake to find that we have lost both our freedom and our security.

STEP 3: REPLACE OTHER DANGEROUS TECHNOLOGIES WITH SAFER ALTERNATIVES

Although all dangerous technologies threaten us with eventual catastrophe, those designed for a beneficial purpose are in a different class from those designed to cause death and destruction on a massive scale. There is no point in looking for alternative technologies capable of doing as much damage as nuclear, chemical, and biological weapons of mass destruction. However, it makes perfect sense to look for safer alternatives to those technologies whose design purpose was benign. After all, nuclear power plants generate energy, the lifeblood of modern society; many dangerous chemical technologies produce materials or fuels that are enormously useful. There is no doubt that we have the talent and creativity to find alternative technologies that provide similar benefits without imposing the threat of disaster. In more than a few cases, safer alternatives already exist. Why then do we so often marginalize or ignore them?

The search for and choice of less risky technologies may appear to be a purely technological problem, but it is not. Powerful economic, social, and political factors have a lot to do with the research directions we follow and the technological choices we make. Safer alternatives may not be used because they are more expensive, more difficult to operate and maintain, harder to monopolize or control, or because businesses are already heavily invested in the technologies they would replace. They may also be pushed to the side by politically influential vested interests that stand to lose power or money if the alternatives are put into use.

Governments sometimes subsidize the development and use of technologies that turn out to be dangerous, even while they withhold subsidy from less favored, but less risky alternatives. Such subsidies distort the technological choices that would otherwise be made in a free market economy. The nuclear power industry is an important case in point. It received massive subsidies from the federal government from the very beginning. Not only was the original development of nuclear technology heavily subsidized (in part, in connection with the nuclear weapons program), but the cost of waste disposal and other operating expenses were also partly funded with taxpayer money.

Even more striking was the huge under-the-table subsidy that resulted from the Price-Anderson Act of 1957, which limited total liability for damages in the event of a major nuclear power accident to $560 million (a small fraction of the estimated multibillion dollar damages) and then used taxpayer money to guarantee $500 million of that amount.[52] In effect, taxpayers were underwriting nuclear power through the guarantee and undercutting their own right to financial compensation in the event of a major accident. Had the federal government simply let the market work, the nuclear power industry might well have died in its infancy.

Not all government subsidies distort choice in the economy. In fact, subsidies can increase economic efficiency when the unaided market does not capture all the benefits to the public generated by the goods or services produced. It makes sense to subsidize public education, for example, because we are all better off living in a more educated society. Everyone gains from everyone else's education as well as from his or her own. If we each had to pay the full cost of our own schooling, many people would get a lot less education, and we would all be deprived of the benefits of their higher skill. This kind of subsidy gives us real value for our money. Subsidizing dangerous technologies, on the other hand, diverts our attention and resources from the search for safer alternatives. This has certainly been true of energy technology.

Alternative Energy Technologies

Oil and coal-fired electric power plants are safer alternatives to nuclear power, and they are actually much more common than nuclear plants. Unfortunately, they are major sources of environmental pollution and major contributors to global warming.[53] They also depend on resources that will eventually run out. It would make more sense to look for more environmentally benign alternative technologies based on renewable resources. Fortunately, we do not have to look far. There is a great deal of energy involved in the natural processes that drive the earth's ecological systems. We can learn to tap it much more efficiently with the right kind of research and development. Solar power, geothermal energy, biomass conversion, hydro power, ocean thermal gradients, the tides, and the wind can all be used together to supply virtually all the energy on which all modern societies depend.[54]

Some of these energy sources are also much more evenly distributed around the earth than oil, coal, natural gas, or uranium. Their development may be the key to widespread national energy independence, as called for by Principle 2 of international economic peacekeeping (discussed earlier). Still, we must take care to use these natural, renewable energy sources wisely. Just because the underlying source of energy is ecologically benign does not mean that the system we use to tap it will be. Suppose we developed a very cheap, technically efficient way of generating electric power from sunlight. We must still make sure that manufacturing the energy conversion devices does not either require large quantities of toxic chemicals as raw materials or result in rivers of toxic waste. Similarly, we must be careful not to overuse natural energy resources to the point where we interfere with critical ecological mechanisms. We must learn to ask whether, for example, locating huge numbers of wind power electric generators in one place might affect airflow enough to alter local or regional rainfall patterns.

We must also avoid the trap of taking projections of energy demand as a given and concentrating all of our attention on the supply side of alternative energy. There is no reason why we cannot find technologies that allow us to maintain or improve our quality of life while using a great deal less energy than we do today. Prompted by the energy crisis more than 30 years ago, I explored a wide variety of possibilities in a book called *The Conservation Response: Strategies for the Design and Operation of Energy-Using Systems.*[55] Without even considering the more advanced technological possibilities of the day, I estimated that total U.S. energy consumption could be cut by 30 percent to 50 percent while maintaining the same standard of living. Many of the energy conservation strategies I recommended were relatively simple, and no single change accounted for more than a small fraction of the total savings.

A serious program of research and development aimed both at exploring key energy conservation technologies and increasing the efficiency of natural energy alternatives to reduce their cost would allow us to move toward a more secure, ecologically sensible energy future. Using public funds to encourage this research and development is easy to justify economically because, like public education, it would generate so many benefits (environmental and otherwise) not captured by the market. It would also help level the playing field by compensating for the enormous subsidies given to fossil fuel and nuclear energy in the past. Ongoing government subsidies that favor nuclear power should also be eliminated.

What is true of energy is also true of many other technologies—there are virtually always alternatives available or within reach. Where the subsidies to dangerous technologies have not been as massive as in the energy business, technologies that are dangerous may still have won out because they were cheaper to make or use, or because they worked better. If that is so, there will be a money and/or performance penalty to pay in the short run when safer alternatives are substituted. In nearly all cases, it will be well worth bearing the extra cost. In rare cases, the advantages of certain key technologies may be so overwhelming and so important that it makes sense to keep using them for the time being, despite the danger. But especially in those cases, high priority should be given to research and development programs aimed at finding less risky replacement technologies so that any necessary compromises will only be very temporary.

STEP IV: FACE UP TO THE LEGACY OF NUCLEAR AND TOXIC CHEMICAL WASTE

Even if we ultimately abandon all dangerous technologies in favor of less socially risky alternatives, we will still have to deal with their deadly legacy. Nowhere is this problem more obvious than with nuclear weapons and nuclear power. Long after nuclear weapons have been dismantled, the vexing problem of safely storing the plutonium and enriched uranium in their cores continues. Plutonium and enriched uranium cannot be neutralized in any practical way.[56] They remain lethally radioactive for millennia.

Both the vitrification and MOX options for disposing of weapons-grade plutonium and the "blending down" option for disposing of highly enriched uranium will generate additional amounts of highly radioactive waste. Enormous quantities of high- and low-level radioactive waste have already been generated in the process of producing nuclear weapons and nuclear power. More is being made every day. All of it will have to be safely stored, isolated, and protected for a very, very long time.[57]

There is still a great deal of controversy among scientists, not to mention the public at large, over the best ways to treat and store this radioactive garbage. We do know that much of what we are doing now is not a viable solution in the long run. As of the late 1990s, close to one-quarter of the U.S. population was living within a 50-mile radius of a storage site for military-related nuclear waste.[58] Storing millions of gallons of liquid nuclear waste in above-ground tanks in places like Hanford, Washington and Savannah River, South Carolina is far too dangerous to be more than a stopgap measure.

For nearly half a century, the managers of the federal government's Hanford site insisted that radioactive waste leaking from their underground tanks (many of which are the size of the Capitol dome) posed no threat to the environment because the waste would be trapped by the surrounding earth. By the mid-1990s, an estimated 900,000 gallons of radioactive waste had leaked from 68 (of 177) tanks and another 1.3 billion cubic meters of liquid radioactive waste and other contaminated fluids had been deliberately pumped into the soil.

They maintained that none of the waste would reach the groundwater for at least 10,000 years. But they were wrong. By November 1997, it was already there.[59]

Every part of the expensive process of dealing with these wastes is fraught with danger and uncertainty. The Department of Energy (DoE) began the first step toward cleanup—"characterizing" (determining the specific contents of)—the waste in Hanford's tanks in 1985. Twenty-four years later in late 2009, GAO pointed out that there were still many "technical uncertainties" remaining in the Hanford cleanup process, not the least of which was the question of whether key waste treatment technologies would even work. That apart, the DoE's estimate of the end date for the cleanup had slipped from 2028 to 2047, and GAO reported that its total cost was likely to increase from the DoE's 2009 estimate of $77 billion to as much as $100 billion, possibly more.[60]

In Russia, the problem is even worse. For example, years after the end of the Cold War the Kola Peninsula had become a graveyard for more than 100 Soviet-era nuclear-powered submarines, rusting away with their nuclear reactors still on board. Decades may be required to transport the more than 50,000 fuel assemblies from those reactors for reprocessing or permanent storage.[61] And in its rush to catch up with the United States, the former Soviet Union spread radioactive waste across the country, "not only near the plutonium–producing reactors in Siberia or the Urals, or on the [nuclear] test range in Kazakhstan . . . [but]also in the midst of daily life in Moscow—near offices, factories, train stations, highways, and homes.[62]

There is no cheap or clever end run around these problems. The longer we put off the day when we abandon dangerous nuclear technologies, the worse the problems become.

Storing Nuclear Waste

Despite decades of effort and billions of dollars of research, there is as yet no generally agreed optimal solution to the problem of storing nuclear waste. Unless we manage to make a spectacular technological breakthrough soon, we are stuck with choosing the best of a lot of unappealing options. The sooner we eliminate the nuclear weapons and power technologies that generate this radioactive refuse, the better. There is no reason to keep adding to the problem.

A variety of options have been proposed, including space disposal, burial by tectonic subduction, disposal under the seabed, and retrievable storage underground on land. The most interesting space disposal option involves launching rockets loaded with radioactive waste on a collision course with the sun. Since the sun is already a gigantic nuclear (fusion) reactor, additional nuclear waste would fit right in. But the number of launches required makes both the cost and the risk of this approach much too great. By one calculation, something like 10,000 spacecraft would have to be launched to carry the waste that the U.S. military alone had accumulated by the mid-1990s.[63] And we have been hard at it since then. This would not only be extraordinarily expensive, it would also divert available launch facilities from more productive uses for quite a long time. More important, the chances of at least one of these rockets exploding, failing to achieve escape velocity, or turning itself around and heading back towards earth is far too high, given the potential consequences.

Burial by tectonic subduction involves placing canisters containing the nuclear waste in deep ocean trenches, in areas where one tectonic plate is sliding beneath another. The idea is that the canisters would be carried deep into the earth by the "diving" plate and thus would be permanently buried far out of reach. The problem is that these subduction zones are geologically unpredictable. We would not know with confidence where the wastes actually went.[64] In addition, for two reasons making the wastes impossible to retrieve is not as good an idea as it sounds. For one, if we succeed in developing breakthrough technology that will somehow allow radioactive waste to be neutralized, it would be good to be able to recover and treat it. In addition, if something went wrong and the waste began to leach into the environment or otherwise go out of control, being able to access it might make it possible to mitigate the problem. While preserving access increases the risk that the waste might someday be recovered and reprocessed for nefarious purposes, storing radioactive waste so that it is difficult but possible to recover seems the least dangerous option.

Disposal under the seabed seems less risky. It could be done at stable locations far from the edges of tectonic plates, using standard deep sea drilling techniques. Drills could bore cylindrical shafts 700 to 1,000 feet deep into the abyssal mud and clays of the seabed (which are already some three miles below the surface of the sea). Canisters loaded with nuclear waste could then be lowered into the shafts, separated from each other by 70 to 100 feet of mud pumped into the borehole. Then the shafts could be backfilled and sealed in such a way that the wastes could be retrieved should that prove necessary or desirable.[65] The seabed disposal option is attractive, though the chances for catastrophic accident at sea, the environmental effects of this proposal, and the implications of storing waste in locations in which surveillance and protection are difficult require further investigation.

Land-based deep underground disposal in specially designed facilities also isolates the waste, while at the same time simplifying surveillance and (if done properly) protection and retrieval. In the United States, two land-based sites have received the most attention for the dubious honor of becoming the nation's long-term nuclear garbage dumps: Yucca Mountain, adjacent to the Nevada Nuclear Test Site (about 100 miles from Las Vegas), and the Waste Isolation Pilot Project (WIPP) site, located in the southeastern corner of New Mexico.

Yucca Mountain is a ridge made of rock formed from volcanic ash. The idea was to create a series of tunnels under the ridge large enough to store 63,000 metric tons of spent fuel from commercial nuclear power plants and 7,000 metric tons of military waste. A huge and expensive undertaking, its planned capacity for military waste storage would not even accommodate all the waste from Hanford alone. If no new commercial nuclear reactors were built and all of those currently in operation in the United States were retired at the end of their licensed operating life, the inventory of spent commercial nuclear fuel (84,000 metric tons) would still be about 30 percent greater than the site's commercial waste storage capacity.[66] Slated to be ready by 1998, an application to actually build the repository was not actually submitted to the NRC until June 2008. In early 2009, after many years of scientific and political controversy and almost $8 billion spent, President Obama essentially defunded the project without suggesting a clear alternative. For the time being, nuclear waste would continue to be stored above ground at 121 sites "within 75 miles of more than 161 million people in 39 states."[67]

Unlike Yucca Mountain, WIPP was developed to store low-level nuclear waste in the same county as the world famous Carlsbad Caverns. WIPP consists of a huge network of vaults that covers almost a square mile, carved out of salt deposits nearly a half-mile below the surface. Because the vaults are carved out of salt rather than rock, government scientists predict that the walls and ceilings of the vaults will slowly engulf and entomb the waste. If so, unlike Yucca Mountain, the buried waste will be extremely difficult if not impossible to retrieve.[68] Able to handle only about half of the low-level waste that has been generated by nuclear weapons production over the past 50 years, the site received the first of an estimated total of 37,000 shipments of transuranic nuclear waste in March 1999.[69] By November 2009, it had received 8,000 more.[70]

Perhaps even more than at Yucca Mountain, WIPP has given rise to serious public opposition. Significant issues have been raised as to the geologic suitability of the site, even as to whether the site meets the DoE's own selection criteria. It is an open question whether the choice of the location may have been driven more by politics than by scientific criteria and just plain common sense.[71]

Meanwhile, scientific research aimed at finding cheaper and safer ways to treat and store nuclear waste has been funded at levels well below the cost of a few modern fighter planes. Funding for this vital research should be sharply increased so that a wider range of alternatives can be thoroughly explored.[72] It is inconceivable that the impact on national security of having two or three fewer fighter planes is anywhere near comparable to the contribution that would be made by real progress in treating nuclear waste.

Treating Chemical Waste

Like chemical weapons, highly toxic chemical waste can be converted into much less dangerous compounds. That is the good news. The bad news is that it is not particularly easy to do, and it can be extremely expensive. There are many different types of toxic chemicals, found in many different forms and combinations. No single approach can deal with them all. Although treatment and chemical detoxification is virtually always possible, it can be so expensive or difficult that it is technically or economically impractical. There are times when, as with nuclear wastes, we are left with simply trying to safely store dangerous chemicals and keep them isolated from the environment.

In addition to its complexity, the problem of chemical toxins is extremely widespread. There are highly toxic chemical waste dumps all over the world. In the United States alone, the Environmental Protection Agency's "Superfund" has located and analyzed tens of thousands of hazardous chemical waste sites since it was established in 1980.[73] As of December 2005, the EPA had completed construction work at 966 private and federal Superfund sites (62%) and begun work at an additional 422 (27%).[74] By the end of 2008, Superfund had completed final assessment decisions on some 40,187 sites.[75] Yet according to the EPA, "[W]ork remains. 1 in 4 Americans lives within 4 miles of a Superfund site. Each year, Superfund assesses potentially hazardous waste sites and finds previously unknown chemicals and wastes that require research and new technologies to properly address ... "[76]

In sum, we still have a long way to go in cleaning up the mess. Meanwhile, we continue to produce lethal chemicals at an alarming rate.

In the mid-1990s, I served under contract as a consultant to the Industrial Partnership Office of the Los Alamos National Laboratories on expanding civilian-oriented research at the Labs. Among other things, we began to explore the possibilities for cooperative research between the chemicals industry and the Labs aimed at dealing with the toxic waste problem by finding alternative chemical processes that would neither require nor produce toxics in the first place. The project was making progress but still in its infancy when, for all practical purposes, it was killed by Congress, which lavishly increased funding for nuclear weapons and related military research and directed the Labs to stay away from more civilian-oriented projects. Yet it is exactly that kind of joint research enterprise with the chemicals industry that has the greatest promise for finding a long-term solution to the toxic chemical threat to our health and security.

WHAT HAVE WE LEARNED?

All human/technical systems are unavoidably subject to failure. Design errors, flaws in manufacturing, mistakes in maintenance, and the complexity of modern technological systems conspire to make them less than perfectly reliable. Substance abuse, mental illness, physical disease, and even the ordinary processes of aging render the people who interact with them unreliable and prone to error as well. So, too, do normal human reactions to what are often stressful, boring, and isolating work environments and to the traumas and transitions that are an ordinary part of everyone's life.

It is impossible to circumvent the problems of human fallibility merely by putting critical decisions and actions in the hands of groups rather than individuals. Groups are subject to a whole set of fallibility problems of their own, problems that sometimes make them even less reliable than individuals. There is the distortion of information and directives in bureaucracies and the bravado of "groupthink." Given the right circumstances, ordinary people can even become so wrapped up in the delusions of a crazy but charismatic leader that they ultimately do terrible things together that none of them would ever have dreamed of doing alone.

The spectacular advance of computer technology has made it seem more feasible than ever to avoid human reliability problems by creating automated systems that leave people "out of the loop." But this too is a false hope. Computers are not only subject to all of the design and hardware failure problems that afflict other equipment, they are subject to an even more perplexing set of problems related to their most important advantage—their programmability. There is plenty of empirical evidence to back up theoretical arguments that flaws are ultimately unavoidable in computer software of any real complexity. Despite the seductive names of such techniques as "automatic programming," "expert systems," and "artificial intelligence," there is no real prospect that they will completely eliminate these problems in the foreseeable future, if ever. Besides, fallible people are always "in the loop" anyway—they are the designers, manufacturers, and programmers behind these impressive machines.

Our imperfectability and the limitations of the devices that embed the technology we have developed create boundaries we must learn to live within. These boundaries only challenge us to channel our creativity. They need not stifle it.

Boundaries have certainly not stifled human creativity in art, music, or literature. Many of the world's most compelling works have been created within the bounds of a structured form and style. The poetry of Wordsworth, the music of Beethoven, and the art of Michelangelo seem so powerful and unconstrained—and yet they are all contained within such limits. All of our advances in science and engineering have been achieved within the boundaries imposed by the biological, chemical, and physical mechanisms of nature. There has been no other choice. If we can respond with such creativity and inventiveness to the external boundaries set by nature, there is no reason we cannot respond with similar ingenuity to the internal boundaries that come from the inherent limitations of being human.

We can no more avoid the boundaries imposed by our fallibility than we can revoke the laws of nature. We cannot allow our fascination with the power of what we can do to blind us to what we cannot do. It is no longer a matter of humility. It is a matter of survival.

NOTES

1. Canberra Commission, *Report on the Elimination of Nuclear Weapons*, 2. At this writing, the full report is only generally available on the World Wide Web at http//www.dfat.gov.au/dfat/cc/cchome.html.

2. Ibid., 2 and 1.

3. J. Diamond, "Air Force General Calls for End to Atomic Arms," *Boston Globe*, July 16, 1994, as cited in Lachlan Forrow and Victor W. Sidel, "Medicine and Nuclear War," *Journal of the American Medical Association*, August 5, 1998, 459.

4. ECAAR Newsletter, volume 9, #1 and #2 (May/June 1996), 4.

5. Lee Butler, "A Voice of Reason," *Bulletin of the Atomic Scientists*, May/June 1998, 59 and 61.

6. "Statesman to Promote Global Nuclear Disarmament," *Global Security Newswire*, December 8, 2008; http://gsn.nti.org/gsn/nw_20081208_6843.php (accessed November 13, 2009); see also Elaine Grossman, "To Nuclear Disarmers, It's Too Early to Worry about Violators," *Global Security Newswire*, December 16, 2008: http://gsn.nti.org/gsn/nw_20081216_9623.php; accessed November 13, 2009).

7. "Abandon Nuclear Weapons, British PM Urges," *Global Security Newswire*, January 23, 2008, http://gsn.nti.org/gsn/GSN_20080123_C6FE1CD2.php (accessed November 13, 2009).

8. John Leicester, Associated Press, "France to Cut Nuclear Arsenal," *USA Today*, March 21, 2008; http://www.usatoday.com/news/world/2008-03-21-3171539782_x.htm (accessed November 13, 2009).

9. Barack Obama, "Remarks by the President at the United Nations Security Council Summit on Nuclear Non-Proliferation and Nuclear Disarmament," New York: United Nations, September 24, 2009; http://www.whitehouse.gov/the_press_office/Remarks-By-The-President-At-the-UN-Security-Council-Summit-On-Nuclear-Non-Proliferation-And-Nuclear-Disarmament (accessed November 13, 2009).

10. William Epstein, *The Last Chance: Nuclear Proliferation and Arms Control* (New York: The Free Press, 1976), Appendix IV, 319.

11. Tim Weiner, "Iraq Spies on UN So It Can Predict Arms Inspections," *New York Times*, November 25, 1997.

12. Kofi Annan, Secretary-General of the UN, "The Unpaid Bill That's Crippling the UN," *New York Times*, OpEd, March 9, 1998.

13. William J. Broad, "Sleuths Patrol Nations for Nuclear Mischief," *New York Times*, December 30, 2003.

14. Office of Technology Assessment, Congress of the United States, *Environmental Monitoring for Nuclear Safeguards* (Washington, DC: U.S. Government Printing Office, September 1995), 1 and 5–6.

15. E. Skindrud, Bomb Testers Beware: Trace Gases Linger," *Science News*, August 10, 1996.

16. Paul G. Richards and Won-Young Kim, "Monitoring For Nuclear Explosions," *Scientific American*, March 2009, 71.

17. Because biological and chemical weapons of mass destruction are much simpler, cheaper, and easier to create, it is harder to achieve this degree of assurance with them than it is with nuclear weapons. But at least

in the case of prohibiting the buildup of national arsenals, there is no reason to believe it cannot be done. Terrorist manufacture of one or a few of these weapons, on the other hand, is much harder to prevent.

18. One of the more interesting and potentially effective proposals for verifying disarmament is "inspection by the people," developed by Seymour Melman and his colleagues more than 50 years ago. Based on the principle that "a common feature of any organized production effort to evade a disarmament inspection system . . . is the participation of a large number of people," inspection by the people would give substantial rewards to anyone with knowledge of illicit activities that revealed them to international authorities, along with safe passage out of the country and protection against future retaliation. See Seymour Melman, ed., *Inspection for Disarmament* (New York: Columbia University Press, 1958), 38.

19. "The Chemical Weapons Convention (CWC) at a Glance, Arms Control Association March 2008; http://www.armscontrol.org/factsheets/cwcglance (accessed November 14, 2009).

20. A number of other processes can also be used to destroy chemical weapons. In a study for the U.S. Army, the National Research Council (of the National Academy of Sciences and National Academy of Engineering) compared three other non-incineration options with chemical neutralization: chemical oxidation using electricity and silver compounds; exposure to high temperature hydrogen and steam; and a process involving a high temperature molten metal bath. National Research Council, *Review and Evaluation of Alternative Chemical Disposal Technologies* (Washington, DC: National Academy Press, 1996).

21. National Research Council, *Review and Evaluation of Alternative Chemical Disposal Technologies* (Washington, DC: National Academy Press, 1996).

22. The Johnston Atoll Chemical Agent Disposal System (JACADS), as the incinerator is known, released VX or GB nerve gas to the environment once in December 1990, once in March 1994, twice in March 1995, and once on the first day of April 1995. Information on the last two of these nerve gas releases is available in Program Manager of Chemical Demilitarization, U.S. Department of Defense, *JACADS 1995 Annual Report of RCRA [Resource Conservation and Recovery Act] Noncompliances* (February 25, 1996). The earlier releases were mentioned in Environmental Protection Agency (Region 9), *Report of JACADS Operational Problems* (August 1994). Information on the explosions and other operational problems at JACADS can be found in Mitre Corporation, *Summary Evaluation of the Jacads Operational Verification Testing* (May 1993). Documents concerning the problems at JACADS have been closely followed by the Chemical Weapons Working Group of the Kentucky Environmental Foundation (P.O. Box 467, Berea, Kentucky 40403).

23. Peter Eisler, "Chemical Weapons Disposal on Fast Track," *USA Today*, May 5, 2009.

24. "Pentagon Boosts Funding for Chemical Weapons Disposal," *Global Security Newswire*, May 6, 2009; http://gsn.nti.org/gsn/nw_20090506_7071.php (accessed November 14, 2009).

25. Op. cit. Peter Eisler and op.cit. Arms Control Association (March 2008).

26. David M. Bearden, "U.S. Disposal of Chemical Weapons in the Ocean: Background and Issues for Congress," *Congressional Research Service*, Updated January 3, 2007; *http:/www.fas.org/sgp/crs/natsec/RL33432.pdf* (accessed November 14, 2009).

27. Ron Chepesiuk, "A Sea of Trouble," *Bulletin of the Atomic Scientists*, September/October 1997.

28. Tim Weiner, "Soviet Defector Warns of Biological Weapons," *New York Times*, February 25, 1998.

29. Ken Alibek, "Russia's Deadly Expertise," *New York Times*, March 27, 1998.

30. Robert Norris and Hans Kristensen, "Dismantling U.S. Nuclear Warheads," *Bulletin of the Atomic Scientists*, January/February 2004, 73.

31. Robert Norris and Hans Kristensen, "U.S. Nuclear Forces, 2007," *Bulletin of the Atomic Scientists*, January/February 2007, 79.

32. U.S. Department of Defense, *Cooperative Threat Reduction Annual Report to the Congress, Fiscal Year 2010*, Information Cutoff Date December 31, 2008; http://www.dtra.mil/documents/oe/ctr/FY10%20CTR%20Annual%20Report %20to%20Congress.pdf (accessed November 14, 2009).

33. U.S. General Accounting Office, Report to the Chairman, Subcommittee on Energy, Committee on Commerce, House of Representatives, *Department of Energy: Problems and Progress in Managing Plutonium* (GAO/RCED-98-68), 4.

34. Matthew L. Wald, "Energy Dept. Announces Dual Plan for Disposal of Plutonium," *New York Times*, January 15, 1997.

35. Andrew E. Kramer, "Power for U.S. From Russia's Old Nuclear Weapons," *New York Times*, November 10, 2009.

36. Weapons-grade plutonium is normally about 93 percent Pu 239, 6 percent Pu 240, and 0.5 percent Pu 241, while typical reactor-grade plutonium is only about 64 percent Pu 239, with 24 percent Pu 240, and 11.5 percent Pu 241. Although neither Pu 240 or Pu 241 is nearly as suitable for weapons purposes as Pu 239, is it a mistake to believe that reactor-grade plutonium could not be used to make a powerful nuclear weapon. The critical mass for reactor-grade plutonium is 6.6 kg, only 40 percent more than for weapons grade. See Brian G. Chow and Kenneth A. Solomon, *Limiting the Spread of Weapon-Usable Fissile Materials* (Santa Monica, CA: National Defense Research Institute, RAND Corporation, 1993), 62–63.

37. Ibid.

38. William J. Weida, "MOX Use and Subsidies," January 28, 1997, unpublished, 4. Weida's calculations are based on Office of Fissile Materials Disposition, Department of Energy, *Technical Summary Report for Surplus Weapons-Usable*

Plutonium Disposition (Washington, DC: DOE/MD-003, 1996) and Revision 1 of that document (issued October 10, 1996). As of early 1997, the Department of Energy was still not being clear about how many reactors would be required to burn all the plutonium coming out of dismantled weapons. It will take at least one, and maybe a few.

An earlier and apparently less comprehensive estimate by the National Defense Research Institute (RAND) put the cost of the MOX option at between $7,600 and $18,000 per kilogram of weapons grade plutonium used. For 50 metric tons, the cost would thus be $380 to $900 million. This study found underground disposal to be a safer and cheaper alternative (op. cit., Brian G. Chow and Kenneth A. Solomon, 66.).

39. Peter Cohn, "Critics of MOX Facility Point to Cheaper Proposal," *Global Security Newswire*, March 16, 2007; http://gsn.nti.org/gsn/GSN_20070316 _0283822F.php (accessed November 14, 2009).

40. James Toevs and Carl A. Beard, "Gallium in Weapons-Grade Plutonium and MOX Fuel Fabrication," Los Alamos National Laboratories document LA-UR-96-4674 (1996).

41. Matthew L. Wald, "Factory Is Set to Process Dangerous Nuclear Waste," *New York Times*, March 13, 1996.

42. For a more complete discussion, see, for example, Dietrich Fischer *Preventing War in the Nuclear Age* (Totowa, NJ: Rowman and Allanheld, 1984), 47–62.

43. Political scientist Gene Sharp has argued persuasively for a completely nonmilitary form he calls "civilian-based defense." See, for example, Gene Sharp, *Civilian-Based Defense: A Post-Military Weapons System* (Princeton, NJ: Princeton University Press, 1990). Dietrich Fischer has also made major contributions to the debate over nonoffensive defense. For an especially clear exposition, see Dietrich Fischer, *Preventing War in the Nuclear Age* (Rowman and Allenheld, 1984).

44. Gary C. Hufbauer, Jeffrey J. Schott, and Kimberly A. Elliott, *Economic Sanctions Reconsidered: History and Current Policy* (Washington, DC: Institute for International Economics, 1990).

45. Ibid., 94–105.

46. David Cortright and George Lopez have done some very interesting work on making sanctions more effective. See, for example, David Cortright and George Lopez, *Sanctions and the Search for Security* (Boulder, CO: Lynne Rienner, 2002).

47. For a more detailed analysis of this proposal and the reasoning behind it, see L. J. Dumas, "A Proposal for a New United Nations Council on Economic Sanctions," in David Cortright and George Lopez, ed., *Economic Sanctions: Panacea or Peacebuilding in a Post-Cold War World* (Boulder, CO: Westview Press, 1995), 187–200.

48. Kenneth Boulding, *Stable Peace* (Austin, TX: University of Texas Press, 1978).

49. Lloyd J. Dumas, *The Peacekeeping Economy: Using Economic Relationships to Build a More Peaceful, Prosperous and Secure World* (forthcoming, Yale University Press, 2010).

50. Ruth L. Sivard, *World Military and Social Expenditures, 1993* (Washington, DC: World Priorities, 1993), 20–21. For total wars and war-related deaths from 1900 to 1995, see Ruth L. Sivard, *World Military and Social Expenditures, 1996* (Washington, DC: World Priorities, 1996), 17–19.

51. See Chapters 3 and 10 of Lloyd J. Dumas, *The Peacekeeping Economy: Using Economic Relationships to Build a More Peaceful, Prosperous and Secure World* (forthcoming 2010).

52. Though these limits of liability have been increased over the years, they remain a small fraction of estimated damages. For example, in 1982 the Sandia National Laboratory prepared a study for the Nuclear Regulatory Commission (NRC), called *Calculation of Reactor Acident Consequences (CRAC2) for U.S. Nuclear Power Plants (Health Effects and Costs)*. The NRC transmitted the CRAC2 study results to Congress later that year. These included estimates of potential damages of $135 billion for a major accident at the Salem Nuclear Plant, Unit 1 reactor (New Jersey), $158 billion for Diablo Canyon, Unit 2 reactor (California), $186 billion for San Onofre, Unit 2 reactor (California), and $314 billion for Indian Point, Unit 3 reactor (New York). Six years later, the 1988 update of Price-Anderson raised the liability limit to $7 billion, only 2 percent to 5 percent of these damage estimates.

53. For an overview of the nature of the global warming problem and a solution-oriented policy analysis, see Lloyd J. Dumas, *Understanding the Challenge of Global Warming* (Newton, MA: Civil Society Institute, 2007; http://www.civilsocietyinstitute.org/reports/GEGWS-DumasChapter.pdf (accessed November 20, 2009).

54. *Solar power* can use the heat of the sun directly, generate electricity with sunlight via solar cells, or even be used to liberate hydrogen that can be cleanly burned from water; *geothermal energy* makes use of the earth's internal heat; *biomass conversion* turns plant or animal waste into clean burning methane gas; turbines that generate electricity can be driven by waterfalls (*hydropower*), the force of the incoming and outgoing tides (*tidal power*), or the wind (*windpower*); and energy can be extracted from the sea by making use of *ocean thermal gradients*, the differences in temperature between different layers of seawater.

55. L. J. Dumas, *The Conservation Response: Strategies for the Design and Operation of Energy-Using Systems* (Lexington, MA: D.C. Heath and Company, 1976).

56. It is theoretically possible to transmute plutonium, uranium, and other dangerous radioactive substances into more benign elements by bombarding them with subatomic particles in particle accelerators. The technology to do this in a practical and economically feasible way with large quantities of radioactive material is nowhere near at hand. While it would be useful to

pursue research in this area, whether or not this can be done must be regarded as a speculative matter for the foreseeable future.

57. Some of the more dangerous radionuclides, like strontium 90 and cesium 137, decay to very low levels within a few centuries. Others, like plutonium 239, americium 241, and neptunium 237, take many thousands to millions of years to become essentially harmless. For an interesting, brief discussion of this issue, see Chris G. Whipple, "Living with High-Level Radioactive Waste," *Scientific American* (June 1996), 78.

58. James Brooke, "Underground Haven, Or a Nuclear Hazard?," *New York Times*, February 6, 1997.

59. Matthew L. Wald, "Admitting Error at a Weapons Plant," *New York Times*, March 23, 1998; Glenn Zorpette, "Hanford's Nuclear Wasteland," *Scientific American*, May 1996, 88–97.

60. U.S. Government Accountability Office, "Nuclear Waste: Uncertainties and Questions about Costs and Risks Persist with DOE's Tank Waste Cleanup Strategy at Hanford" (Washington, DC: September 2009, GAO-09-913), "GAO Highlights."

61. David Hoffman, "Rotting Nuclear Subs Pose Threat in Russia: Moscow Lacks Funds for Disposal," *Washington Post*, November 16, 1998.

62. C. J. Chivers, "Moscow's Nuclear Past Is Breeding Perils Today," *New York Times*, August 10, 2004.

63. James C. Warf and Sheldon C. Plotkin, "Disposal of High Level Nuclear Waste," Global Security Study No.23 (Santa Barbara, CA: Nuclear Age Peace Foundation, September 1996), 6.

64. Charles D. Hollister and Steven Nadis, "Burial of Radioactive Waste Under the Seabed," *Scientific American*, January 1998), 61–62.

65. Ibid., 62–65.

66. Chris G. Whipple, "Can Nuclear Waste Be Stored Safely at Yucca Mountain," *Scientific American*, June 1996, 72–76.

67. Editorial, "Mountain of Trouble: Mr. Obama Defunds the Nuclear Depository at Yucca Mountain," *Washington Post*, March 8, 2009.

68. Op. cit. James Brooke.

69. "Nation's First Nuclear Waste Depository Opens," Associated Press, March 26, 1999.

70. "WIPP Receives 8000th TRU Waste Shipment," *TRU TeamWorks*, November 6, 2009; http://www.wipp.energy.gov/TeamWorks/index.htm (accessed November 21, 2009).

71. For a particularly interesting and erudite analysis of the problems with the WIPP site, see Citizens for Alternatives to Radioactive Dumping, "Greetings from WIPP, the Hot Spot: Everything You Always Wanted to Know About WIPP" (CARD, 144 Harvard SE, Albuquerque, NM 87106: 1997).

72. One interesting alternative is bioremediation, though much careful study is necessary before living organisms can be seriously considered as agents for degrading nuclear waste. See John Travis, "Meet the Superbug:

Radiation-Resistant Bacteria May Clean Up the Nation's Worst Waste Sites," *Science News*, December 12, 1998.

73. U.S. Environmental Protection Agency, "Superfund: Basic Information—How Superfund Works"; http://www.epa.gov/superfund/about.htm (accessed November 21, 2009).

74. U.S. Environmental Protection Agency, "Superfund's 25th Anniversary"; http://www.epa.gov/superfund/25anniversary/index.htm; updated June 3, 2009 (accessed November 21, 2009).

75. U.S. Environmental Protection Agency, "Superfund National Accomplishments Summary Fiscal Year 2008"; http://www.epa.gov/superfund/accomp/numbers08.htm (accessed November 21, 2009).

76. U.S. EPA, "Superfund's 25th Anniversary"; http://www.epa.gov/superfund/25anniversary/index.htm.

Index

accidental war, 145–167. *See also* accidents, false warnings; communications, command and control problems, 157–162, 229, 236; incidents of failure, 146–147, 151; crises and the generation of, 17, 145, 147–149; India and Pakistan and, 145–146; risk of, 149–151, 166; triggering events for, 148–150. *See also* launch-on-warning; World War I as, 146–147

accidents: aviation, 108–110, 123, 133, 137, 149, 198 n.9, 206, 210, 265, 283, 291, 332; Freudian Slips, 210–211; biological weapons related, 105–107, 112–113, 149; dangerous materials, 116–117; nuclear waste, 115–116, 149; nuclear weapons, 104–105. *See also* "broken arrows;" nuclear powered satellites, 113–115; triggers of accidental war: weapons of mass destruction, 147–149; consequences of, 148; likelihood of, 107–108; public record of, 108–112

Acheson, Dean, 241

Afghanistan War: Soviet, 35,188, 242, 351; United States, 189–190, 213

Albright, Joe, 86

alcohol: abuse by airline pilots, 176, 198 n.9; average consumption of, in Soviet Union/Russia, 179; average consumption of, in US, 176–177; in Soviet/Russian military, 178–179, 199 n.20; in U.S. military, 83, 177–178, 192–194, 208. *See also* personnel reliability

Alcohol, Tobacco and Firearms, Bureau of, 76, 246

Aloha Airlines, 283

Alibekov, Kanatjan ("Ken Alibek"), 357

Allen, Robert, 261

Alternative energy technologies, 367–369

Al Qaeda, 34, 38, 46, 56–57, 207; attacks by. *See* September 11th; previous U.S. support for, 35

Amagasaki, Japan, 206

American Airlines, 283, 292, 331; Flight 11, 30; Flight 77, 30; Flight 965, 205; flight groundings due to computer problems, 293; reservation system failure, 293

anthrax, 10, 20, 25 n.5, 39, 63 n.14, 78, 106, 107, 112, 119 n.19, 121 n.40, 139, 141, 357

Apollo: Pennsylvania, 80; Space mission, 265

Applewhite, Marshall Herff, 247–248

Argyris, Chris, 229

About the Author

LLOYD J. DUMAS is Professor of Political Economy, Economics, and Public Policy at the University of Texas at Dallas. Trained both as a social scientist and an engineer, he was formerly Associate Professor of Industrial and Management Engineering at Columbia University and served as a consultant to the Los Alamos National Laboratories. He has published six books and more than 120 articles in 11 languages in books and journals of economics, engineering, sociology, history, public policy, military studies, and peace science, as well as in such periodicals as the *New York Times*, *Los Angeles Times*, *International Herald Tribune*, *Science*, *Technology Review*, and *Defense News*. He has discussed his work on more than 300 television and radio programs in the United States, the former Soviet Union, Russia, Canada, Europe, Latin America, and the Pacific. His books include *The Conservation Response: Strategies for the Design and Operation of Energy-Using Systems*, *The Overburdened Economy: Uncovering the Causes of Chronic Unemployment, Inflation and National Decline* and *The Peacekeeping Economy: Using Economic Relationships to Build a More Peaceful, Prosperous and Secure World*.